Adsorption–Desorption Phenomena

International Symposium on

Adsorption–Desorption Phenomena *, 2d, Florence, 1971*

edited by F. Ricca
University of Turin, Italy

Proceedings of the
II International
Conference
Held at Florence
in April 1971

1972

Ⓐ𝐏 Academic Press
London and New York

Academic Press Inc (London) Ltd
24-28 Oval Road
London, NW1

US edition published by
Academic Press Inc
111 Fifth Avenue
New York, New York 10003

Library of Congress Catalog Card number: 73-172370

ISBN: 0.12.587750.1

Printed in Great Britain by William Clowes & Sons Ltd, London, Colchester and Beccles

CONTRIBUTORS

M ABON *Départment de Chimie-Physique, Institut de Recherches sur la Catalyse, CNRS, Villeurbanne, France.*

T G ANDRONIKASHVILI *Institute of Physical and Organic Chemistry, Academy of Sciences of the Georgian SSR, Tbilisi, USSR.*

Y BALLU *Centre d'Études Nucléaires de Saclay, Section d'Études des Interactions Gaz-Solides, Gif-sur-Yvette, France.*

Y BARBAUX *Laboratoire de Catalyse, Université des Sciences et Techniques de Lille, France.*

J-P A BEAUFILS *Laboratoire de Catalyse, Université des Sciences et Techniques de Lille, France.*

R CHAPMAN *National Research Council of Canada, Ottawa, Canada.*

R CHAPPELL *Department of Chemistry, Imperial College, London, England.*

M A CHESTERS *Department of Chemistry, Queen Mary College, London, England.*

T A CHUMBURIDZE *Institute of Physical and Organic Chemistry, Academy of Sciences of the Georgian SSR, Tbilisi, USSR.*

R CLAMPITT *Culham Laboratory, Abingdon, England.*

R C A CONTAMINARD *Chemistry Department, Imperial College of Science and Technology, London, England.*

R C COSSER *Chemistry Department, Imperial College of Science and Technology, London, England.*

R CREASEY *School of Chemistry, The University of Leeds, England.*

G DALMAI-IMELIK *Institut de Recherches sur la Catalyse, CNRS, Villeurbanne, France.*

J G DAUNT *Stevens Institute of Technology, Hoboken, New Jersey, USA.*

D A DEGRAS *Centre d'Études Nucléaires de Saclay, Section d'Études des Interactions Gaz-Solides, Gif-sur-Yvette, France.*

E J DERDERIAN *Department of Chemistry, The Pennsylvania State University, Pennsylvania, USA.*

G J DOOLEY *Aerospace Research Laboratories (LJ), Wright-Patterson AFB, Ohio, USA.*

M M DUBININ *Institute of Physical Chemistry, Academy of Sciences of the USSR, Moscow, USSR.*

G ERTL *Institut für Physikalische Chemie und Elektrochemie, Technische Universität, Hannover, Germany.*

E FERRONI *Institute of Physical Chemistry, University of Florence, Italy.*

O FRANK *Fritz-Haber-Institut der Max-Plank-Gesellschaft, Berlin-Dahlem, Germany.*

E GARRONE   *University of Turin, Italy.*

B A GOTTWALD   *Technische Universität, Hannover, Germany.*

L GOWLAND   *UKAEA Culham Laboratory, Abingdon, England.*

J T GRANT   *Aerospace Research Laboratories (LJ), Wright-Patterson AFB, Ohio, USA.*

T B GRIMLEY   *The Donnan Laboratories, The University of Liverpool, England.*

T W HAAS   *Aerospace Research Laboratories (LJ), Wright-Patterson AFB, Ohio, USA.*

D E HAGEN   *Battelle Memorial Institute, Columbus, Ohio, USA.*

D O HAYWARD   *Department of Chemistry, Imperial College, London, England.*

J P HOBSON   *National Research Council of Canada, Ottawa, Canada.*

A M HORGAN   *School of Chemical Sciences, University of East Anglia, Norwich, England.*

B IMELIK   *Institut de Recherches sur la Catalyse CNRS, Villeurbanne, France.*

D A KING   *School of Chemical Sciences, University of East Anglia, Norwich, England.*

R KLEIN   *National Bureau of Standards, Washington, USA.*

J KOCH   *Institut für Physikalische Chemie und Elektrochemie, Technische Universität, Hannover, Germany.*

P KRONAUER   *Lehrstuhl für Physikalische Chemie, Technische Hochschule Darmstadt, West Germany.*

R M LAMBERT   *Department of Physical Chemistry, University of Cambridge, England.*

J LECANTE   *Centre d'Études Nucléaires de Saclay, Section d'Études des Interactions Gaz-Solides, Gif-sur-Yvette, France.*

T J LEE   *Royal Observatory, Edinburgh, Scotland.*

E LERNER   *Stevens Institute of Technology, Hoboken, New Jersey, USA.*

J W LINNETT   *Department of Physical Chemistry, University of Cambridge, England.*

M MAGLIETTA   *Institute of Physical Chemistry, University of Florence, Italy.*

G-A MARTIN   *Institut de Recherches sur la Catalyse, CNRS, Villeurbanne, France.*

J D McKINLEY   *National Bureau of Standards, Washington, USA.*

D MENZEL   *Institut für Physikalische Chemie und Elektrochemie, Technische Universität, München, West Germany.*

R P MERRILL   *Department of Chemical Engineering, University of California, Berkeley, California, USA.*

F J MILFORD   *Battelle Memorial Institute, Columbus, Ohio, USA.*

D MOUROT   *Centre d'Études Nucléaires de Saclay, Section d'Études Interactions Gaz-Solides, Gif-sur-Yvette, France.*

A D NOVACO   *Battelle Memorial Institute, Columbus, Ohio, USA.*

J PATIGNY   *Laboratoire de Catalyse, Université des Sciences et Techniques de Lille, France.*

L A PETERMANN   *Battelle Institute, Geneva Research Centre, Switzerland.*

C PISANI   *University of Turin, Italy.*

F PRATESI   *Institute of Physical Chemistry, University of Florence, Italy.*

J PRITCHARD   *Department of Chemistry, Queen Mary College, London, England.*

F RICCA   *University of Turin, Italy.*

G ROVIDA   *Institute of Physical Chemistry, University of Florence, Italy.*

Sh D SABELASHVILI   *Institute of Physical and Organic Chemistry, Academy of Sciences of the Georgian SSR, Tbilisi, USSR.*

M D SCHEER   *National Bureau of Standards, Washington, USA.*

L D SCHMIDT   *Department of Chemical Engineering and Materials Science, University of Minnesota, Minneapolis, USA.*

W A SCHMIDT   *Fritz-Haber-Institut der Max-Planck-Gesellschaft, Berlin-Dahlem, Germany.*

A SCHRAM   *Centre d'Études Nucléaires de Saclay, Section d'Études des Interactions Gaz-Solides, Gif-sur-Yvette, France.*

J A SCHWARZ   *Department of Physical Chemistry, University of Cambridge, England.*

M L SIMS   *Department of Chemistry, Queen Mary College, London, England.*

W A STEELE   *Department of Chemistry, The Pennsylvania State University, Pennsylvania, USA.*

B TARDY   *Départment de Chimie-Physique, Institut de Recherches sur la Catalyse, CNRS, Villeurbanne, France.*

N TAYLOR   *School of Chemistry, The University of Leeds, England.*

S J TEICHNER   *Départment de Chimie-Physique, Institut de Recherches sur la Catalyse, CNRS, Villeurbanne, France.*

F C TOMPKINS   *Chemistry Department, Imperial College of Science and Technology, London, England.*

G V TSITSISHVILI   *Institute of Physical and Organic Chemistry, Academy of Sciences of the Georgian SSR, Tbilisi, USSR.*

W H WEINBERG   *Department of Chemical Engineering, University of California, Berkeley, California, USA.*

FOREWORD

These are the proceedings of one of two International Symposia held in 1971, at Florence. The other, "Residual Gases in Electron Tubes", edited by Dr. T. A. Giorgi and Dr. P. della Porta, is also published by Academic Press.

PREFACE

The present volume is comprised of the papers presented at the II International Symposium on Adsorption–Desorption Phenomena, held at the Palazzo dei Congressi, Florence from 14 to 17 April, 1971. This symposium was organized by the Italian Association of Physical Chemistry and was sponsored by the Consiglio Nazionale delle Ricerche. It was held in connection with the IV International Symposium on Residual Gases organized by the Italian Vacuum Society, because experience at the preceding symposia proved the value of combining science and technology on this subject.

The meeting was prepared by an International scientific committee in which Professor Dubinin, of the Institute of Physical Chemistry of the USSR Academy of Sciences; Dr. Hobson, of the Canadian National Research Council; Professor Steele, of Pennsylvania State University; Professor Teichner, of the Institute of Catalysis of the French National Council for Scientific Research; Professor Tompkins, of the Imperial College of Science and Technology; and Professor Ferroni, of the University of Florence and president of the Italian Association of Physical Chemistry, participated. Such a committee guaranteed the quality of contributions and ensured the participation of scientists from so many countries and I wish to warmly express, here, my sincere thanks for their invaluable help.

The themes chosen for the meeting were intended to cover fundamental problems in adsorption as distinct from catalysis and solid reactivity, and were divided into three sub-sections which allowed particular attention to be given to some of the novel aspects of research in these fields. Devoted to theoretical studies, to particle beams and to chemisorption of gases by metals, these sub-sections offered an opportunity for stimulating discussions at the meeting. It is hoped that the present volume will now prove useful in extending information and discussion to a much greater number of scientists working in the field of adsorption and desorption phenomena.

Turin, *February, 1972*

F. Ricca

CONTENTS

I. THEORETICAL STUDIES ON PHYSICAL ADSORPTION

FUNDAMENTALS OF THE THEORY OF PHYSICAL ADSORPTION OF GASES AND VAPOURS IN MICROPORES

M. M. DUBININ

Institute of Physical Chemistry, Academy of Sciences of the USSR, Moscow, USSR

I. LIMITING CASES OF PHYSICAL ADSORPTION

Regardless of their chemical nature, adsorbents are usually divided into non-porous and porous. The non-porous adsorbents include those whose surface curvature radii are very wide and, in principle, tend to infinity. The porous adsorbents contain different pore varieties whose role in adsorption is characterized, depending on their parameters, either by exclusively quantitative or qualitative differences of radical nature.

In general, one can use a rather far-reaching analogy between disperse and porous solids, regarding the latter as disperse systems; in the first case the solid forms a disperse substance while, in the second case, the solid itself is the disperse medium. Thus, disperse and porous solids are a sort of transformed system. In this scheme, communicating pores are analogues of branching aggregations formed by colloid particles as a result of coagulation.

A characteristic parameter of the disperse systems under study are the linear dimensions of the disperse particles or pores. For particles such a parameter may be, for instance, the effective radius, equal to a cubic root of the ratio of the particles volume to $4\pi/3$, while the effective radius of a pore of arbitrary shape is equal to the doubled ratio of the area of the normal cross-section of the pore to its perimeter. Let us now turn our attention to the variation in the general properties of the disperse systems under review with changes in the linear dimensions or particles or pores.

It may be considered a generally accepted standpoint that a coarse suspension is a two-phase system. In this case the surface of the solid phase is the most important thermodynamic parameter characterizing the behaviour of the system. With a sufficiently increased degree of dispersity, i.e. with reduced linear dimensions of the particles, we pass over from the region of suspensions to that of colloidal solutions, in which the surface still plays the determining role. With a further increase in the degree of disperity, however, we inevitably find ourselves in the region directly adjoining true molecular solutions for which the concept, surface of the solid, loses its meaning because the system becomes

3

one-phase. Thus, with an increasing degree of dispersity, the quantitative changes in the properties of the system, consisting in an increase of the interface, will inevitably result in qualitative differences: i.e. the transformation of the two-phase system into one-phase.

The analogy between the disperse and porous systems leads to a conclusion of importance. Transition from non-porous adsorbents through wide-pore ones to the limiting case of porosity, when the pores are commensurate with the sizes of the molecules adsorbed, is at first associated with the accumulation of quantitative differences which inevitably lead to qualitative differences. Most important is the fact that the concept of surface loses its physical meaning and the adsorbent-adsorbate system becomes in a sense single-phase.

Our conception of the varieties of adsorbent pores (Dubinin, 1968) are based on the difference in the mechanisms of the adsorption and capillary phenomena taking place in them. The effective radii of the largest variety of adsorbent pores—macropores—exceed 1000-2000 Å and their specific surface area ranges between 0.5 and 2 m^2/g. Adsorption on the surface of macropores is usually negligible and capillary condensation is unfeasible for a number of reasons. Therefore, the macropores act exclusively the role of transport pores. The effective radii of a smaller variety of adsorbent pores—intermediate pores—greatly exceed the sizes of the molecules adsorbed. On the surface of intermediate pores there occurs monomolecular and polymolecular adsorption of vapours: i.e. the formation of successive adsorption layers leading to the volume—filling of this variety of pores by the capillary condensation mechanism. The effective radii of intermediate pores range from 18-19 Å to 1000-2000 Å. As we have shown previously (Bering *et al.,* 1966) the lower boundary, which corresponds to meniscus curvature radii of 15-16 Å, corresponds, in the pores of indicated dimensions, to the limit of applicability of the Kelvin equation. Depending on the development of the volume of intermediate pores and their predominant radii, the specific surface areas of intermediate pores may range from 10 to 400 m_2/g. On the whole, for the non-porous, macroporous, and intermediate-pore adsorbents of the same chemical nature the difference in vapour adsorption is basically quantitative due to the different specific surface areas since the surface curvature (prior to capillary condensation) in practice exerts only a slight effect on adsorption. In all these cases the concept of adsorbent surface has a clear-cut physical meaning, and vapour adsorption amounts to the formation of successive adsorption layers.

The effective radii of the smallest variety of pores—micropores—are substantially below the lower boundary of sizes of intermediate pores. Small-angle X-ray scattering data (Dubinin *et al.,* 1964; Dubinin and Plavnik, 1968) show that the principal volume of micropores ordinarily lies within the range of effective radii (inertia radii) from 5 to 10 Å. Thus, the micropores are commensurate with the sizes of the molecules adsorbed. In particular, this range includes the sizes of zeolite voids. One of the main parameters of micropores is their total volume per unit mass of adsorbent, whose maximal value generally does not exceed 0.5 cm^3/g by any considerable extent.

II. THERMODYNAMIC DIFFERENCE BETWEEN LIMITING CASES OF ADSORPTION

At present one cannot ignore the fact that the term physical adsorption refers to many phenomena widely differing both in the nature of molecular interactions and in their mechanism, which require different models and the application of different mathematical language for their description. From the thermodynamic standpoint the whole diversity of physical adsorption phenomena lies between two limiting cases of adsorption on non-porous, homogeneous, surfaces and adsorption in micropores. All real cases of adsorption represent a more or less complex superposition of these two adsorption phenomena.

In adsorption on non-porous, and also on wide-pore adsorbents, a pictorial molecular model is the coverage of the adsorbent surface with the formation of successive adsorption layers. The most important parameter which characterizes adsorption equilibrium is the surface area of the adsorbent. On the other hand, a vivid molecular model of adsorption in micropores is their volume filling. This adsorption mechanism is the consequence of the relative smallness of the sizes of micropores in whose entire space an adsorption field is set up, in adsorption interactions of almost any sort. For the same reason the substance adsorbed in micropores cannot be regarded as a separate phase.

Bering *et al.* (1970) formulated, in rigorously thermodynamic language, the qualitative differences between adsorption on non-porous and microporous adsorbents. It was shown that, from the thermodynamic point of view, the basic equation describing adsorption on non-porous or relatively wide-pore adsorbents is the well-known Gibbs equation, while the equation describing adsorption in micropores is the Gibbs-Duhem equation. In other words, from the thermodynamic viewpoint adsorption in micropores is a process similar to the formation of a sort of solution resulting in a change in the chemical potentials of both components of the system.

III. EQUILIBRIA IN ADSORPTION IN MICROPORES

Equilibrium states in gas or vapour adsorption are expressed by the thermal equation of adsorption which establishes the relationship between the adsorption value a, pressure p, and temperature T. Any one of these values may be regarded as a function of the other two. One of the principal tasks of the theory is to obtain this equation in an explicit form.

The concept of volume filling of micropores leads to a clear-cut notion of the limiting adsorption value a_0 corresponding to the filling of the whole adsorption space of micropores by the molecules adsorbed. The dependence of a_0 on temperature is determined by the thermal coefficient of limiting adsorption α

$$\alpha = -\frac{1}{a_0}\frac{da_0}{dT} = -\frac{d\ln a_0}{dT} \tag{1}$$

which is practically constant over a wide range of temperature. The coefficient α can be calculated to a good approximation from the physical constants of the adsorptive (Nikolaev and Dubinin, 1958; Bering et al., 1966). If the limiting adsorption value a_0° is determined experimentally for a certain temperature T_0, then, according to (1), the limiting adsorption values a_0 for other temperatures will be expressed thus

$$a_0 = a_0^\circ \exp\left[-\alpha(T - T_0)\right] \tag{2}$$

Sufficiently reliably computable limiting adsorption values a_0 make it possible to use, in place of the adsorption value a, the dimensionless parameter θ expressing the degree of filling of micropores. By definition

$$\theta = a/a_0 \tag{3}$$

The theory of volume filling of micropores bears a thermodynamic character and, therefore, in describing adsorption equilibria, this theory uses such thermodynamic functions as enthalpy, entropy, and free energy. To calculate variations in these functions, it is customary to adopt as the standard reference state at the temperature under consideration, the bulk liquid phase, which corresponds to the adsorptive and is in equilibrium with its saturated vapour at a pressure p_s or fugacity f_s.

The principal thermodynamic function in this section is the differential maximal molar work of adsorption A equal, with a minus sign, to the variation in Gibbs' free energy of adsorption ΔG

$$A = -\Delta G = RT\ln p_s/p \quad \text{or} \quad A = RT\ln f_s/f \tag{4}$$

where p is the equilibrium pressure or fugacity f of a vapour at a temperature T. The introduction of fugacities in place of pressures enables the non-ideal nature of the gas phase to be taken into account. If adsorption is expressed in dimensionless units, then the differential molar work of adsorption is expedient to express in a similar manner in the form of the dimensionless ratio A/E, where E is characteristic free energy of adsorption whose physical meaning will be explained below. Then the thermal equation of adsorption can be represented in the general form

$$\theta = f\left(\frac{A}{E}, n\right) \tag{5}$$

Equation (5) expresses the distribution function of filling of micropores, θ, over the differential molar work of adsorption A; E is one of the parameters of this function. Since most of the distribution functions in the normalized form are characterized by two parameters, the second parameter, which we will denote conventionally by n, enters as a constant parameter into the analytic expression for the function (5).

According to equation (5) we obtain the expression for the so-called characteristic curve

$$A = E\varphi(\theta, n) \tag{6}$$

If the function φ and parameter n remain invariable for different vapours, then

$$\left(\frac{A}{A_0}\right)_\theta = \frac{E}{E_0} = \beta \tag{7}$$

i.e. the characteristic curves in the coordinates $A \div \theta$ show an affinity relationship. In other words, the ratios of the ordinates of A taken at equal θ's are constant and equal to the coefficient of affinity β, in the range of variation of filling θ, in which the initial assumptions of invariability of the function φ, and of constancy of its parameter n, hold good. In equation (7) the subscript 0 denotes the value for a standard vapour. The fulfilment of the affinity of the characteristic curves was first substantiated in investigations of the example of microporous carbonaceous adsorbents (Dubinin, 1966).

From equation (6) it follows that $E = A$ for a certain filling $\theta°$ or for a characteristic point determined, in the general case, from the condition

$$\varphi(\theta°, n) = 1 \tag{8}$$

With an invariable function φ the filling $\theta°$ will be the same for different vapours. The role of n will be considered below. The aforesaid is the basis for experimental determination of the characteristic free energy of adsorption from a single point of the adsorption isotherm corresponding to filling θ° and expressed by equation (8). Naturally, the absolute value of θ° depends on the type of the function φ.

By regarding the function f in the thermal equation as a distribution function, we assumed the temperature invariance of this function, supposing that its parameters E and n are constant values for the adsorption system under consideration. Since, as has just been said, $E = A$ for filling θ° (i.e. E is one of the points of the characteristic curve) the assumption of temperature invariance automatically leads to temperature independence of the characteristic free energy of adsorption E and, as a consequence, of the parameter n.

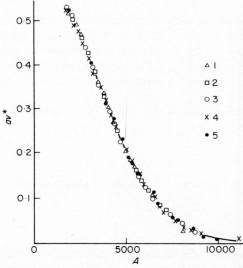

Fig. 1. Characteristic curve of adsorption of benzene on active carbon CK. 1-20°C. 2-50°, 3-80°, 4-110°, 5-140° (av^*, cm^3/g, A, cal/mole, $av^* \sim \theta$).

In principle, the temperature invariance condition is not mandatory for the theory of adsorption in micropores. However, the case when temperature invariance is not observed is only in the initial stage of investigation. Experience shows that the conception of temperature invariance of the characteristic curve is a rational approximation for the numerous adsorption systems studied. This was first shown experimentally by Polanyi on several examples for relatively narrow temperature ranges. The interpretations of this fact offered by Polanyi, however, followed from entirely improbable assumptions in calculating the adsorption potential (Polanyi, 1932).

The expediency of the assumption of temperature invariance of characteristic curves in adsorption of various vapours on microporous adsorbents of different nature is testified by the graphs on which points calculated from experimental isotherms for different temperatures are denoted by different symbols. Figure 1 depicts the characteristic curve of adsorption of benzene on active carbon CK for the temperature range from 20 to 140°C (Dubinin and Polstyanov, 1966). Figure 2 illustrates the observance of temperature invariance for adsorption of carbon dioxide on zeolite LiX in the temperature range from 0 to 90°C (Avgul *et al.*, 1968). Investigation of many adsorption systems has led to the conclusion that the condition of temperature invariance of characteristic curves is well observed for different kinds of interactions causing physical adsorption.

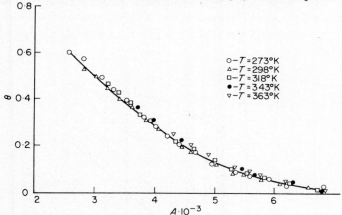

Fig. 2. Characteristic curve of adsorption of carbon dioxide on zeolite LiX (A, cal/mole).

The generalization of the computational methods of the volume filling of micropores theory was based on applying a more general Weibull distribution function (Dubinin and Astakhov, 1971), which made it possible to express the thermal equation of adsorption (5) in an explicit form

$$\theta = \exp\left[-\left(\frac{A}{E}\right)^n\right] \qquad (9)$$

where the exponent n is a small integer. Expressing A from equation (9) as a function of θ, we obtain the characteristic curve equation, which satisfies all the above-described properties of the similar equation (6), represented in the general form. Thus, according to (9) $E = A$ for $\theta° = 1/e = 0.368$, where e is the base of natural logarithms.

Substituting the expression for θ according to (3) and for a_0 according to (2) into equation (9), we obtain the thermal equation of adsorption in the form:

$$a = a_0° \exp\left[-\left(\frac{A}{E}\right)^n - \alpha(T - T_0)\right] \qquad (10)$$

The parameters of this equation—the limiting adsorption value a_0° and the characteristic free energy E—are determined from a single experimental adsorption isotherm for a temperature T_0. In connection with the integral value of the parameter n only its tentative estimate is required, which is readily obtainable from the same initial adsorption isotherm.

As has been shown in our investigations (Dubinin and Astakhov, 1971), the thermal equation of adsorption (10) is widely applied for different microporous adsorbents, including zeolites. The only exception is the case of adsorption of relatively small molecules, for instance CO_2, H_2O, and others on zeolites with relatively large voids; for instance faujasites, when equation (9) becomes a two-term one. In this case, different values of the parameters E and n correspond to adsorption on cations and to the filling of the remaining adsorption space of micropores.

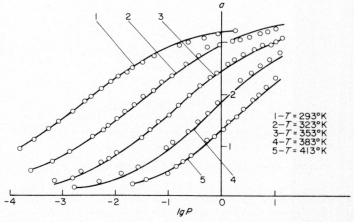

Fig. 3. Calculated and experimental absorption isotherms of cyclohexane on active carbon CA (a, mmole/g, p, Torr).

In the graphs given above the continuous curves denote adsorption isotherms calculated from equation (10), while the circles indicate experimental points. Only new data, corresponding to $n \neq 2$, will be presented here since equation (9) in connection with (3) and (4) at $n = 2$ transforms into the Dubinin-Radushkevitch equation, the wide applicability of which for adsorbents with not-too-small micropores has already been described (Dubinin, 1966). Figure 3 displays calculated the experimental adsorption isotherms of cyclohexane on active carbon from polyvinyl-idenchloride with very fine micropores for the temperature range from 20 to 140°C (Dubinin and Polstyanov, 1966). In this case $n = 3$ and $E = 6930$ cal/mole were determined from an isotherm at 80°C. The family of adsorption isotherms of methane on zeolite L given in figure 4 for the temperature range from $-117°C$ to $-30°C$ was calculated on the basis of

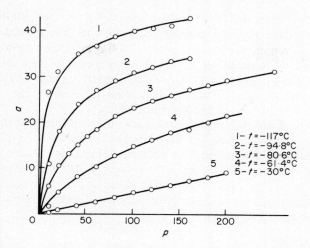

Fig. 4. Calculated and experimental adsorption isotherms of methane on zeolite L (a, cm³ NTP/g, p, Torr).

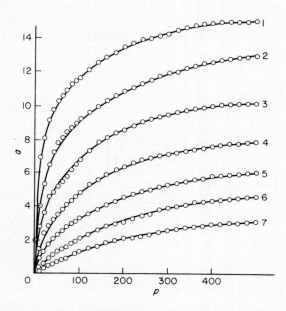

Fig. 5. Calculated and experimental adsorption isotherms of carbon dioxide on Na, K—erionite: 1-20°C, 2-40°, 3-60°, 4-80°, 5-100°, 6-120°, and 7-140° (a, %, p, Torr).

experiments by Barrer and Lee (1968) from the initial isotherm at $-80°C$ ($n = 3$, $E = 2350$ cal/mole). Figure 5 exhibits calculated and experimental adsorption isotherms of carbon dioxide on erionite, following experiments by L. A. Meyerson and G. A. Stepanov for the temperature range from 20 to 140°C. From the initial isotherm for 80°C we obtained $n = 3$ and $E = 5350$ cal/mole. For a number of cases of vapour adsorption on zeolites, the exponent n amounted to 4 or 5 (Dubinin and Astakhov, 1971).

It should be noted that for calculating adsorption isotherms at above-critical temperatures use was made of the effective magnitudes of the limiting adsorption value a_0 determined from equation (2) and the effective values of p_s extrapolated by an equation of the form $\log p_s = C - D/T$ into the above-critical region. In this case the values of p_s usually served to determine the fugacities f_s, which were then used for calculating the differential molar works of adsorption (Dubinin and Astakhov, 1971).

IV. DIFFERENTIAL HEATS AND ENTROPIES OF ADSORPTION

On the basis of general thermodynamic considerations it is possible to write quite rigorous expressions for the differential molar entropy of adsorption ΔS and for the net differential heat of adsorption q equal to the variation in the differential molar enthalpy of adsorption ΔH with a minus sign (Bering *et al.*, 1966)

$$\Delta S = \left(\frac{\partial A}{\partial T} \right)_\theta + \alpha \left(\frac{\partial A}{\partial \ln a} \right)_T \tag{11}$$

$$q = -\Delta H = A - T \left[\left(\frac{\partial A}{\partial T} \right)_\theta + \alpha \left(\frac{\partial A}{\partial \ln a} \right)_T \right] \tag{12}$$

These equations are valid also for the absence of temperature invariance of the characteristic curves. However, for the case in hand, when this condition is observed

$$\left(\frac{\partial A}{\partial T} \right)_\theta = 0 \tag{13}$$

and equations (11) and (12) take the form

$$\Delta S = \alpha \left(\frac{\partial A}{\partial \ln a} \right)_T \tag{14}$$

$$q = A - \alpha T \left(\frac{\partial A}{\partial \ln a} \right)_T \tag{15}$$

From equation (14) there follows a radically important conclusion. Since the coefficient α is a positive value, while the differential molar work of adsorption A decreases with increasing adsorption, the thermodynamic criterion of the upper limit of A at which temperature invariance is rigorously possible is

$$\Delta S < 0 \tag{16}$$

This condition can be expressed, according to Bering and Serpinsky, through the lower boundary of filling θ_l, which determines the limits of rigorous applicability of the thermal equation of adsorption (9)

$$\log \theta_l = -0.434 \left(\frac{\alpha E}{nR} \right)^{n/(2n-1)} \tag{17}$$

For the example of adsorption of benzene on active carbon from saran, $\theta_l = 0.31$. However, taking into account the deviations acceptable in the calculation of adsorption equilibria, the practical value of θ_l reduces to 0.1-0.20.

Finding from the thermal equation of adsorption (9), taking into consideration (3) the derivative entering into the expressions for ΔS and q is

$$\left(\frac{\partial A}{\partial \ln a} \right)_T = \frac{E}{n} \left(\ln \frac{a_0}{a} \right)^{1/n-1} \tag{18}$$

and substituting it into equations (14) and (15), we obtain the analytic expression for the differential molar entropy of adsorption ΔS and the net differential heat of adsorption q whose range of applicability is determined by the thermodynamic criteria (16) and (17) (Bering *et al.*, 1971):

$$\Delta S = -\frac{\alpha E}{n} \left(\ln \frac{a_0}{a} \right)^{1/n-1} \tag{19}$$

$$q = E \left[\left(\ln \frac{a_0}{a} \right)^{1/n} + \frac{\alpha T}{n} \left(\ln \frac{a_0}{a} \right)^{1/n-1} \right] \tag{20}$$

For the characteristic point $a_0/a = e$ and according to (20)

$$q_0 = E \left[1 + \frac{\alpha}{n} T \right] \tag{21}$$

M. M. DUBININ

The contribution of the second term of equation (21) to the value q_0 usually does not exceed 10-13%. Thus, the characteristic free energy of adsorption E is close to the differential heat of adsorption for the characteristic point.

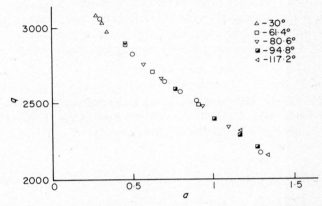

Fig. 6. Dependence of net differential heat of adsorption of methane on zeolite L on adsorption value. Circles denote experimental isosteric heats of adsorption (q, cal/mole, a, mmole/g).

By way of example, figure 6 compares the experimental isosteric differential heats of adsorption of methane on zeolite L (denoted by circles) with those calculated from equation (20) for different temperatures and adsorption values.

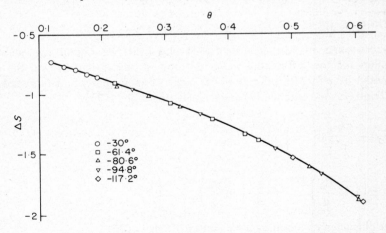

Fig. 7. Calculated dependence of differential molar entropy of adsorption of methane on zeolite L on filling. The different symbols denote the points corresponding experimental isotherms of adsorption for temperatures in the range from −117 to −30°C (ΔS, cal/mole, deg.).

The calculated and experimental results are in close agreement. Figure 7 shows the dependence of the differential molar entropy of adsorption ΔS on filling θ as calculated from equation (19). The points corresponding to experimental adsorption isotherm in the temperature range from -30 to $-117°C$ are denoted by different symbols. From equation (19) it follows that the dependence of ΔS on θ is independent of temperature.

Thus, the equations of the theory of volume filling of micropores which correspond to temperature invariance in the characteristic equation make it possible to describe adsorption equilibria over wide ranges of temperatures and pressures, and to calculate thermodynamic functions from the parameters of the thermal equation of adsorption for many adsorption systems of practical importance. The theory, however, is not universal and is restricted by the above-mentioned limits of applicability. In particular, its basic propositions cease to be sufficiently rigorous for filling below 0.1-0.2.

REFERENCES

Avgul, N. N., Aristov, B. G., Kiselev, A. V., Kurdjukova, L. N., and Frolova, N. V. (1968). *Zh. Fiz. Khim.* **42**, 2682.

Barrer, R. M. and Lee, J. A. (1968). *Surface Sci.* **12**, 341.

Bering, B. P. Dubinin, M. M. and Serpinsky, V. V. (1966). *J. Colloid Interface Sci.* **21**, 373.

Bering, B. P., Myers, A. and Serpinsky, V. V. (1970). *Dokl. Akad. Nauk SSSR* **193**, 119.

Bering, B. P., Gordeeva, V. A., Dubinin, M. M., Efimova, L. I. and Serpinsky, V. V. (1971). *Izv. Akad. Nauk SSSR, Ser. Khim.* **N1**, 22.

Dubinin, M. M. (1966). *In* "Chemistry and Physics of Carbon". (P. L. Walker, ed.) Vol. 2, pp. 54-120. Marcel Dekker, New York.

Dubinin, M. M. (1968). *Adv. Coll. Interface Sci.* **2**, 217.

Dubinin, M. M. and Astakhov, V. A. (1971). *In* "Molecular Sieve Zeolites-II", Amer. Chem. Soc., Washington.

Dubinin, M. M. and Plavnik, G. M. (1968). *Carbon* **6**, 183.

Dubinin, M. M. and Polstyanov, E. F. (1966). *Izv. Akad. Nauk SSSR, Ser. Khim.* **N4**, 610.

Dubinin, M. M., Plavnik, G. M., and Zaverina, E. D. (1964). *Carbon* **2**, 261.

Nikolaev, K. M. and Dubinin, M. M. (1958). *Izv. Akad. Nauk SSSR, Otd. Khim. Nauk* **N10**, 1165.

Polanyi, M. (1932). *Trans. Faraday Soc.* **28**, 316.

SUPPLEMENT TO M. M. DUBININ'S PAPER

In conclusion I should like to mention the application of a type (9) equation for the description of monomolecular adsorption of vapours on non-porous adsorbents; for instance on the surface of glass, in the region of very low equilibrium pressures. According to (9) and (3), at $n = 2$ we obtain the Dubinin-Radushkevich equation

$$a = a_0 \exp\left[-\left(\frac{A}{E}\right)^2\right] \tag{22}$$

in which, for the case of adsorption in micropores, the limiting adsorption value a_0 is a function of temperature and is expressed by equation (2). For the case of monomolecular adsorption, a_0 is assumed constant (Kaganer, 1957; Hobson and Armstrong, 1963; Ricca *et al.*, 1966) and, according to (1), we obtain $\alpha = 0$.

On the basis of direct experimental data on adsorption equilibria of nitrogen and argon on Pyrex glass in the low-temperature range (Hobson and Armstrong, 1963) in the region of adsorption values a ($\theta = 10^{-6} - 0.3$), equation (22) is in agreement with the experiment and variation in the differential molar entropy of adsorption $\Delta S = S_{ads} - S_{liq}$ within experimental error is equal to zero. In line with the rigorous thermodynamic expression (11), which is valid for any cases of equilibrium physical adsorption, at $\Delta S = 0$ and $\alpha = 0$ (a_0 = const) we set $(\partial A/\partial T)_\theta = 0$, i.e. temperature invariance of equation (22). Since $E = A$ at $\theta° = 1/e$, the parameter E in equation (22) is temperature invariant. According to (22)

$$A = E\sqrt{\ln a_0/a} \tag{23}$$

and for the case in hand we have, from (23), $\Delta S = (\partial A/\partial T)_a = 0$ because E and a_0 are constants. Note that the same conclusions follow from the more general form of equation (6), provided that E, a_0, and n are constants and φ is temperature invariant.

In general the limiting adsorption values a_0 are substantially lower than the "monolayer capacities" a_m and they correspond to adsorption on active centres of the adsorbent surface, with differential heats of adsorption exceeding the latent heats of condensation of adsorptives several times over. A type (9) equation has been applied (on condition that a_0 is equal to the number of adsorption centres) for adsorption of relatively small molecules on active centres (cations), of faujasite-type zeolites, in the initial region of coverages (Dubinin and Astakhov, 1971).

The case of applicability, discussed above, of a type (22) equation in the region of low coverages, indicates most clearly that the function $A = RT\ln p_s/p$ is not the adsorption potential as Polanyi (1932) stated. The conception that saturated pressure p_s exists over a separate molecule localized on the adsorbent's active centre has no physical meaning. This function is variation in Gibbs' free energy taken with a minus sign, and p_s corresponds to the selected standard reference state—the bulk liquid phase in equilibrium with its saturated vapour at the temperature under review.

ADDITIONAL REFERENCES

Hobson, J. P. and Armstrong, R. A. (1963). *J. Phys. Chem.* **67**, 2000.
Kaganer, M. G. (1957). *Dokl. Akad. Nauk SSSR* **116**, 251.
Ricca, F., Bellardo, A. and Medana, R. (1966). *Ric. Sci.* **36**, 460.

THE INFLUENCE OF ENERGETIC HETEROGENEITY ON THE ADSORPTION ISOTHERM AT LOW COVERAGES

B. A. GOTTWALD

Technische Universität Hannover, Germany

I. INTRODUCTION

The method at present most widely used for giving an analytic expression for adsorption isotherms at low coverages is the Dubinin-Kaganer-Radushkevich (DKR) isotherm equation. According to Dubinin and Radushkevich (1947) and Kaganer (1957) it has the form

$$n_\sigma = n_{\sigma m} \cdot \exp(-B \cdot \epsilon^2) \qquad (1)$$

n_σ is the number of molecules adsorbed per cm^2, $n_{\sigma m}$ is the number of atoms adsorbed per cm^2 in the DKR monolayer, B is a constant, and $\epsilon = -R \cdot T \cdot \ln p/p_0$ is the Polanyi potential. The DKR equation (1) is assumed to show a transition into the linear Henry adsorption isotherm where its slope in a double-logarithmic plot of amount-adsorbed-versus-pressure reaches unity, as quoted by Hobson and Armstrong (1963).

Another concept for describing physical adsorption is that of Ross and Olivier (1964). They form the adsorption isotherm on a heterogeneous surface by integration over the local isotherms on homotattic patches. Ross and Olivier (RO) present their isotherms in the form of tables. To use their method at very low surface coverages their isotherms had to be recalculated. Ross and Olivier by their choice of the upper and lower limit for integration have neglected about 0.4% of the surface. This might influence the isotherm significantly at very low surface coverages, especially the transition into Henry's law.

19

II. THE SLOPE α OF THE DOUBLE-LOGARITHMIC ADSORPTION ISOTHERM

Gottwald and Haul (1968) have shown that the discussion of the slope α of a double-logarithmic plot of the adsorption isotherm is a very useful approach to adsorption problems. The parameter

$$
\alpha = \left(\frac{\partial \ln n_\sigma}{\partial \ln p} \right)_T = \left(\frac{\partial \ln v}{\partial \ln p} \right)_T
$$

$$
= \left(\frac{\partial \ln \theta}{\partial \ln p} \right)_T = \left(\frac{\partial \ln \theta}{\partial \ln p/p_0} \right)_T \tag{2}
$$

is especially suited for characterizing adsorption isotherms. p is the equilibrium pressure, v is the amount adsorbed in moles per gram adsorbent, θ is the surface

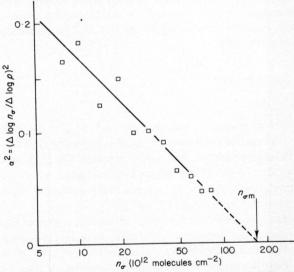

Fig. 1. Comparison of equation (3) with the experimental data of Troy and Wightman (1970) for nitrogen adsorbed on stainless steel at 77°K.

coverage, and p_0 is the saturation vapor pressure of the adsorptive. The parameter α is independent of the choice of the units for the amount adsorbed and the pressure. Plots of the slope α versus the amount adsorbed have the advantage that they are independent of the right choice of the vapor pressure at low temperatures, which might be that of either the solid or the liquid adsorptive. This problem is discussed in detail by Ricca *et al.* (1967).

For the DKR isotherm equation (1) follows by application of (2) the equation

$$\alpha^2 = -4 \cdot B \cdot (R \cdot T)^2 \cdot \ln \theta \tag{3}$$

as shown by Gottwald and Haul (1968). That means that for adsorption isotherms, following the DKR equation, a plot of α^2 versus the logarithm of the amount adsorbed must give a straight line with an intercept on the abscissa axis at the monolayer capacity. This is shown in figure 1 for the data of Troy and Wightman (1970) on the adsorption of nitrogen on stainless steel at $77°$K.

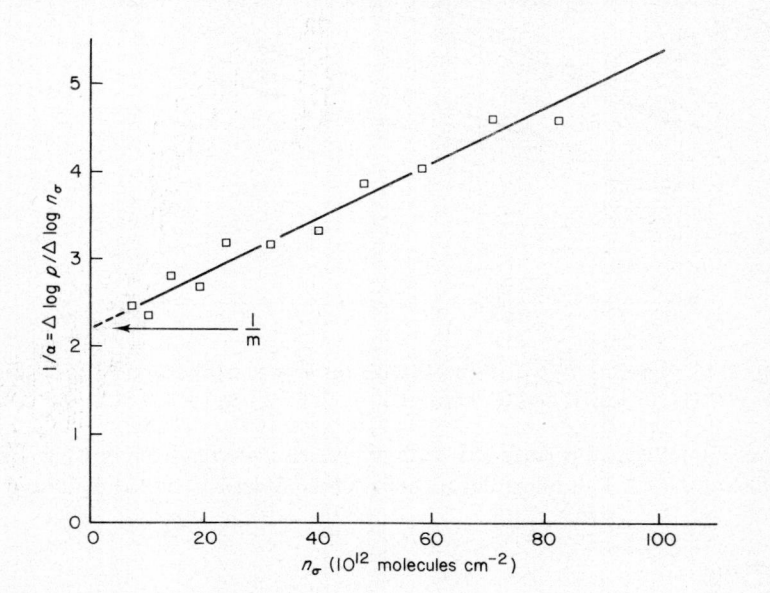

Fig. 2. Comparison of equation (4) with the experimental data of Troy and Wightman (1970) for nitrogen adsorbed on stainless steel at $77°$K.

III. MODIFIED VIRIAL ADSORPTION ISOTHERM

On the other hand many experimental adsorption isotherms at low surface coverages show a linear dependence of $1/\alpha = d \ln p/d \ln n_\sigma$ versus amount adsorbed. This is shown in figure 2 for the data of Troy and Wightman (1970) on the adsorption of nitrogen on stainless steel at $77°$K. The data of Hobson and Armstrong (1963) on the adsorption of nitrogen and argon on pyrex glass may be plotted in a similar way In all these cases the extrapolated intercept on the ordinate axis is not equal to unity, as would be expected for the onset of Henry's law. Even the RO model isotherms show a linear dependence of $1/\alpha$ versus surface coverage θ for low values of the heterogeneity parameter γ (i.e.

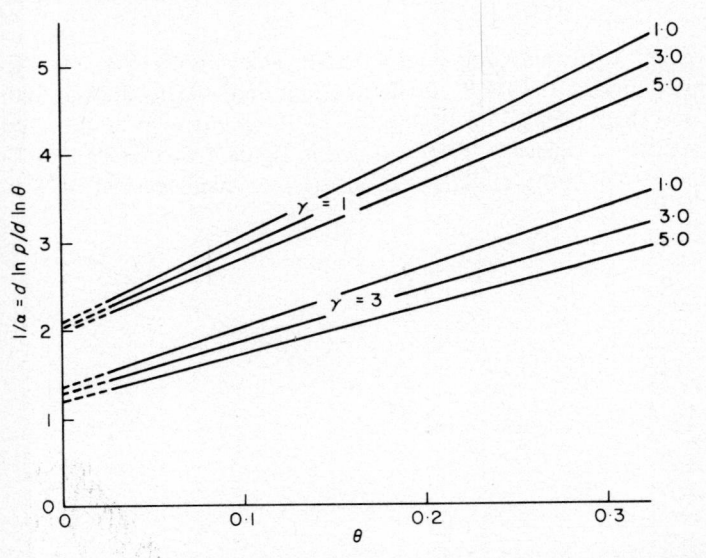

Fig. 3. Comparison of equation (4) with the model isotherm data of Ross and Olivier (1964) for low values of γ with $(2 \cdot a_2)/(k \cdot T \cdot b_2) = 1.0, 3.0,$ and 5.0.

relatively heterogeneous surfaces), with intercepts differing from unity as can be seen from figure 3. This linear dependence can be described by the equation

$$\frac{1}{\alpha} = \frac{1}{m} + C \cdot n_\sigma \tag{4}$$

$1/m$ is the intercept on the ordinate axis and C is a constant. With

$$\frac{1}{\alpha} = \frac{d\ln p}{d\ln n_\sigma} = n_\sigma \cdot \frac{d\ln p}{dn_\sigma} \tag{5}$$

the differential form of the Gibbs adsorption isotherm

$$d\pi = k \cdot T \cdot n_\sigma \cdot d\ln p \tag{6}$$

takes the form

$$\pi = k \cdot T \cdot \int_0^{n_\sigma} \frac{1}{\alpha} \cdot dn_\sigma \tag{7}$$

π is the two-dimensional spreading pressure. By comparing (7) with the most frequently used expression for the integrated form of the Gibbs adsorption isotherm

$$\pi = k \cdot T \cdot \int_0^p n_\sigma \cdot d\ln p \tag{8}$$

it is obvious that the integration (7) is quite useful: The integration of n_δ over the logarithm of the pressure is replaced by an integration of $1/\alpha$ over the amount adsorbed. From figure 2 and figure 3 it may be seen that this is especially advantageous at low surface coverages. Applying the integration (7) on (4) results in

$$\pi = \frac{k \cdot T}{m} \cdot n_\delta + k \cdot T \cdot C \cdot \frac{1}{2} \cdot n^2{}_\delta \tag{9}$$

By rearranging and making use of $n_\sigma \cdot \sigma = 1$ we get

$$\pi \cdot \sigma = \frac{k \cdot T}{m} + k \cdot T \cdot C \cdot \frac{1}{2} \cdot \frac{1}{\sigma} \tag{10}$$

σ is the area per molecule. For small values of π and large values of σ equation (10) may be replaced by

$$\pi \cdot \delta = \frac{k \cdot T}{m} + m \cdot \frac{1}{2} \cdot C \cdot \pi \tag{11}$$

With the definition of the two-dimensional (distinguished by the subscript 2 from the corresponding three-dimensional variable) virial coefficient

$$B_2 = \lim_{\pi \to 0} \frac{d(\pi \cdot \sigma)}{d\pi} \tag{12}$$

we get $\qquad \pi \cdot \sigma = k \cdot T \cdot \dfrac{1}{m} + B_2 \cdot \pi \tag{13}$

and thus $\qquad C = \dfrac{2}{m} \cdot B_2 \tag{14}$

Introducing this into equation (4) gives

$$\frac{1}{\alpha} = \frac{1}{m} \cdot (1 + 2 \cdot B_2 \cdot n_\sigma) \tag{15}$$

Upon integration we get the corresponding adsorption isotherm equation

$$p^m = \frac{K}{n_{\sigma m}} \cdot n_\sigma \cdot \exp(2 \cdot B_2 \cdot n_\sigma) \tag{16}$$

$K/n_{\sigma m}$ is the appropriate form of the integration constant.

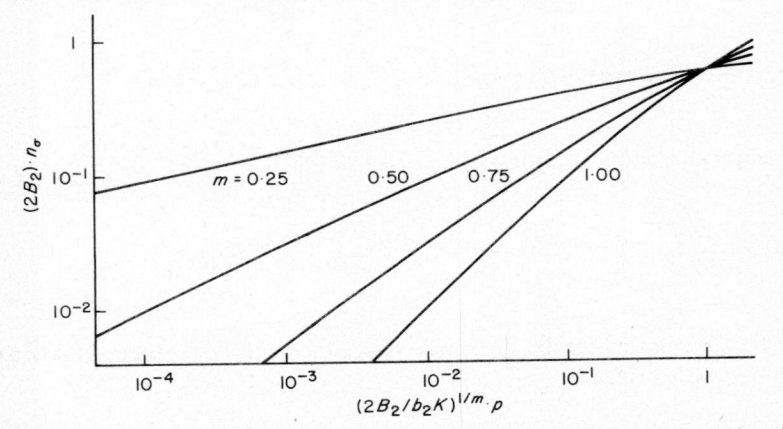

Fig. 4. Adsorption isotherms calculated from the modified viral equation (16) for temperatures above the two-dimensional Boyle temperature ($B_2 > 0$).

The modified virial adsorption isotherm equation (16) is plotted in a reduced form in figure 4 for $B_2 > 0$. For $B_2 > 0$ (i.e. above the two-dimensional Boyle temperature) the adsorption isotherms are concave versus the pressure axis in a double-logarithmic plot. A shape similar to figure 4 is found experimentally quite often. This is reflected even by the fact that many authors describe their experimental findings by the DKR equation, which necessarily must be concave versus the pressure axis in double-logarithmic plots.

The modified virial adsorption isotherm equation (16) is plotted for $B_2 < 0$ (i.e. below the two-dimensional Boyle temperature) similar to figure 4 in figure 5. For $B_2 < 0$ the adsorption isotherms are convex versus the pressure axis in a double-logarithmic plot. Some experimental isotherms on heterogeneous surfaces as, for example, those of Baker and Fox (1965) for the adsorption of xenon on a nickel film show a shape similar to figure 5. Such isotherms cannot be described by the DKR equation.

For experimentally observed isotherm data appropriate values of m and B_2 may be evaluated in two ways: (i) by plotting $1/\alpha$ vs. n_σ and analyzing intercept and slope according to equation (15), or (ii) by comparing a double-logarithmic plot of the isotherm data with a plot analogous to figure 4 or figure 5, respectively: $2 \cdot B_2$ appears as a scale factor for the ordinate axis. $^m\sqrt{(2 \cdot B_2)/(b_2 \cdot K)}$ is a scale factor for the pressure axis. Both parameters, as well as the parameter m are determined, when the experimental and computed isotherms are brought into coincidence. This method for obtaining the appro-

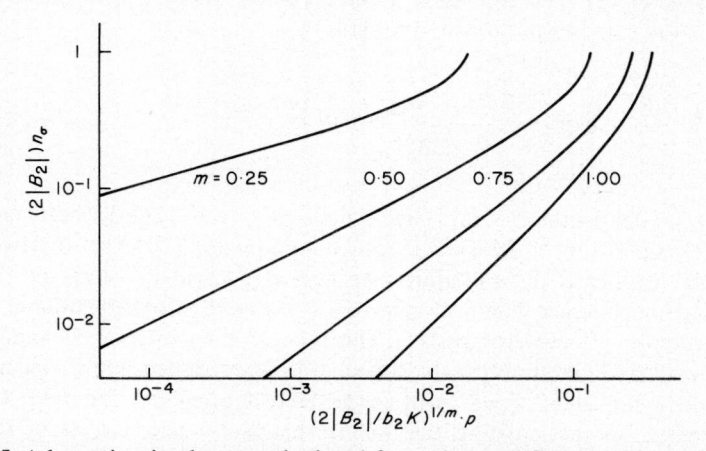

Fig. 5. Adsorption isotherms calculated from the modified virial equation (16) for temperatures below the two-dimensional Boyle temperature ($B_2 < 0$).

priate values of the parameters is similar to that proposed by Ross and Olivier (1964). The temperature dependence of the initial slope m and the two-dimensional virial coefficient B_2 is discussed below.

IV. MODIFIED HILL-deBOER ADSORPTION ISOTHERM: TWO-DIMENSIONAL CRITICAL TEMPERATURE AND TWO-DIMENSIONAL BOYLE TEMPERATURE

On comparing the equation of state (13) with that generated from the equation of state of a completely mobile two-dimensional gas

$$\left(\pi + \frac{a_2}{\sigma^2}\right) \cdot (\sigma - b_2) = k \cdot T \tag{17}$$

for low pressures

$$\pi \cdot \sigma = k \cdot T + B_2 \cdot \pi \tag{18}$$

the only modification is the introduction of the parameter m. Therefore it is obvious to modify even equation (17) by introducing m on the right side and to convert the resulting equation of state

$$\left(\pi + \frac{a_2}{\sigma^2}\right) \cdot (\sigma - b_2) = \frac{k \cdot T}{m} \tag{19}$$

into the corresponding adsorption isotherm

$$p^m = K \cdot \frac{\theta}{1 - \theta} \cdot \exp\left(\frac{\theta}{1 - \theta} - \frac{2a_2}{kTb_2} \cdot m \cdot \theta\right) \tag{20}$$

The RO model isotherms for large values of γ (i.e. slightly heterogeneous surfaces) qualitatively show a shape similar to equation (20). Quantitatively the RO model isotherms do not show a vertical step even for values of $(2 \cdot a_2)/(k \cdot T \cdot b_2)$ much larger than 6.75: very much below the two-dimensional critical temperature for a homotattic surface. This is because the integral over an infinite sum of small step functions (i.e. local isotherms with condensation), occurring at slightly different values of p/K', will not generate another step function. This has been verified by recalculating the RO isotherms for large values of $(2 \cdot a_2)/(k \cdot T \cdot b_2)$ with very small intervals in p/K'. Following these arguments a real two-dimensional condensation must not occur on heterogeneous surfaces.

Nevertheless, an apparent two-dimensional critical temperature frequently is calculated in the following way: The Hill-deBoer adsorption isotherm equation

$$p = K \cdot \frac{\theta}{1 - \theta} \cdot \exp\left(\frac{\theta}{1 - \theta} - \frac{2a_2}{kTb_2} \cdot \theta\right) \tag{21}$$

is rearranged as

$$W = \ln \frac{\theta}{1 - \theta} + \frac{\theta}{1 - \theta} - \ln p = \frac{2a_2}{kTb_2} \cdot \theta - \ln K \tag{22}$$

An apparent value of $2 \cdot a_2/b_2$ is then derived from

$$\left(\frac{2a_2}{b_2}\right)_{app} = k \cdot T \cdot \frac{dW}{d\theta} \tag{23}$$

On heterogeneous surfaces a plot of W vs. θ is not linear over the entire range $0 \leqslant \theta \leqslant 1$. In this case the slope of the approximately straight portion in the neighbourhood of $\theta = 0.33$ is given by

$$\frac{dW}{d\theta} = \frac{d}{d\theta}\left\{\ln\frac{\theta}{1-\theta} + \frac{\theta}{1-\theta}\right\} - \frac{d\ln p}{d\theta} \approx 6.75 - \frac{d\ln p}{d\theta} \tag{24}$$

Therefore it is convenient to replace equation (23) by

$$\left(\frac{2a_2}{b_2}\right)_{\text{app}} = kT\cdot\left(6.75 - \frac{d\ln p}{d\theta}\bigg|_{\theta=0.33}\right) \tag{25}$$

Applying (25) to the modified Hill-deBoer isotherm equation (20) results in

$$\left(\frac{2a_2}{b_2}\right)_{\text{app}} = \left(\frac{2a_2}{b_2}\right)_{\text{ideal}} - k\cdot T\cdot\left(\frac{1}{m} - 1\right)\cdot 6.75 \tag{26}$$

The apparent value of $2\cdot a_2/b_2$ thus is lower than the ideal value for $0 < m < 1$. The relation between the apparent two-dimensional critical temperature and its ideal value is similar to (26).

Another interesting subject is the two-dimensional virial coefficient. Its ideal value can be derived from the two-dimensional van der Waals equation (17)

$$B_2 = b_2 - \frac{a_2}{k\cdot T} \tag{27}$$

This gives a two-dimensional Boyle temperature ($B_2 = 0$) of

$$(T_{2\text{ Boyle}})_{\text{ideal}} = \frac{a_2}{k\cdot b_2} \tag{28}$$

Similarly the modified van der Waals equation (19) for the heterogeneous surface gives a two-dimensional virial coefficient of

$$B_2 = b_2 - m\cdot\frac{a_2}{k\cdot T} \tag{29}$$

As m is dependent upon temperature the two-dimensional Boyle temperature cannot be derived from (29) without further assumptions.

From the definition (2) of the parameter α follows

$$\left(\frac{\partial 1/\alpha}{\partial 1/RT}\right)_\theta = \frac{\partial}{\partial 1/RT}\left(\frac{\partial \ln p}{\partial \ln \theta}\right)_T = \frac{\partial}{\partial \ln \theta}\left(\frac{\partial \ln p}{\partial 1/RT}\right)_\theta$$

$$\left(\frac{\partial 1/\alpha}{\partial 1/RT}\right)_\theta = -\frac{\partial}{\partial \ln \theta} q_{st} \tag{30}$$

q_{st} is the isosteric heat of adsorption. According to equation (30) the dependence of q_{st} upon surface coverage determines the dependence of $1/\alpha$ upon temperature. For low coverages ($\alpha \approx m$) equation (30) may be replaced by

$$\frac{d 1/m}{d 1/RT} = -\frac{dq_{st}}{d\ln \theta} \tag{31}$$

Experimentally it is found quite often that the relation between the isosteric heat of adsorption q_{st} and $\ln n_\sigma$ is linear for low surface coverages. Thus equation (31) takes the form

$$\frac{d 1/m}{d 1/RT} = E \tag{32}$$

Upon integration

$$\frac{1}{m} = \frac{E}{RT} + 1 \tag{33}$$

where the integration constant is chosen equal to 1, so that for high temperatures Henry's law ($m = 1$) is reached. Introducing (33) into (26) gives

$$\left(\frac{2a_2}{b_2}\right)_{app} = \left(\frac{2a_2}{b_2}\right)_{ideal} - E\cdot 6.75 \tag{34}$$

and similarly for the two-dimensional critical temperature

$$(T_{2\,crit})_{app} = (T_{2\,crit})_{ideal} - E/k \tag{35}$$

This is similar to the linear relation of figure V-4 of Ross and Olivier (1964). Equation (35) makes it possible to relate the energy parameter E to the width of the distribution function.

Introducing (33) into (29) gives for the two-dimensional Boyle temperature on a heterogeneous surface finally the expression

$$T_{2\,Boyle} = \frac{a_2/b_2}{k} \cdot \left(1 - \frac{E}{a_2/b_2}\right) \tag{36}$$

A three-dimensional van der Waals equation similar to (19) has been suggested by Haward (1966) for liquids: By introducing an interaction coefficient B into the three-dimensional van der Waals equation, $R \cdot T$ is replaced by $B \cdot R \cdot T$. From the values of $B \cdot R$ given for some liquids in Table 1 of Haward (1966) the following values of $1/B$ can be calculated: 0.75 for argon, 0.52 for oxygen, 0.58 for nitrogen and 0.47 for methane. These values of $1/B$ thus correspond to the values of m, calculated from adsorption isotherms according to equation (16). From this point of view the energy parameter E is determined by the temperature dependence of the interaction constant: i.e. by the temperature dependence of the accessible degrees of freedom.

V. ISOSTERIC HEAT OF ADSORPTION

For comparing the proposed isotherm equations (16) and (20) with experimental data even the variation of q_{st} with θ in the intermediate surface coverage region is to be considered. By application of the definition of the isosteric heat of adsorption

$$q_{st} = -\left(\frac{\partial \ln p}{\partial 1/RT}\right)_{n_\sigma} \tag{37}$$

on the modified Hill-deBoer equation (20) follows

$$q_{st} = -\frac{d\,1/m}{d\,1/RT} \cdot \ln K - \frac{1}{m} \cdot \frac{d\ln K}{d\,1/RT}$$

$$-\frac{d\,1/m}{d\,1/RT} \cdot \left\{\ln \frac{\theta}{1-\theta} + \frac{\theta}{1-\theta}\right\} + \frac{2a_2}{b_2} \cdot \theta \tag{38}$$

The coverage dependent contribution to q_{st} is plotted in figure 6 as a function of surface coverage for several values of $E/(2 \cdot a_2/b_2)$. For $0 \leqslant E/(2 \cdot a_2/b_2) \leqslant$ 0.148 the isosteric heat of adsorption passes through a minimum value, with increasing θ until finally going through a maximum. This shape of q_{st} vs. surface coverage is frequently observed on not too heterogeneous surfaces. These isosteric heat curves have a point of common intersection at $\theta = 0.36$ with a slope of

$$\frac{dq_{st}}{d\theta}\bigg|_{\theta=0\cdot36} = \frac{2a_2}{b_2} - E\cdot6.78 \tag{39}$$

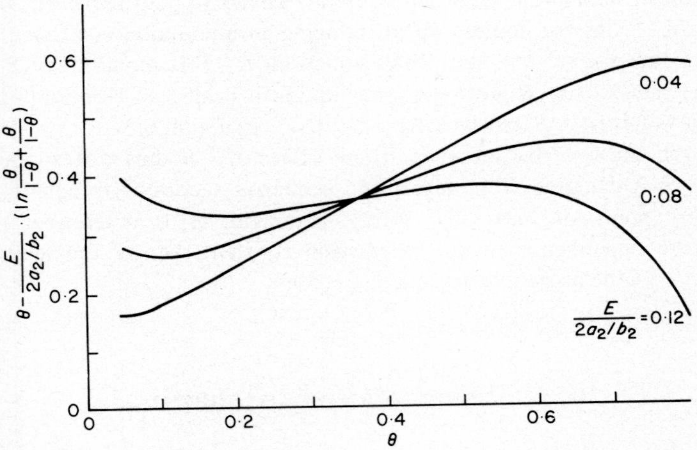

Fig. 6. Isosteric heat curves computed from equation (39) for typical values of the parameter $E/(2 \cdot a_2/b_2)$.

VI. APPLICATION TO EXPERIMENTAL DATA

Application of the modified virial isotherm equation (16) to the data of Hobson and Armstrong (1963) on the adsorption of nitrogen on pyrex glass results in values of 0.25 to 0.50 for m in the temperature range between 63.3°K and 90.2°K. For all temperatures the minimum sum of squares of the residuals

$$Q = \sum_i [\ln p_i - \ln p(\theta_i)]^2 \tag{40}$$

is considerably smaller if the experimental data are described by the modified virial isotherm than for the DKR equation, as shown in table 1. This means that the modified virial adsorption isotherm is a better approximation to the experimental data than the DKR equation. Inspection of the temperature dependence of the m-values suggests that the choice of the integration constant equal to 1 in equation (33) might be too stringent. Unfortunately the uncertainty in the m-values caused by the experimental scatter is too large for calculating a suitable value for E. The values of B_2 are found in the range 20 to

30 $Å^2$/molecule. Even the uncertainty in the B_2-values is too large for evaluating the two-dimensional van der Waals constants a_2 and b_2 by making use of equation (29).

The xenon adsorption isotherms on graphitized carbon of Cochrane et al. (1967) give variations of the isosteric heat of adsorption which are quite similar to figure 6. From figure 7 of Cochrane et al. (1967) a value of $dq_{st}/d\theta$ = 1.37 kcal/mole for xenon on Sterling MT 3100 is derived for the intermediate part of the curve. By use of (39) this results in a value of E = 84 cal/mol. $(2 \cdot a_2/b_2)_{ideal}$ for xenon is assumed to be equal to 1.94 kcal/mole, as

Table 1. *Minimum sum of squares of the residuals*

Dubinin-Kaganer-Radushkevich equation:

$$Q = \sum_i \left[\ln p_i - \ln p_0 + \sqrt{\frac{\ln n_{\sigma m} - \ln n_{\sigma i}}{B \cdot (R \cdot T)^2}} \right]^2$$

Modified virial isotherm equation:

$$Q = \sum_i \left[\ln p_i - \frac{1}{m} \cdot \left(\ln \frac{K}{n_{\sigma m}} + \ln n_{\sigma i} + 2 \cdot B_2 \cdot n_{\sigma i} \right) \right]^2$$

Hobson and Armstrong (1963): N_2/pyrex glass

$T[°K]$	63.1	67.1	72.6	77.4	81.0	84.9	90.2
DKR : Q_{min} =	0.881	0.426	0.315	0.893	0.145	0.904	1.138
Mod. vir. : Q_{min} =	0.637	0.304	0.282	0.703	0.112	0.236	0.200

calculated from the three-dimensional van der Waals constants according to Hill (1946). The corresponding value for xenon on Graphon from figure 8 of Cochrane et al. (1967) is E = 149 cal/mole with $dq_{st}/d\theta$ = 0.93 kcal/mole. The lower value of E on Sterling MT 3100 than on Graphon corresponds to the smaller amount of heterogeneity in the former. Calculating E according to equation (34) from the values for $(2 \cdot a_2/b_2)_{app}$, as given in table 3 of Cochrane et al. (1967), results in somewhat lower values for both adsorbents. This again supports the assumption that equation (33) might be too stringent.

A more detailed analysis of experimental data by the present theory will be published elsewhere.

VII. CONCLUSION

An analytic expression for physical adsorption isotherms on heterogeneous surfaces is proposed, taking into account that many experimental adsorption isotherms do not show a transition into the linear Henry isotherm at low coverages. A modified virial adsorption isotherm equation is derived from the linear dependence of $1/\alpha$ (α: slope of double-logarithmic adsorption isotherm) upon surface coverage. Similarly a modified Hill-de Boer adsorption isotherm equation is proposed and discussed with respect to the two-dimensional critical temperature and the two-dimensional Boyle temperature. A suitable energy parameter is introduced into the corresponding equations of state. This energy parameter stands for the activation energy for the surface diffusion or for the width of the distribution function for the adsorption energies, which are related to each other. Finally, the proposed modified adsorption isotherm equations are compared with the experimental data of Hobson and Armstrong (1963) and Cochrane et al. (1967).

ACKNOWLEDGEMENT

The author is indebted to Prof. Dr. R. Haul for stimulating discussions and encouraging support during the course of this work.

REFERENCES

Baker, B. G. and Fox, P. G. (1965). *Trans. Faraday Soc.* **61**, 2001.
Cochrane, H., Walker, P. L. Jr., Diethorn, W. S., and Friedman, H. C. (1967). *J. Coll. Interface Sci.* **24**, 405.
Dubinin, M. M. and Radushkevich, L. V. (1947). *Proc. Acad. Sci. USSR, Phys. Chem. Sect.* **55**, 327.
Gottwald, B. A. and Haul, R. (1968). *Proc. 4th Int. Vac. Congr. Manchester* 96.
Haward, R. N. (1966). *Trans. Faraday Soc.* **62**, 828.
Hill, T. L. (1946). *J. Chem. Phys.* **14**, 441.
Hobson, J. P. and Armstrong, R. A. (1963). *J. Phys. Chem.* **67**, 2000.
Kaganer, M. G. (1957). *Proc. Acad. Sci. USSR, Phys. Chem. Sect.* **116**, 603.
Ricca, F., Medana, R., and Bellardo, A. (1967). *Z. Phys. Chemie NF* **52**, 276.
Ross, S. and Olivier, J. P. (1964). "On Physical Adsorption". Interscience, New York.
Troy, M. and Wightman, J. P. (1970). *J. Vac. Sci. Technol.* **7**, 429.

THE ONSET OF HENRY'S LAW FOR PHYSICAL ADSORPTION OF A VAPOR ON A HETEROGENEOUS SURFACE

J. P. HOBSON and R. CHAPMAN

National Research Council of Canada, Ottawa, Canada

I. INTRODUCTION

Experimental data obtained during the last decade in many laboratories has demonstrated that the physical adsorption isotherm of a vapor on a heterogeneous surface covers many orders of magnitude in pressure and coverage, between the expected limits of Henry's Law at low pressures and the vapor pressure of the adsorbate at high pressures. The span between these limits for adsorption of a vapor on a homogeneous surface, in orders of magnitude of pressure and coverage, is roughly half of that for a heterogeneous surface, as shown by the results of Cochrane *et al.* (1967) for the adsorption of xenon on graphitized carbon blacks. Recently Hobson (1969) measured the adsorption isotherms of argon, krypton, and xenon on the heterogeneous adsorbent porous silver at $T = 77.4°K$ over a pressure range from vapor pressure to ultrahigh vacuum and did not find Henry's Law at the lowest pressures measured. In particular the measurements for argon covered the range from $p_0 = 205$ Torr (vapor pressure) to a pressure above the adsorbed layer of $p = 10^{-9}$ Torr. A proposed isotherm equation gave good agreement with all measured data and predicted the onset of Henry's Law for $T = 77.4°K$ at $p = 10^{-14}$ Torr, which was a pressure too low for practical measurements. However, for $T = 110°K$ the predicted onset for Henry's Law was $p = 10^{-4}$ Torr, which was well within the range of practical measurement. It was decided to calculate isotherms at various temperatures from the proposed equation, and to calculate as well the simpler thermodynamic quantities (isosteric heat and differential molar entropy), and simultaneously to perform experiments to test the calculations. A report on these two objectives represents the purpose of the present paper.

Schram (1967a, 1967b) has reported considerations not unlike those described below, but Schram did not explicitly calculate isotherms, and his experimental system (argon on nickel) was subject to large corrections for adsorption on the Pyrex walls. These corrections became worse as the region of

Henry's Law was approached. Further Schram's adsorption vessel was not isothermal, and the effect of this is not known.

It is emphasized that at the present stage of studies like those of Schram and of the present authors, the primary objective is not the study of the specific adsorbate-adsorbent system, but rather a study of the form of the isotherm as it approaches the low pressure limit.

II. CALCULATIONS BASED ON THE ISOTHERM EQUATION

The isotherm equation proposed by Hobson (1969) for argon on porous silver was:

$$N = N_m \left[1 + \frac{E_3}{\epsilon} \right] \exp \left(- \frac{\epsilon^2}{2E_1{}^2} \right) \text{ for } \epsilon < \frac{E_1{}^2}{RT} \tag{1}$$

$$N = N_m \left[1 + \frac{E_3}{\epsilon} \right] \exp \left(- \frac{\epsilon}{RT} + \frac{E_1{}^2}{2R^2T^2} \right) \quad \text{for } \epsilon > \frac{E_1{}^2}{RT} \tag{2}$$

where N = number of atoms adsorbed, N_m = number of atoms adsorbed in a Dubinin-Radushkevich (DR) monolayer, $\epsilon = RT \ln p/p_0$, p = pressure above adsorbed layer, p_0 = vapor pressure of adsorbate. For the system argon-porous silver the following constants were found: $N_m = 1.9 \times 10^{19}$, $E_1 = 933$ cal mole^{-1}, $E_3 = 882$ cal mole^{-1}. A constant E_2 which appeared in the isotherm equations for krypton and xenon on porous silver was found to be zero for argon and does not appear in the above equation.

Equations (1) and (2) describe the entire isotherm from vapor pressure ($\epsilon = 0$) to indefinitely low pressures ($\epsilon \to \infty$). Figure 1 shows calculated graphs of these equations for five temperatures $T = 77.4, 85, 90, 100$ and $110°$K. Experimental data obtained by Hobson (1969) at $T = 77.4°$K are shown as circles. The equation clearly gives results in general agreement with experiment over the whole experimental range at $T = 77.4°$K. Discussion was given previously of the relationship between the BET and DR monolayers and nothing further will be added here. The DR monolayer will be used below to define the relative coverage θ

$$\theta = \frac{N}{N_m} \tag{3}$$

Equations (1) and (2) are characterized by three regions at any given temperature

(a) A multilayer region at high pressures limiting at the bulk condensate ($\epsilon = 0$ or $p/p_0 = 1$).

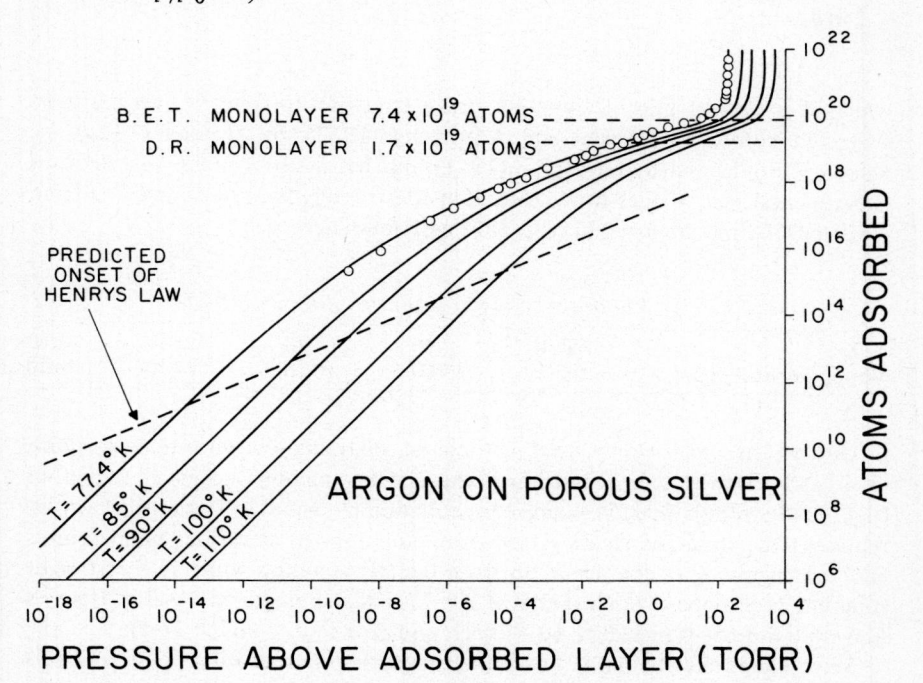

Fig. 1. Calculated graphs of equations (1) and (2) with $\log_{10} p$ as abscissa for constants $N_m = 1.7 \times 10^{19}$, $E_1 = 933$ cal mole^{-1}, $E_3 = 882$ cal mole^{-1}. Values of p_0 from Weast (1970) (table 1). Circles are experimental points of Hobson (1969).

(b) A submonolayer region characterized by the Dubinin-Radushkevich (1947) isotherm equation, often written

$$\ln \theta = -B\epsilon^2 \tag{4}$$

with $B = \dfrac{1}{2E_1^{\,2}}$ \hfill (5)

(c) A Henry's Law region with onset marked by

$$\epsilon = \frac{E_1^{\,2}}{RT} = \frac{1}{2BRT} \tag{6}$$

The onset of Henry's Law described by equation (6) is not a function of θ alone, but is described to a good approximation in figure 1 by the relation

$$2 \ln \theta = \ln \frac{p}{p_0} \tag{7}$$

which is a straight line of slope $\frac{1}{2}$. Such a transition to Henry's Law has been suggested before by Hobson and Armstrong (1963), by Hobson (1966), by Schram (1967b) and by Cerofolini (1971), but has never been put to a rigorous experimental test. Values of p_0 for use in constructing figure 1 were taken from Weast (1970) and are listed in table I for easy reference.

Table 1. *Vapor pressures of argon from Weast (1970)*

T°K	77.4	85	90	100	110
Vapor pressure Torr	205	540	1000	2400	4800

One of the most widely used methods of analysing low pressure adsorption data has been to plot the logarithm of the adsorbed amount against ϵ^2 (equation (4)). Such a plot is given in figure 2 for equation (1) and (2). The vertical arrows represent the onset of Henry's Law from equation (6) at various temperatures and the beginning of the deviation from the DR equation which is the straight solid line with slope $-B$. Note that at $T = 77.4^\circ$K the predicted onset of Henry's Law does not occur until $\theta \sim 10^{-8}$ (corresponding to $p \sim 10^{-14}$ Torr).

Since equations (1) and (2) describe the isotherm at all values of θ, p, T it is possible to calculate the isosteric heat (fig. 3) and the differential molar entropy (fig. 4) as a function of θ. The equations used for calculating figures 3 and 4 were

$$q_{st} = -R \left(\frac{\partial \ln p}{\partial \frac{1}{T}} \right)_N \tag{8}$$

$$T(\bar{S}_s - \bar{S}_L) = -RT \ln \frac{p}{p_0} - q_{st} + q_1 \tag{9}$$

where q_1 is the heat of vaporization of liquefaction of the bulk adsorbate and has been taken as the heat of liquefaction (1558 cal mole^{-1}) throughout the present calculations.

In keeping with the result that the onset of Henry's Law is not a function of θ alone in the Henry's Law range, q_{st} depends upon T, increasing as T becomes smaller. The envelope of q_{st} in the intermediate range of figure 3 is

$$q_{st} = q_1 + B^{-1/2} \left(\ln \frac{1}{\theta} \right)^{1/2} \tag{10}$$

a result which is readily derived from equation (4). Experimental results qualitatively similar to the calculated results of figure 4 have been reported by Schram (1967a).

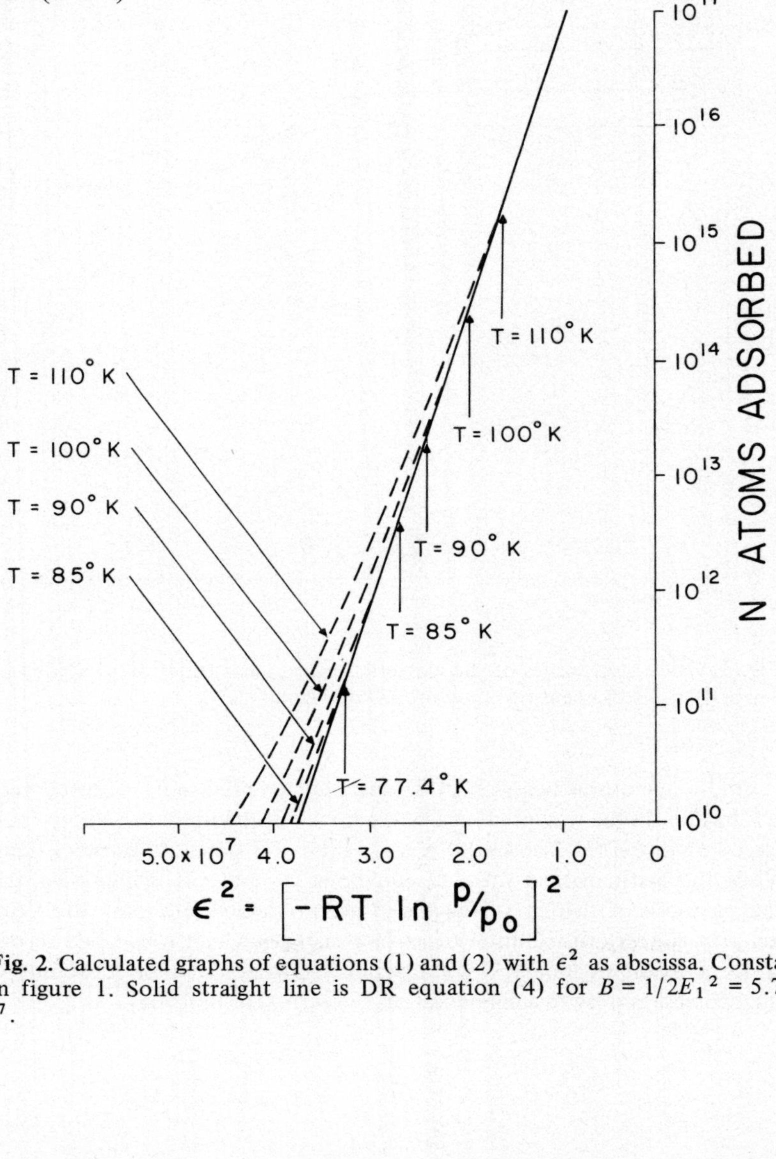

Fig. 2. Calculated graphs of equations (1) and (2) with ϵ^2 as abscissa. Constants as in figure 1. Solid straight line is DR equation (4) for $B = 1/2E_1{}^2 = 5.74 \times 10^{-7}$.

Similarly the differential molar entropy in figure 4 remains at zero until the onset of Henry's Law at which time it soon establishes a slope the same as that of the BET model but displaced in coverage by an amount depending on the temperature. A very small differential molar entropy was reported experimentally by Hobson and Armstrong (1963) for argon on Pyrex for $10^{-5} < \theta <$

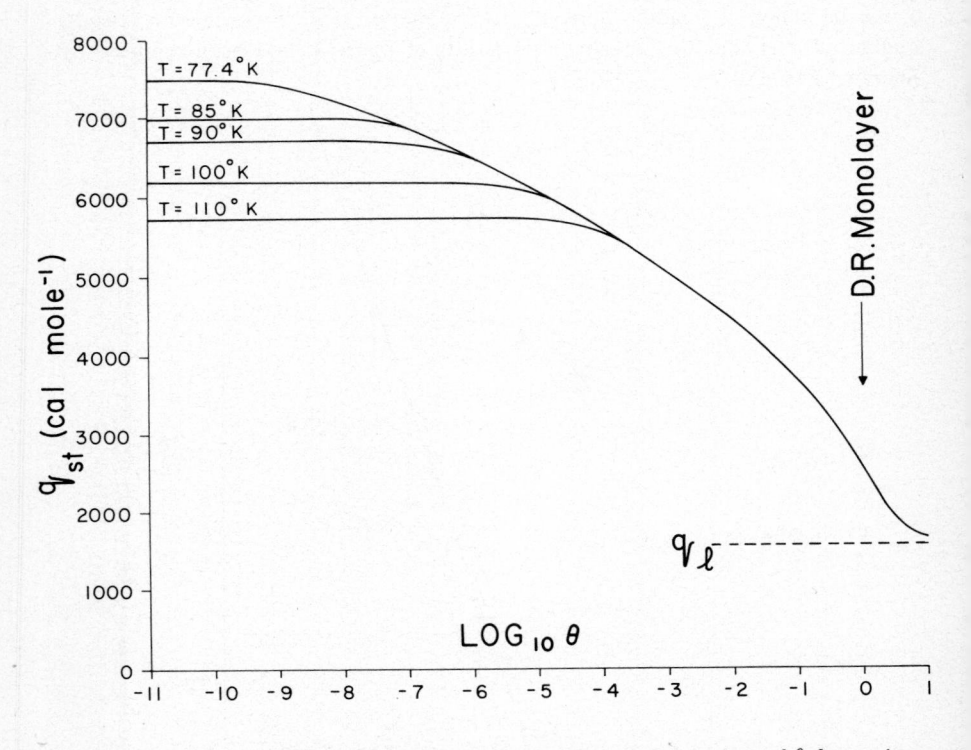

Fig. 3. Calculated values of the isosteric heat q_{st} as a function of θ for various values of T. q_1 is the heat of liquefaction of argon.

10^{-1} for temperatures below 77.4°K, and a differential molar entropy varying with T but not in the way predicted by figure 4 was reported by Schram (1967a) for argon on nickel for T above 80°K and values of θ similar to those of figure 4.

While the mathematical form of equations (1) and (2) is empirical, results similar to those of figures 1-4 have also been obtained (Hobson 1966) from a model postulating a distribution of surface energies which were filled according to a local isotherm similar to a Langmuir isotherm. Other local isotherms calculated were similar to a condensation isotherm (Harris 1968).

III. EXPERIMENTAL APPARATUS AND PROCEDURE

The experimental procedure was in general similar to that used earlier by Hobson (1969). Several improvements in the ultrahigh vacuum apparatus were incorporated in addition to the use of a variable temperature dewar. A cooled radiation shield was added to the pot containing the porous silver to prevent

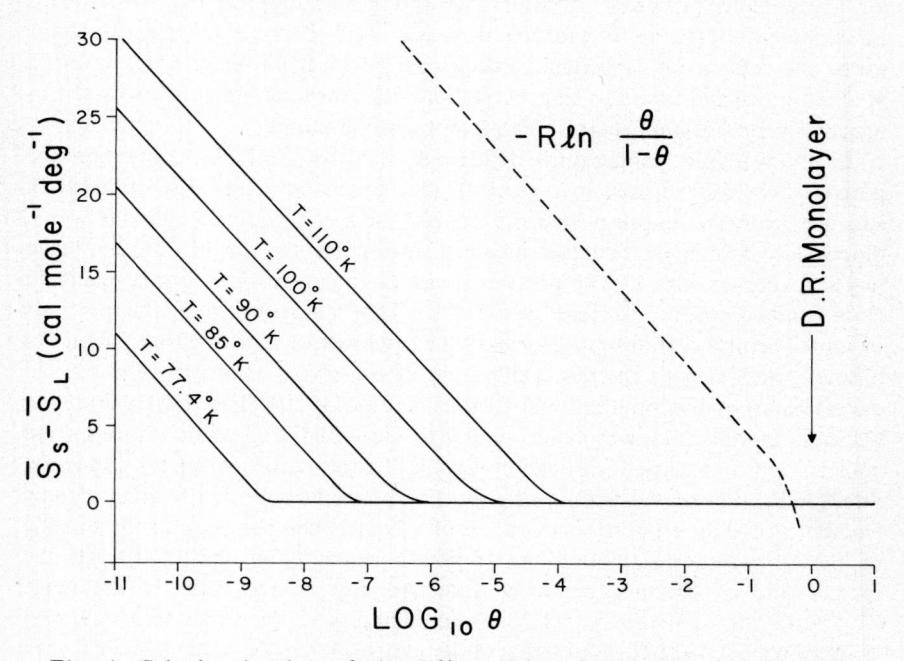

Fig. 4. Calculated value of the differential molar entropy (expressed as a difference between that of the adsorbed phase $\bar{S}_S$ and that of the Liquid $\bar{S}_L$) as a function of θ for various values of T. Shown dashed is value of $\bar{S}_S - \bar{S}_L$ for the BET model.

possible errors in adsorbent temperature caused by radiation entering through the pipe to the adsorbent. The results at $T = 77.4°K$ were in close agreement with those obtained previously and it is thought that radiation errors were negligible in both experiments. Argon pressures were measured with a mass spectrometer (90° Magnetic Sector). This permitted measurement of equilibrium argon pressures as low as 10^{-11} Torr. While there seems little doubt that some adsorption of the background gases of the system took place continuously during the measurements, no evidence was found that this process effected the isotherm measurements. Total background pressures at room temperature before adsorption measurements were typically 4×10^{-10} Torr, with values of 3×10^{-10} Torr after the porous silver had been cooled to 77.4°K. The system contained a Bayard-Alpert gauge which was used as a check on the mass spectrometer at all pressures, and was used alone at pressures above 10^{-5} Torr, when the mass spectrometer was switched off. Linearity measurements with steady argon leaks were carried out with the mass spectrometer and the Bayard-Alpert gauge, and corrections made where necessary to convert observed ion currents to pressures. Absolute calibration was supplied by simultaneous measurement of an argon pressure of 2.5×10^{-4} Torr by the Bayard-Alpert gauge and a Baratron capacitance manometer which had been checked against a McLeod gauge and found to be accurate to $\pm 5\%$. Absolute pressure errors of this magnitude are not important in the present considerations.

In the variable temperature dewar the stainless steel pot containing the porous silver was encased in a close fitting copper cover ($\frac{1}{8}''$ wall) filled with silicone grease to improve thermal contact. A second copper cylinder which dipped into a reservoir of liquid nitrogen surrounded the pot but did not make physical contact with it. The pot was thus bathed in nitrogen vapor in relatively close contact with a surface at 77.4°K. The temperature of the pot was measured with a platinum resistor having a total surface area of about 15 cm^2 in intimate contact with the pot. Calibration of the platinum resistance was based on values from the International Critical Tables (1929). The resistance of the platinum thermometer was monitored by the potential drop produced by a small constant current passed through it. A five figure digital voltmeter allowed a direct indication of the developed potential, which was also fed to an electronic circuit to provide proportional control of a heater wrapped around the outside of the adsorbent pot. With a heater power capability of about 10 watts the temperature of the pot could be fixed to any desired value in the range 77.4°K-130°K to within $\pm .05^\circ$K. Reproducibility of the Pt thermometer was checked almost daily at room temperature and 77.4°K.

Two copper-constantin thermocouples at the top and bottom of the pot respectively, arranged as a differential thermocouple, measured ΔT over the pot. After some experience it was found possible to make pressure measurements with the mean temperature of the pot within $\pm .05^\circ$K of that desired and with $\Delta T \leqslant .1^\circ$K. Such conditions could not be maintained continously but would occur sufficiently frequently to make the above limits practical. Pressures never rose to a level where thermal transpiration corrections other than limiting values determined by $(T_1/T_2)^{\frac{1}{2}}$ were required. Ion pumping by the gauges was a minor problem at electron currents of 100 μa in the mass spectrometer and 80 μa in the Bayard-Alpert gauge, and was reduced to negligible levels if necessary, by intermittent gauge operation.

No adsorption took place other than in the porous silver pot and the porous silver container was an isothermal system.

IV. EXPERIMENTAL RESULTS AND COMPARISON WITH
CALCULATIONS

Schram (1968) has reported some time dependence in the pressure above an argon layer adsorbed on nickel. Periodically during our measurements we checked the approach to equilibrium with the results shown in figure 5. It is

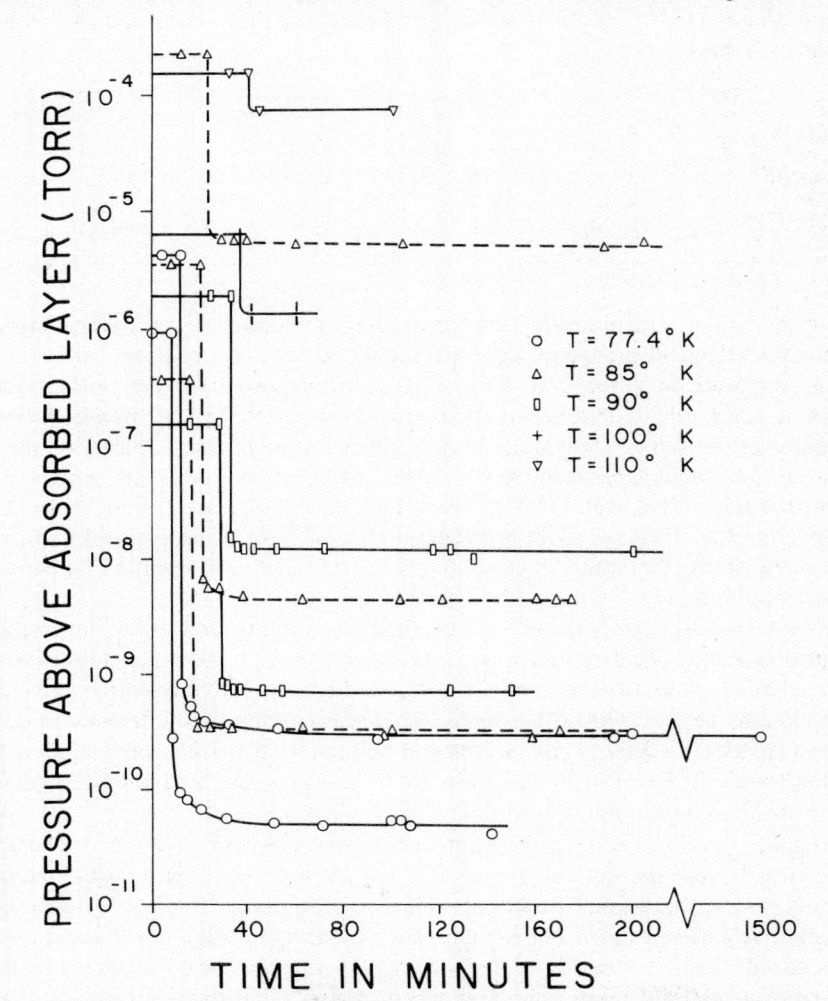

Fig. 5. Time dependence of argon pressure following commencement of physical adsorption at various pressure levels. Zeros of time have been displaced for clarity.

clear that some time dependence is present. The approach to equilibrium is slower the lower the pressure and the lower the temperature. The typical results of the run at 77.4°K extending over 1500 min were the most extensive and are collected in table 2.

The final approach to equilibrium was independent of whether the adsorbent was cooled in the presence of the adsorbate, or whether the adsorbate was admitted to the already cooled adsorbent. This proved that the final approach to equilibrium was not due to the lack of thermal equilibrium, as found also by Schram (1968).

Table 2. *Time dependence of Argon pressure over adsorbed layer at T = 77.4°K*

Time in mins. after start of physical adsorption	1	4	6	11	21	41	82	180	246	1186	1282	1472	1500
p/p_∞	3.00	1.80	1.56	1.42	1.33	1.22	1.02	1.11	1.13	1.02	1.00	1.02	1.00

p_∞ = Pressure after 1500 mins = 2.7×10^{-10} Torr

Because of the requirements of controlling temperature and temperature gradients of the adsorbent pressure measurements were not made at fixed times after the start of adsorption. Of the 20 points reported for the isotherm at 77.4°K the mean time of observation was 45 min after the start of adsorption (excluding the run of 1500 min). Table 2 indicates that $p/p_\infty = 1.2$ after 45 min. Scatter in individual readings is also indicated in table 2. Errors at temperatures other than 77.4°K were less than those at $T = 77.4$°K. In estimating the effect of errors in the results below it has been assumed that the pressures above the adsorbed layer might be 20% high as a result of failure to reach equilibrium.

Measured isotherms are shown in figure 6 (points and solid lines) and compared with calculations (dotted) i.e. with figure 1. Two conclusions are immediately clear: (a) the good fit achieved between the experiment and the calculations at $T = 77.4$°K becomes progressively worse as the temperature is raised (b) Henry's Law is not measured at any temperature the closest approach being found at $T = 110$°K with $N \propto p^{.92}$ corresponding to a line inclined at 42.5° to the x-axis in figure 6.

Figure 7 reflects the same results when the adsorbed amount is plotted against ϵ^2. Here the data at $T = 77.4$°K fall close to the D-R straight line of figure 2 but deviations occur at higher temperatures. The deviations are qualitatively as calculated but the experimental deviations are much larger than calculated. Figure 8 shows the experimental isosteric heat calculated from equation 8. Here the experimental results decisively point to the conclusion that q_{st} is a function of θ alone and not of T, although it should be remembered that data at $T = 77.4$°K were not available below $\theta = 10^{-6}$ for the measurement of

q_{st}. The differential molar entropy calculated from equation (9) is shown in figure 9. Once again it would appear that $\bar{S}_S - \bar{S}_L$ is a function of θ alone and not of T as predicted by the calculations. The limiting slope between $\theta = 10^{-6}$ and $\theta = 10^{-8}$ comes close to the limiting slope predicted by all the calculations.

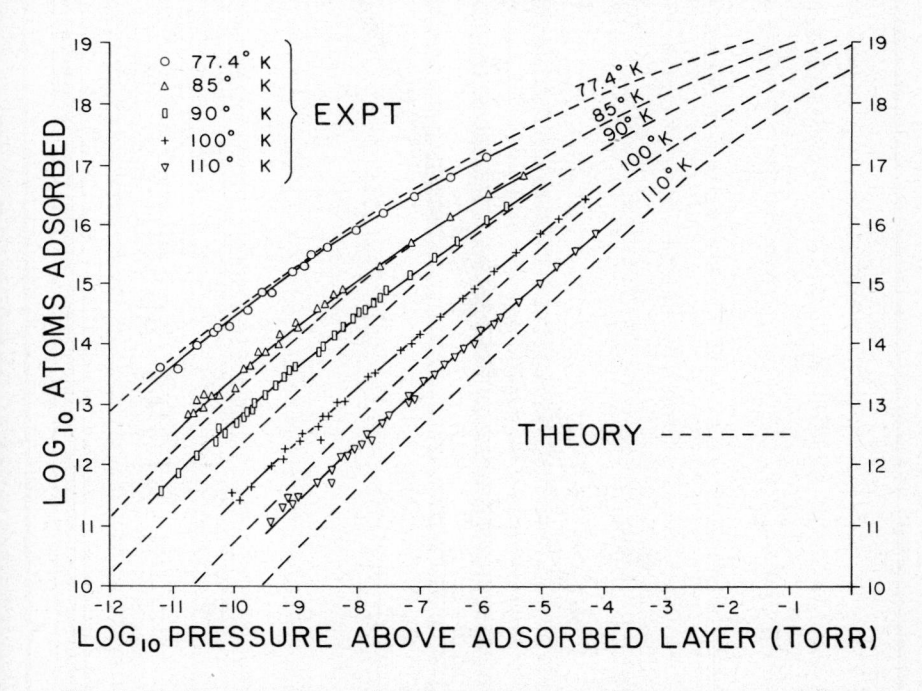

Fig. 6. Measured isotherms for the adsorption of argon on porous silver at various temperatures. Solid lines have been passed through the data visually. Dashed lines are the calculations of figure 1.

Before acceptance of these experimental conclusions it is necessary to ask whether the experimental error, in particular an error of 20% in the pressure, could nullify them. In figure 6 an error of 20% in p has no hope of changing any important result. In figure 7 errors in ϵ^2 might be 1% or an error of 2.5 x 10^5—too small to be significant. In figure 8 errors in q_{st} might be 100 cals mole^{-1} but are probably much smaller because the same % error in p at two adjacent temperatures does not result in any error in q_{st}. For figure 9 the possible errors are the largest being 20% at most, but once again would be reduced if the same % error in p occurred at adjacent temperatures.

Figures 8 and 9 suggest that Henry's Law might be present below $\theta = 10^{-6}$, i.e. in the range where q_{st} is constant and the slope of the entropy is nearly that of the calculations. In figure 6 $\theta = 10^{-6}$ corresponds to 1.7×10^{13} atoms

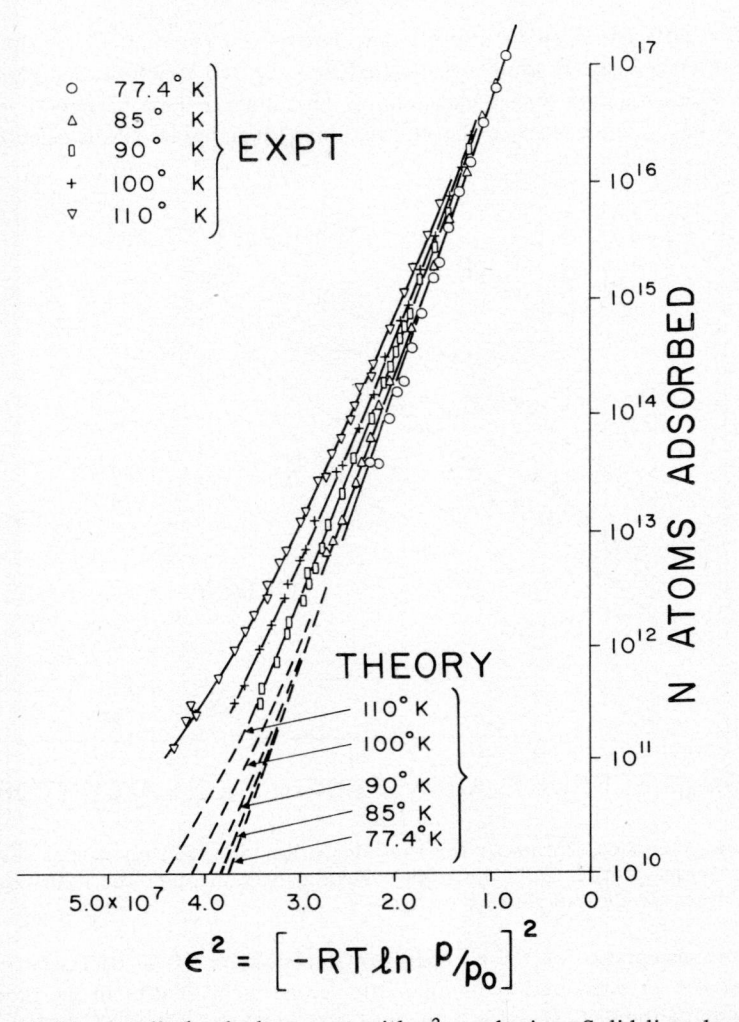

Fig. 7. Measured adsorbed amount with ϵ^2 as abscissa. Solid lines have been passed through the data visually. Dashed lines are the calculations of figure 2.

adsorbed and the question arises as to whether Henry's Law could be present within the experimental error over this range only. It is our conclusion that it is not although it must be admitted that a marginal area is being approached.

Our main conclusions are therefore:

(a) While experiment confirmed the calculations in a general way, adequate for many practical purposes, there were important deviations.

(b) Henry's Law was not found exactly at any temperature.

(c) Department from the D-R equation were found to be in qualitative agreement with the calculations, but with the following particular deviations.

(d) The isosteric heat was only a function of coverage and not of temperature, and reached a limiting value of 5600 cals mole^{-1} below $\theta = 10^{-6}$ maintaining this value to $\theta = 10^{-8}$.

(e) The differential molar entropy was only a function of coverage and not of temperature but reached a limiting slope

$$\left(\frac{\partial(\bar{S}_S - \bar{S}_L)}{\partial \ln \theta}\right)_T \simeq -R \qquad (11)$$

between $10^{-8} < \theta < 10^{-6}$.

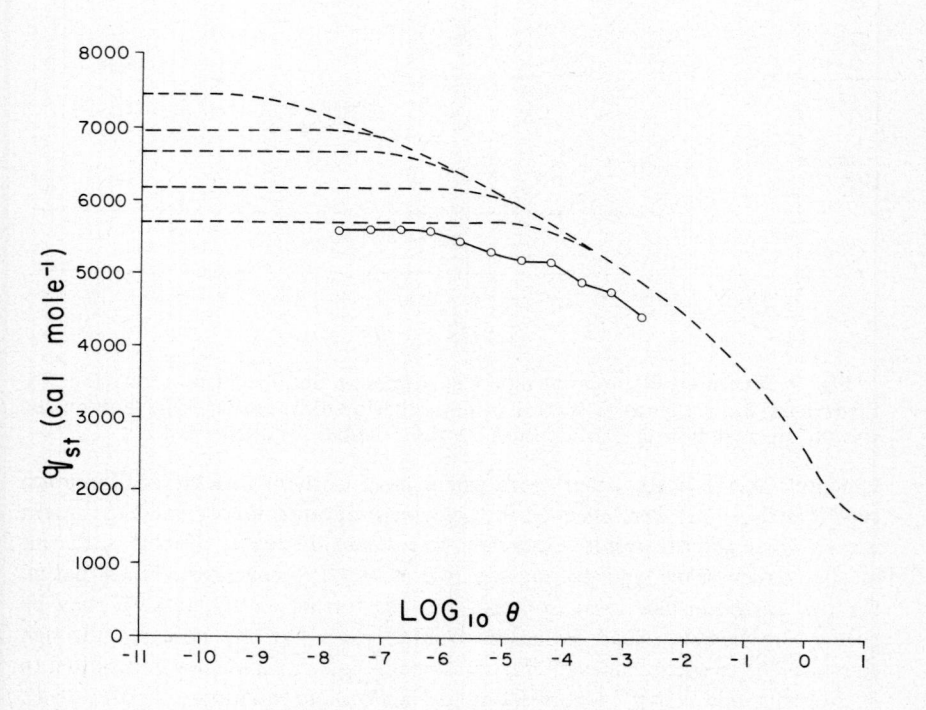

Fig. 8. Measured isosteric heats obtained from equation (8). Solid line passed visually through points. Dashed lines are the calculations of figure 3.

V. DISCUSSION

The prediction of the onset of Henry's Law which has been proposed and tested in this work is clearly not adequate. Since the same general form for the onset was calculated by Hobson (1966) for a heterogeneous surface using conventional assumptions it follows that at least one of Hobson's earlier results is not valid. This applies even for the case where the assumed local isotherm was

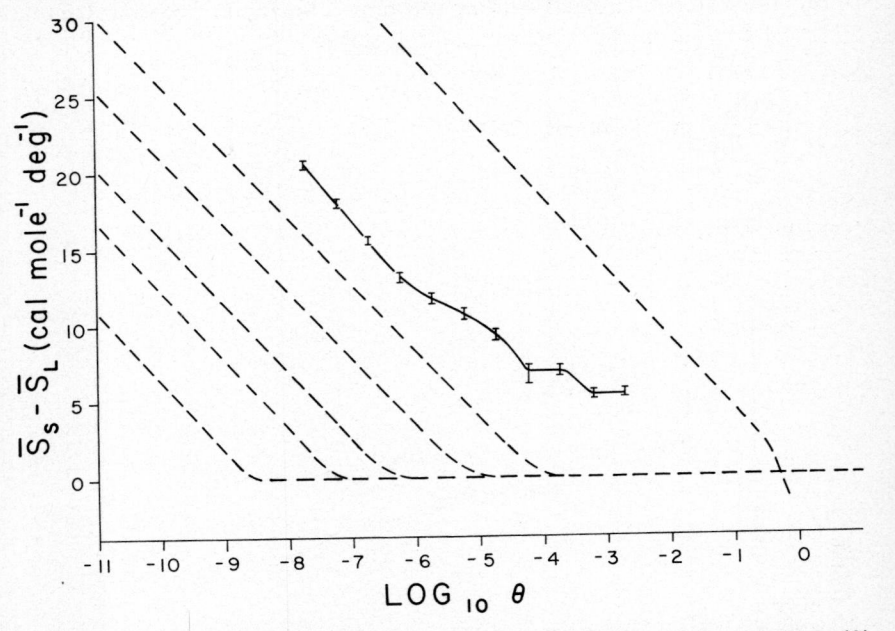

Fig. 9. Measured differential molar entropies as obtained from equation (9). Error bars show the total scatter in the experimental results. Solid line passed visually through points. Dashed lines are the calculations of figure 4.

Langmuir-like. For the other cases where local isotherms of the condensation type (Harris 1968) were assumed the agreement between theory and experiment is even worse and this result would seem to rule out the reality of local isotherms of the condensation type playing any role at very low coverages. The situation for the Langmuir-like local isotherm requires further clarification. It may be noted here however that deviations from Henry's Law are reflected through equation (9) to yield values of entropy slope in equation (11) not quite equal to R. For example $N \propto p^{.92}$ corresponds to a slope in equation (11) of 1.08 R. Cochrane et al. (1967) could still measure small deviations from Henry's Law at $\theta = 10^{-10}$ even for a surface much more homogeneous than porous silver.

ACKNOWLEDGEMENTS

We would like to thank R. D. Cottee who assisted with the construction of the apparatus and B. G. Baker who made valuable comments about the manuscript.

REFERENCES

Cerofolini, G. F. (1971). *Surface Sci.* **24**, 391.
Cochrane, H., Walker, P. L. Jr., Diethorn, W. S. and Friedman, H. C. (1967). *J. Colloid Interface Sci.* **24**, 405.
Dubinin, M. M. and Radushkevich, L. V. (1947). *Proc. Acad. Sci. USSR,* **55**, 331.
Harris, L. B. (1968). *Surface Sci.* **10**, 129.
Hobson, J. P. and Armstrong, R. A. (1963). *J. Phys. Chem.* **67**, 2000.
Hobson, J. P. (1966). *J. Vac. Sci. Technol.* **3**, 281.
Hobson, J. P. (1969). *J. Phys. Chem.* **73**, 2720.
International Critical Tables (1929). Vol. VI, p. 130, McGraw-Hill, New York.
Schram, A. (1967a). *Nuovo Cimento, Suppl.* **5**, 291.
Schram, A. (1967b). *Nuovo Cimento, Suppl.* **5**, 309.
Schram, A. (1968). "Proc. 4th Int. Vac. Congr." p. 106, Institute of Physics and the Physical Society, London.
Weast, R. C. (1970). p. 141, Chemical Rubber Handbook, Cleveland.

DEFINITION AND MEASUREMENT OF AN ELECTRIC FIELD AT THE SURFACE OF A SOLID

J. PATIGNY, Y. BARBAUX and J.-P. A. BEAUFILS

Laboratoire de Catalyse, Université des Sciences et Techniques de Lille, France

I. INTRODUCTION

Several authors (Mignollet, 1950; Suhrmann, 1956) have observed that the adsorption of a rare gas on the surface of a solid produced a variation of its workfunction ΔW.

The surface is only slightly and reversibly altered by the adsorption of the rare gas, so that ΔW can be considered as information about the surface as it was before adsorption. The rare gas atom plays the role of a gauge. However, the information obtained is more or less useful depending on its interpretation. It is generally assumed that a dipole moment is associated to the adsorbed atom and that this dipole moment μ can be calculated by the formula

$$\Delta W = 4\pi N\mu$$

Where N is the number of atoms adsorbed by unit area. In the case of multilayer adsorption, only the first layer has to be taken into account.

For the dipole moment two types of interpretation have been proposed:

(1) Mignollet (1950) has suggested that there exists an electric field E_S at the surface of the solid and that the adsorbed atom is polarized by it. The surface electric field can therefore be calculated by the formula

$$\mu = \alpha E_S$$

where α is the polarizability of the adsorbed atom.

(2) Mignollet (1953) later suggested that the dipole moment was due to a partial charge transfer between the atom and the solid, described quantum mechanically by the no-bond-charge transfer (NBCT) model of Mulliken (1952).

If hypothesis (1) is correct the calculated surface field should be the same for different rare gases: in other words the dipole moments should be in the ratio of polarizabilities, whereas there is no reason for such a relationship to be verified if hypothesis (2) is correct. The comparison of the effect of different rare gases is therefore a crucial test (Gundry and Tompkins, 1960; Suhrmann, 1956) and the conditions of this comparison will be discussed in what follows.

49

II. CORRECTIONS TO THE DIPOLE MOMENT

Whatever may be the cause of the dipoles, their presence will produce electric fields which will contribute to the dipole moment. We think that two corrections ought to be considered:

(1) The dipole moment will polarize the solid near it. This polarization will in turn create a field E_I acting on the dipole. This effect will greatly depend on the properties of the solid and a macroscopic description of these properties is a very rough basis for an evaluation of this effect. Assuming however that the solid is a perfectly plane conducting body, the field E_I will be the field produced by the dipole image of the absorbed dipole. The field E_I will increase the dipole moment from μ_0 to μ. This field will be approximately

$$E_I = \frac{2\mu}{(2R)^3}$$

R radius of the atom. From this equation and $\mu = \mu_0 + \alpha E_I$ we obtain

$$\mu = \frac{\mu_0}{1 - \alpha/4R^3}$$

$\alpha/4R^3$, independent of the nature of the rare gas, is of the order of 0.12 so that the correction is not negligible. This formula is true only when the surface coverage is small. At high surface coverage the contributions of all image dipoles add up and this sum tends towards zero: when the surface is completely covered the image dipoles form a quasi-continuous double layer and it is well known that the field is zero outside such a layer.

(2) The dipoles around a given dipole produce on it a depolarizing field E_D. We have proposed (Beaufils and Ranchoux, 1968) a way of computing E_D leading to the expression

$$E_D = \frac{2\pi N\mu}{R}$$

The correction will be important when the surface coverage $\theta = N/N_0$ is great enough and, therefore, is calculated separately from the first corrections. One obtains

$$\mu = \frac{\mu_0}{1 + N/N_c}$$

$$N_c = \frac{R}{2\pi\alpha}$$

N/N_c is smaller than N_0/N_c, which is of the order of 0.5, independently of the nature of the rare gas, the depolarization correction is therefore rather small except near saturation.

To compare with the experiment we assumed that the coverage of the first layer on the solid is

$$\frac{N}{N_0} = \frac{CP}{1 + CP}$$

where P is the pressure of the gas, obtaining thus

$$\Delta W = 4\pi\mu_0 \frac{CN_0 P}{1 + CP(1 + N_0/N_c)}$$

The experiment effectively showed a law of this type for coverages which were not too small. For small coverages the image field correction must be included, but there remains a steep variation of ΔW for very small pressures, which may be due to the adsorption of impurities or to the heterogeneity of the surface.

III. EXPERIMENTAL MEASUREMENTS

We have just seen that a comparison of the effect of different rare gases on the work function, necessitates a knowledge of surface coverage. In particular, the comparison by Suhrmann (1956) between measurements at low coverage for argon and saturation for krypton involves a correction factor greater than the one tentatively proposed by him.

The surface coverage must be either directly measured or deduced from pressure measurements. We adopted the latter. The pressure measurement has a meaning only if the thermomolecular effects are absent: this is achieved if the pressure is not too low. Consequently, methods of work function measurement, involving the detection of free electrons and low pressure operation, had to be discarded and we have adopted the Kelvin-Zisman method of measurement of contact potential differences (Zisman, 1932) by vibrating condensator.

In the absence of heating, the variation of the contact potential difference owing to the introduction of rare gas is the difference between contributions of the two electrodes. This variation increases, by heating, to a saturation value. If the heating is too strong the temperature of the other electrode also increases and the variation of contact potential difference decreases. The optimum value of heating depends on other experimental conditions. The apparatus is schematically represented in figure 1.

When the two electrodes are not at the same temperature, the thermoelectric power contributes to the contact potential difference. This contribution has been evaluated from hand-book data and from direct measurements. The maximum contribution between room temperature and liquid nitrogen temperature is 3.5 mV. The following data are corrected accordingly.

The main experimental problem is obviously the elimination of any contribution to adsorption due to other gases. High purity argon (99.995%) and krypton (spectroscopically pure) are used and gas from all apparatus is carefully removed.

However cooling of the studied electrode from room temperature to liquid nitrogen temperature in the vacuum produces a variation of the corrected

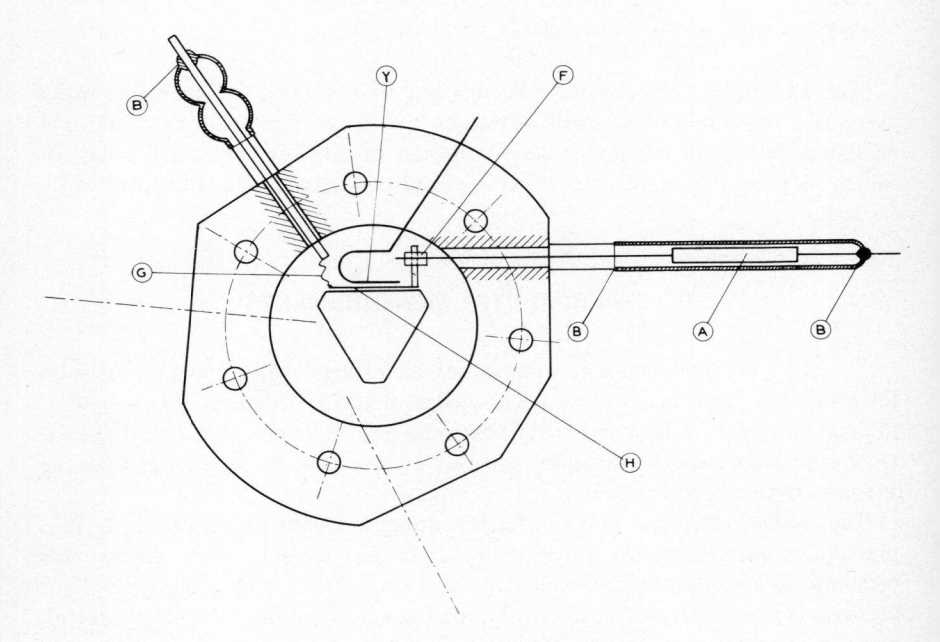

Fig. 1. Apparatus:

A: Magnet for acting the vibrating reed.
B: Glass—metal seal.
F: Insulator.
D-G: Flexible wires.
H: Reference electrode.
Y: Heater.

contact potential difference of the order of 10 to 20 mV. This variation is reduced to 3.5 mV by cooling the cell walls to liquid nitrogen temperature. The experiments are perfectly reproducible. However molecular sieve pumping of a long duration must be avoided because it produces a drift of contact potential difference. Under these conditions we have compared contact potential difference variations from vacuum to saturation pressure of argon and krypton, for the stainless steel wall of our cell. Values of 24 and 34 mV respectively have been obtained. The expected ratio of these results, if the electric field hypothesis is correct, is

$$\frac{\Delta W}{\Delta W'} = \frac{\alpha N_0}{\alpha' N_0'} = 1.12$$

to be compared with the experimental ratio: 1.41.

IV. DISCUSSION OF RESULTS

Gundry and Tompkins (1960) computing surface electric fields from adsorption heat data obtained comparable values with different rare gases. Our own results indicate that the dipole moment increases more rapidly from argon to krypton than expected, on the basis of the electric field hypothesis. This is also the case in Suhrmann's experiments if we interpret them with the necessary corrections.

To discuss the surface field hypothesis we shall consider four models, two electrostatic and two quantum mechanical. In the electrostatic models it is assumed that there is no penetration between the electron densities of the atom and the solid:

(1) Because the electric field is zero outside a double layer of plus and minus charges, it is necessary that the atom penetrates inside the double layer.

(2) The surface can also be considered as a perfect plane. To study how an electric field can however be produced we must define the boundary conditions: assume that in front of the studied surface is a second surface, the potential of which is adjusted so that the charge of the condensator formed by the two solids is zero, then this is precisely the situation realized in the Kelvin method. The electric field between the two surfaces is equal to zero and any line of force of the field coming outside of the studied surface must go back to it. The corresponding distribution of lines of force is represented in figure 2. To any region of the surface where the electric field is turned toward the outside corresponds a region where the electric field is turned toward the inside. We can say that there are on the surface positive and negative sites. A variation of work function on adsorption of rare gas will be observed if the adsorption coefficients of positive and negative sites are fairly different. In these two models the field is highly inhomogeneous. When the size of the atom increases from argon to krypton, a greater part of the atom escapes to the action of the field. We can expect that the dipole moment will accordingly increase less than proportionally to polarizability. This is contrary to present experimental data.

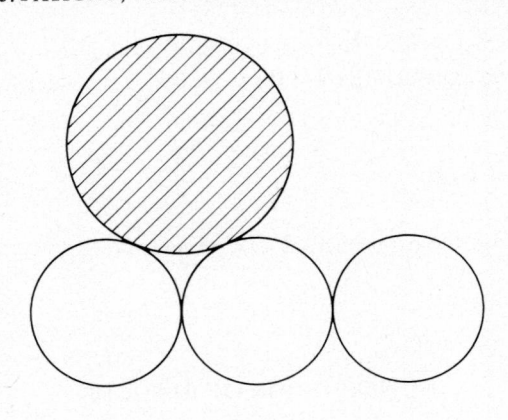

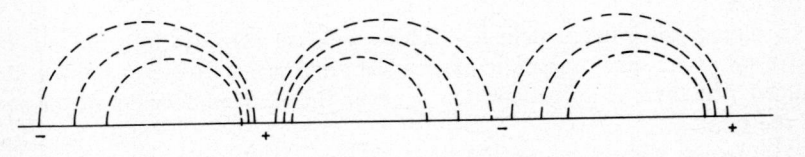

Fig. 2. Electrostatic models

Let us now consider quantum mechanical models. In these models it is assumed that the wave function is

$$\varphi = a\varphi_0 + b\varphi_1$$

where φ_0 is the wave function of the fundamental state of the unperturbed system, and φ_1 is the wave function of an excited state of the unperturbed system. Mulliken (1952) obtains

$$\lambda = -\frac{H_{01} - SW_0}{W_1 - W_0} = \frac{b}{a}$$

where

$$H_{01} = \langle \varphi_0 | H | \varphi_1 \rangle$$
$$S = \langle \varphi_0 | \varphi_1 \rangle$$
$$W_0 = \langle \varphi_0 | H | \varphi_0 \rangle$$
$$W_1 = \langle \varphi_1 | H | \varphi_1 \rangle$$

(3) In the model of Mulliken the excited state φ_1 corresponds to the transfer of one electron in one or the other direction between metal and adsorbate. Then the dipole moment is essentially due to the excess charge on the rare gas atom which Gundry and Tompkins (1960) calculate as

$$\mu \simeq (2\lambda S + \lambda^2) eR$$

They also reformulate the evaluation by Brodd (1958) of $W_1 - W_0$

$$W_1 - W_0 = \begin{cases} \Phi - A - \dfrac{e^2}{4R} \\[2mm] I - \Phi - \dfrac{e^2}{4R} \end{cases}$$

according to the direction of the transfer, where Φ is the work function of the metal, A the electron affinity and I the ionization energy of the rare gas.

(4) Other excited states can be envisaged, provided the excitation energy is not greater than $W_1 - W_0$. This will be the case for excited states without charge transfer, in which the metal is unchanged and rare gas electron is excited to a higher level. In this case the dipole moment will be entirely localized on the atom, the excitation energy $W_1 - W_0$ will be characteristic only of the rare gas atom, and matrix elements involved will be essentially of the form $\langle A' \mid V_M \mid A \rangle$ where $|A\rangle$ and $|A'\rangle$ are atom states, and V_M is the potential of the metal in the region of the atom, owing to leakage of electron density.

The problem studied in this way is formally identical to the problem of polarization of a rare gas atom by an electric field E: the main difference being the replacement of $\langle A' | V_M | A \rangle$ by $\langle A' | E_z | A \rangle$. Consequently a model of this type essentially predicts effects proportional to the polarizability of the atom.

V. CONCLUSIONS

The real situation on a surface is probably described by taking simultaneously into account the features of the four models. The variation of the dipole moment with size of the atoms should be studied experimentally with more detail, especially with regard to the effect of coverage. The present results however suggest that the contribution of the charge transfer effect is important and that the proper electrostatic effects are of minor importance.

From this we can say more precisely what information can be expected from experiments of this type on different types of surfaces. Because the charge transfer is rather important, the interpretation of the results will be difficult. But the knowledge of the physical adsorption preceding chemisorption will be of great importance for understanding the latter, by giving some insight into the beginning of the reaction path towards chemisorption. The experimentally measured "electric field" will give an idea of the tendency of the surface to promote charge transfer, and correlations between this "field" and activation energy of chemisorption are to be expected.

This would not be the case if the field was purely electrostatic: the interpretation would be simpler but the effect of the field would be the same for a chemisorbed atom or a rare gas; fairly small in most cases.

REFERENCES

Beaufils, J. P. and Ranchoux, R. (1968). *C.R. Acad. Sc.* **267**, 1089.
Brodd, R. J. (1958). *J. Phys. Chem.* **62**, 54.
Gundry, P. M. and Tompkins, F. C. (1960). *Trans. Faraday Soc.* 846.
Mignollet, J. C. P. (1950). *Discuss. Faraday Soc.* **8**, 105.
Mignollet, J. C. P. (1953). *J. Chem. Physics.* **21**, 1298.
Mulliken, R. S. (1952). *J. Amer. Chem. Soc.* **74**, 811.
Suhrmann, R. (1956). *J. Amer. Chem. Soc.* **53**, 15.
Zisman, W. A. (1932). *Rev. Sci. Instrum.* **3**, 367.

ELECTROSTATIC FORCES IN PHYSICAL ADSORPTION OF RARE GASES ON TRANSITION METALS

A. SCHRAM

Centre d'Études Nucléaires de Saclay, Section d'Études des Interactions Gaz-Solides, Gif-sur-Yvette, France

I. INTRODUCTION

Only a few years ago it was still a common statement that physical adsorption was a rather non-specific process in relation to the nature of the solid adsorbent. In fact physical adsorption was considered as mainly a condensation phenomenon, dependent principally upon gas species, pressure, temperature and microgeometry (pores) of the adsorbent. The decrease, often observed, of adsorption heat with increasing coverage (below the monolayer) was generally explained by an intrinsic surface heterogeneity.

Today increasing experimental evidence, mostly from FEM or FIM experiments, suggests that at least for transition metals other forces than those of dispersion alone are involved, and that physical adsorption of rare gases depends strongly upon the bandstructure of the metal. Thus we have here a process highly relevant to the nature of the solid adsorbent.

Extensive experimental results will not be given in this paper, because accurate results, at low coverage for rare gases on transition metal monocrystals, are not yet known to be available. We intend only to formulate the problem from generally accepted experimental evidence and to discuss qualitatively some simple physical interpretations. Moreover a few suggestions will be made for future experimental work.

EXPERIMENTAL EVIDENCE

The following facts have to be explained in the case of rare gases adsorbed on metal surfaces. Below monolayer coverage, the adsorption heat increases regularly with decreasing coverage, at constant temperature. Often this behaviour follows the Dubinin-Radushkevitch isotherm equation, as discussed by the author previously (Schram, 1967). At constant (low) coverage,

57

the heat decreases with increasing temperature by an amount much larger than kT, as shown in our earlier work (1967). On transition metals, adsorption of rare gases has been shown by Engel and Gomer (1970) and by Mignolet (1953) to produce high negative work function variations, while Van Oirschot and Sachtler (1970)report that on alkali metals such variations seem to be absent; the noble metals probably occupying a medium position.

A very strong cooperative effect is observed between argon, introduced as low energy ions on molybdenum surface, and preadsorbed carbon monoxide. When heated the argon leaves the surface with the chemisorbed CO, and not before. (Guernigou, 1971). At very low coverage, approaching zero, high adsorption heats are measured, about three times the condensation heat.

THE VARIATIONS OF THE ADSORPTION HEAT

It is always possible to interpret decreasing adsorption heat with increasing coverages as essentially due to surface heterogeneity. So far practically all isotherms have been taken on non-uniform (metal) surfaces; the only exception being Rhodin's early results (1950) concerning nitrogen physisorbed on copper monocrystal surface planes. He found there constant heat for $\theta < 0.2$ rising to a maximum for $\theta \to 1$. The heterogeneity hypothesis has been widely accepted. But applying, for instance, the procedure of Ross and Olivier (1967) or analysis based on the Dubinin-Radushkevitch equation, very large and similar "site distributions" are often found for different gas-metal systems. Moreover the very high low-coverage heats cannot be readily explained by classical dispersion forces only.

The rather high heats found for rare gases on transition metals at low coverages suggests the need, as in chemisorption, for a more detailed investigation of the exact nature of the forces involved, and the mutual interaction forces in the adsorbate itself, rather than invoking site distribution of energies. In Rhodin's work only about 20% difference between (110) and (111) faces for nitrogen on copper with (100) intermediate is reported. In the following we shall consider only uniform surfaces, making the bold hypothesis that, just as in chemisorption, the polycrystalline surface, which is formed by small monocrystal facets, will not show an essentially different behaviour.

STATISTICAL MECHANICS

An adsorption isotherm on a uniform surface may be expanded in the following way (see Steele 1967)

$$n_a = B_{AS} \left(\frac{p}{kT} \right) + C_{AAS} \left(\frac{p}{kT} \right)^2 + D_{AAAS} \left(\frac{p}{kT} \right)^3 + \cdots \qquad (1)$$

n_a being the number of gas particles adsorbed on the surface area A at pressure p and temperature T, in equilibrium conditions. The "mixed" virial coefficients B_{AS}, C_{AAS}, D_{AAAS}, functions of T, contain respectively the interaction potential of one particle with the surface, of two particles with the surface and with one another, and so on; they may be determined experimentally (at least in principle). These mixed virial coefficients are simply related to the virial coefficients B_{2D}, C_{2D}, ... of the two-dimensional equation of state

$$\frac{\varphi A}{n_a kT} = 1 + B_{2D}\frac{n_s}{A}\theta + C_{2D}\frac{n_s^2}{A^2}\theta^2 + \ldots \tag{2}$$

$$\frac{B_{2D}}{A} = -\frac{C_{AAS}}{2B_{AS}^2} \tag{3}$$

φ being the spreading pressure, θ the coverage and n_s the number of particles to form a complete monolayer on a unit surface area. Here B_{2D} contain the two-body interaction in the adsorbate, C_{2D} the three-body interaction, and so on. Negative B_{2D} indicates attractive forces, positive B_{2D}, repulsive forces.

The virial coefficients are closely related to the cluster integrals

$$B_{AS} = \int \{\exp[-u_S(r)/kT] - 1\}\,d\mathbf{r} \tag{4}$$

$$B_{2D} = -\pi \int_0^\infty \{\exp[-u(\tau)/kT] - 1\}\tau\,d\tau \tag{5}$$

where $u_s(\mathbf{r})$ is the interaction potential between a particle at $\mathbf{r}$ and the surface, and $u(\tau)$ is the interaction between two adparticles separated by a distance τ.

While, according to Steele, this approach is valid for both localized and mobile adsorption at low temperatures, another two-dimensional equation of state has been found by Armand (1971) for the case of localized adsorption and a given intrinsic defect distribution (heterogeneity)

$$\frac{A\varphi}{kT} = -\log(1-\theta) + \frac{1}{kT}\sum_{r=2}^{m}\frac{r-1}{r}\theta_r V_r - \theta^2 \frac{\frac{\partial}{\partial\theta}\sum_{2}^{\infty}}{1+\sum_{2}^{\infty}}$$

$$\sum_{2}^{\infty} = \sum_{2}^{\infty}\frac{(-1)^p}{p!}\frac{\overline{(w-\bar{w})^p}}{(kT)^p} \tag{6}$$

where V_r represents the r-body interaction energy in the adsorbate for $\theta = 1$, and $\overline{(w-\bar{w})^2} = \sigma^2$ the width of the combined intrinsic and induced energy distribution. For low values of θ the V_r are related to the virial coefficients in a very simple manner:

$$n_s B_{2D} = \frac{1}{2} \left(1 + \frac{V_2}{kT} \right)$$

$$n_s^2 C_{2D} = \frac{1}{3} \left(1 + 2 \frac{V_3}{kT} \right) \tag{7}$$

The limiting isosteric heat of adsorption, for θ approaching zero, may be written

$$\operatorname*{Lim}_{\theta \to 0} q_a^{is} = kT + K \frac{\partial \log B_{AS/A}}{\partial (1/T)} \tag{8}$$

Armand's theory gives for the isosteric heat as function of θ and T

$$q_a^{is} = \operatorname*{Lim}_{\theta \to 0} q_a^{is} - \sum_{r=2}^{m} \theta^{r-1} \cdot V_r + f(\theta, T) \tag{9}$$

where $f(\theta, T)$ is a complicated function containing the combined intrinsic and induced energy distribution. It is clear that high $\operatorname*{Lim}_{\theta \to 0} q_a^{is}$ and strong θ dependency of q_a^{is} at constant T requires a high slope of the $\log B_{AS}/A = f(1/T)$ curve, and a high positive V_2 (or B_{2D}) value. The temperature dependency of q_a^{is} at constant θ is not immediately evident and will be considered later.

Now it is customary in physisorption to introduce Lennard-Jones type potentials in the cluster integrals representing the virial coefficients. For B_{AS} this means a z^{-3} attractive, z^{-9} repulsive potential, and for B_{2D} a τ^{-6} attractive, τ^{-12} repulsive potential. Steele (1967) has calculated the reduced virial coefficient B_{AS}/Az_0 as function of u_{min}/kT, z_0 being the distance where u_s is zero, and u_{min} the minimum value of the potential (equilibrium distance). The experimental results of our previous work (1967) with argon on nickel yield B_{AS}/Az_0 values at least two orders of magnitude higher than given by the extrapolated curve of Steele for a reasonable value of u_{min}/kT value. In this case $\operatorname*{Lim}_{\theta \to 0} q_a^{is}$ is of course equal to $u_{min} + \frac{5}{2}kT$. For B_{2D} the situation is still worse, because the Lennard-Jones potential yields a negative virial, while the θ dependency of q_a^{is} requires a high positive value. The London dispersion forces alone in the cluster integrals are thus unable to explain the experimental facts. Because of the important work function changes upon adsorption it seems reasonable to introduce explicitly electrostatic forces.

POLARISATION OF THE ADATOMS BY A SURFACE ELECTRIC FIELD

This idea is not new. Mignolet (1953) and Gundry and Tompkins (1960), among others have already discussed the possibility that a surface electric field of the order of 10^8 V/cm might polarize the adatoms, explaining by the Helmholtz equation the observed $\Delta\varphi$ values

$$\Delta\varphi = 4\pi n_s \mu\theta \tag{10}$$

where $\Delta\varphi$ is the work function variation and μ the dipole moment. The dipole moment is related to the polarizing field F and the polarizability α by the relation $\mu = \alpha F$. If F does not vary too rapidly in the interaction region an extra

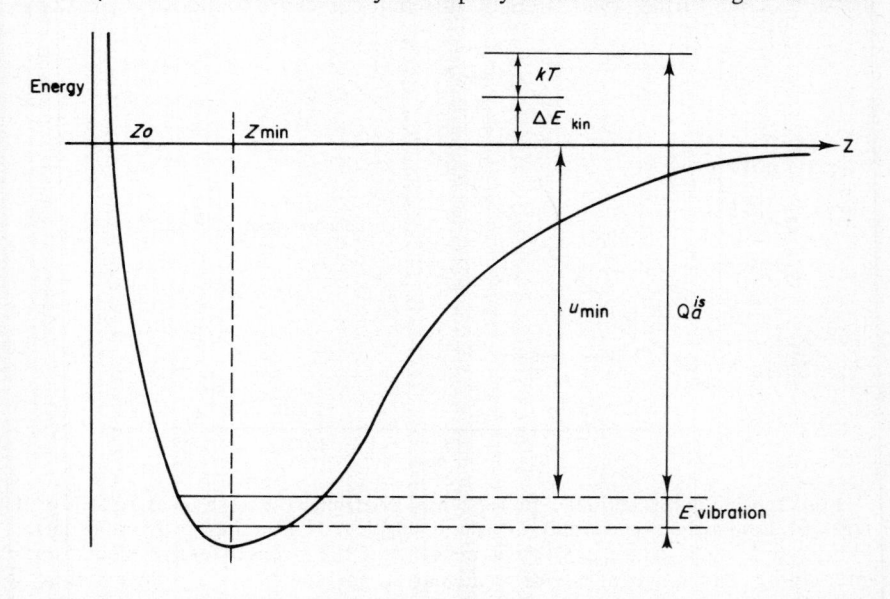

Fig. 1. Schematic Lennard-Jones type potential well diagram for a physisorbed atom, at equilibrium distance z_{min} from the surface. The relation of the isosteric heat q_a^{is} to the minimum u_{min} of the well is shown. ΔE_{kin} is the loss of kinetic energy on adsorption and E_{vibr} is the vibration energy of the adatom on the surface.

attractive term equal to $\frac{1}{2}\alpha F^2$ will arise in the interaction potential between adatom and surface. We will not discuss here the physical nature of this electric field, but suppose that such an intense field perpendicular to the metal surface exists. This will greatly modify the values of the virial coefficients B_{AS} and B_{2D}. In the cluster integral corresponding to B_{AS} we introduce the attractive term $\frac{1}{2}\alpha F^2$ in the following way. F is assumed to decay as z^{-2} like the image force law. The new term is put in the form

$$\frac{1}{2}\alpha F^2 = Cu_{min}\left(\frac{z}{z_{min}}\right)^{-4} \tag{11}$$

u_{min} being the classical Lennard-Jones potential minimum which permits easy numerical integration of the cluster integral

$$\frac{B_{AS}}{Az_0} = \int_0^\infty \left\{ \exp\left[\frac{u_{min}}{kT}\left(\frac{3}{2}x^{-3} + Cx^{-4} - \frac{1}{2}x^{-9}\right)\right] - 1 \right\} dx \qquad (12)$$

According to the value of the adjustable parameter C, which is essentially a measure of the surface field strength, different curves are found for B_{AS}/Az_0 as

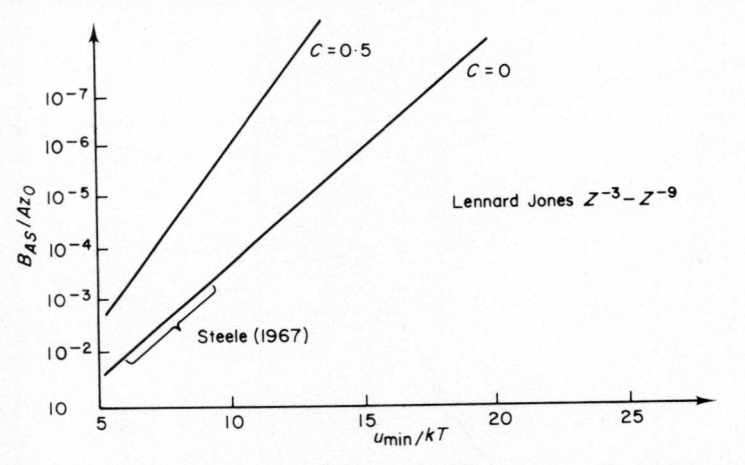

Fig. 2. Calculated reduced mixed virial coefficient B_{AS}/Az_0 as function of reduced Lennard-Jones minimum energy u_{min}/kT. Parameter C, from equation (11): $\frac{1}{2}\alpha F^2 = Cu_{min}(z/z_{min})^{-4}$ is a measure of the strength of the electrostatic interaction. The range of Steele's calculation for the $z^{-3} - z^{-9}$ interaction is indicated.

function of u_{min}/kT. Figure 1 shows schematically the potential curves, while figure 2 gives the reduced virial coefficient B_{AS}/Az_0 as function of u_{min}/kT. It seems that for an electrostatic attraction which is half the dispersion attraction, much higher values are obtained. $\underset{\theta \to 0}{\text{Lim}} q_a^{is}$ becomes about $\frac{3}{2}u_{min}$. Both the high limiting adsorption heats in the Henry's law region and the high experimental B_{AS} values may thus be explained by such surface field, which for our case of $C = 0.5$ amounts to about 10^8 V/cm.

In the adsorbate repulsive dipole interactions will counteract the normal dispersion forces attraction, in order to obtain a net repulsion and hence a positive B_{2D} virial coefficient. The potential $u(\tau)$ of equation (5) becomes thus

$$u(\tau) = 4\epsilon\left[\left(\frac{\tau}{\sigma}\right)^{-12} - \left(\frac{\tau}{\sigma}\right)^{-6}\right] + \frac{\mu^2}{\sigma^3}\left(\frac{\tau}{\sigma}\right)^{-3} \qquad (13)$$

τ being the distance between the two adatoms, σ the distance where the Lennard-Jones potential goes to zero, and ϵ the minimum of this potential. With $4\epsilon/kT$ as parameter some numerical integrations were carried out which yielded n_sB_{2D} as a function of

$$b = \frac{1}{kT} \cdot \frac{\mu^2}{\sigma^3}$$

or

$$\frac{1}{2}\alpha F^2 \cdot \frac{1}{kT} \cdot \frac{2\alpha}{\sigma^3} \approx \frac{1}{2}\alpha F^2/kT$$

as shown in figure 3. These curves clearly indicate that as the electrostatic attraction energy of the adatoms by the surface exceeds a few kT, the second

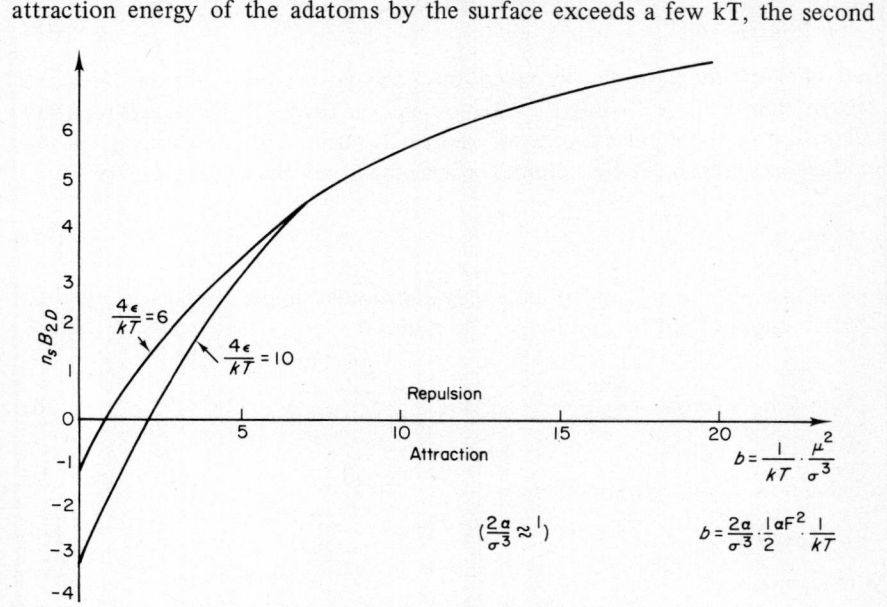

Fig. 3. Calculated reduced two-dimensional virial coefficient, n_sB_{2D} as function of reduced dipole strength for two different lateral Lennard-Jones attractive potentials.

two-dimensional coefficient becomes positive. For $\frac{1}{2}\alpha F^2 > 10$ kT, the Lennard-Jones potential no longer influences the result, but finally the n_sB_{2D} value, or the two-body interaction potential V_2, will remain rather limited to about a few tens of kT. This is less satisfying than the results for B_{AS}, because the variation of heat with θ suggests stronger repulsive interaction, while a crude estimate of C_{AAS} and B_{AS} from experimental isotherms yield very high B_{2D} values. If this should be confirmed in the near future, a stronger interaction than that according with the dipole r^{-3} interaction law should be used. This, of course, will cause the introduction of a screening length in order to avoid divergence of the integral. Physically such an interaction implies that the dipole character of the adatoms should be more or less lost, because a very short range electric field acts upon the outer electron cloud of the atom, very near to the surface electron sea. But this picture leads us to quite another approach to the problem: discussed later as the charge-transfer no-bond theory.

A more direct and very simple evaluation of the variation of the isosteric heat with coverage will now be given. The isosteric heat is the sum of the constant dispersion energy term and the part due to induced dipoles by the surface electric field

$$q_a^{is} = q_{disp} + \tfrac{1}{2}\alpha F^2 \tag{14}$$

But two effects have to be taken into account: (i) the dipoles create a depolarizing field, decreasing the effective electric field; (ii) the mutual repulsive interaction of the dipoles decreases the isosteric heat. The depolarizing field F_d produced at a distance r by a dipole μ perpendicular to the surface is given by

$$F_d = \frac{\mu}{r^3} \tag{15}$$

For an assembly of n_a dipoles uniformly distributed simple integration gives the depolarizing field felt by any dipole with radius d

$$F_d = \int_d^\infty n_a \frac{\mu}{r^3} 2\pi r\, dr = \frac{4\pi n_a \mu}{2d} \tag{16}$$

μ being now the effective moment

$$\mu = \alpha(F - F_d) \tag{17}$$

With $n_a = n_s \theta$

$$F_d = \frac{4\pi n_s \alpha \theta F}{2d + 4\pi n_s \alpha \theta} \tag{18}$$

and instead of the energy term $\tfrac{1}{2}\alpha F^2$ we have now

$$\tfrac{1}{2}\alpha F^2 \left(\frac{1}{1 + a\theta} \right)^2 \approx \tfrac{1}{2}\alpha F^2 (1 - 2a\theta) \tag{19}$$

with $a = 4\pi n_s \alpha/2d$ (a is of the order of unity).

Combining this effective field with the Helmholtz equation, we obtain

$$\frac{1}{\Delta\varphi} = \frac{1}{2d \cdot F} + \frac{1}{4\pi n_s \alpha} \cdot \frac{1}{\theta} \tag{20}$$

or

$$\Delta\varphi = 4\pi n_s \alpha F \cdot \frac{\theta}{1 + a\theta} \tag{20a}$$

which is the equation used by Beaufils and Ranchoux (1968) to evaluate F from measurements of $\Delta\varphi$ as a function of θ. The $\Delta\varphi$ versus θ data recorded by Engel and Gomer (1970) for different surface crystal planes of tungsten seem to follow this $\theta(1 + a\theta)^{-1}$ proportionality fairly well. The mutual dipole repulsion can be accounted for in several ways. The most simple is to integrate, as in equation (14) the interaction potential between two dipoles, yielding

$$q_{rep} = \int_{d}^{\infty} 2\pi n_s \theta \frac{\mu^2}{r^3} r \, dr = 2\pi n_s \theta \frac{\mu^2}{d} \tag{21}$$

With (15) and (16) this becomes

$$q_{rep} = \alpha F^2 \frac{a\theta}{(1 + a\theta)^2} \tag{22}$$

Finally the total isosteric heat of equation (14) will now be given by

$$q_a^{is} = q_{disp} + \tfrac{1}{2}\alpha F^2 \frac{1}{(1 + a\theta)^2} - \alpha F^2 \frac{a\theta}{(1 + a\theta)^2}$$

or

$$q_a^{is} = q_{disp} + \tfrac{1}{2}\alpha F^2 \frac{1 - 2a\theta}{(1 + a\theta)^2} \approx q_{disp} + \tfrac{1}{2}\alpha F^2(1 - 4a\theta) \tag{23}$$

for $\theta \ll 1$.

This result is a reasonable representation of the true variation of the adsorption heat for coverages between about $\theta = 10^{-2}$ and a monolayer, but at lower coverages it fails to reproduce the heat curves derived from the Dubinin-Radushkevitch isotherms. We saw already that the r^{-3} dipole repulsion law gave too low a value for B_{2D}, and therefore a stronger interaction with longer range can not be excluded. If, for instance, we crudely suppose a direct electrostatic repulsion, restricted to a certain number of neighbours (screening length), a $-\theta^{\frac{1}{2}}$ variation results for low coverages, which is much nearer to the experiment.

The variation of the isosteric heat with temperature at constant coverage can be explained qualitatively in the case of an appreciable $\frac{1}{2}\alpha F^2$ term, by examining figure 4, which shows schematically the potential well for an adatom. This potential well being very dissymmetric, and the adatom virbating on its energy level around an equilibrium position midway between the well walls, a rise in

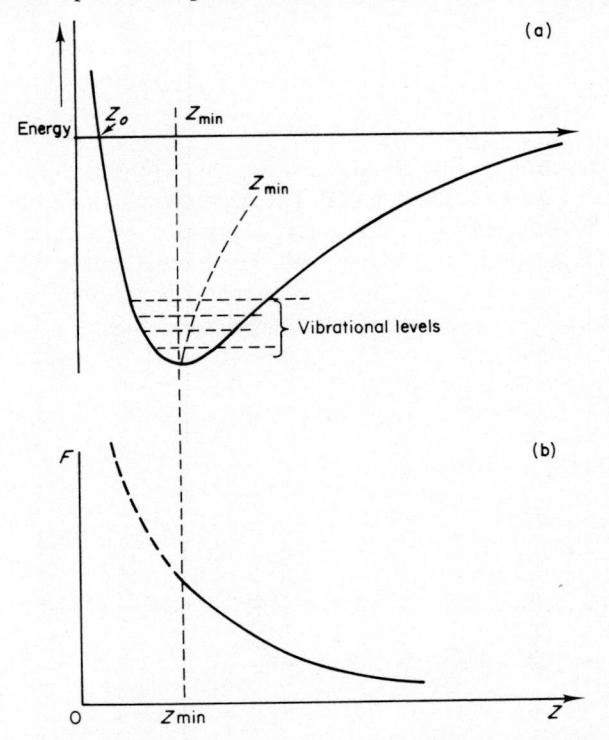

Fig. 4. (a) Schematic potential well diagram. When the temperature rises the adatom vibrates midway between the well walls, on a higher vibration level. Curve z_{min} shows the locus of the equilibrium distance. (b) The electric field F decreases parabolically with increasing equilibrium distance.

temperature brings it to a higher level and the equilibrium position will be shifted slightly to the right. The electric field being strongly dependent on this distance, at least as a z^{-2} law, the adsorption heat will decrease rather rapidly. A strong electric surface field may thus explain, at least qualitatively, many experimental facts in adsorption of rare gases on metals. But there remain some problems; e.g. why should this field apparently not exist (or be very weak) at alkali metal surfaces, as shown by Van Oirschot and Sachtler (1970)? And, of course the very important question of the existence and the nature of this field!

THE CHARGE TRANSFER NO-BOND THEORY

Another theory was advanced long ago, which avoids these difficulties: the "charge transfer no-bond theory" (CTNB), which may be traced back to Mulliken (1952) and which was already envisaged by Mignolet (1953). More recently Tompkins and Gundry (1960), Engel and Gomer (1970) and Van Oirschot and Sachtler (1970) discussed the applicability of this theory to our problems. We shall give here only a very short outline of this theory, before comparing it with the surface field theory. The additional energy term may be written here

$$q_{CTNB} = \frac{\beta^2}{W_1 - W_0} \tag{24}$$

where $\beta = H_{01} - SW_0$

H_{01} being the resonance integral $\langle \psi_0 | H | \psi_1 \rangle$, S the overlap integral $\langle \psi_0 | \psi_1 \rangle$, with ψ_0 and ψ_1 respectively the wavefunctions of the system in the no-bond neutral state and the ionic state, and W_0 and W_1 the energies of the system in these states.

With $W_0 = E_{metal} - I$ and $W_1 = E_{metal} - \varphi - e^2/4z$ we have:

$$W_1 - W_0 = I - \varphi - \frac{e^2}{4z} \tag{25}$$

I being the ionisation energy of the atom, φ the workfunction and the last term representing the image potential, the whole problem lies in the values of the overlap and resonance integrals. As stated by Engel and Gomer, the new energy term depends on a small difference between rather large terms. Actually β of equation (24) is an adjustable parameter. As to the physics underlying this quantum mechanical description, the necessary overlapping of the wavefunctions ψ_0 and ψ_1 suggests, as stated by Van Oirschot and Sachtler, that the d-orbitals of the metal should play an important role. Thus the high adsorption energy of rare gas atoms on transition metals with empty d-levels may be explained by the low position of these levels, compared to the electron level of the adatom. For instance strong overlapping may be expected between xenon and nickel: the bandstructure of the latter showing a level at about the same depth as the ionization energy of the gas atom. These authors attribute the very low adsorption energy for these atoms on potassium to the much higher position of the empty d-levels for this metal. Resonance may shift and broaden the atomic level thus facilitating the tunneling of electron charge through the barrier. Noble metals should thus occupy an intermediate position in their adsorptive behaviour towards rare gases.

II. DISCUSSION

The electric field theory is able to give a relatively easy and simple explanation of many experimental facts in a classical way. If we accept the existence of these fields self-consistent calculations are straightforward. More over Beaufils and Ranchoux (1968) has claimed the possibility of measuring these fields in a simple way. But against these positive arguments one may oppose:

(i) The measurements made by Beaufils are based on equation (20) which seems to be only valid in a rather short range of values. The extrapolation to $1/\theta = 0$ to yield $d \cdot F$ appears to us somewhat hazardous. By using equation (20a), applied to Engel and Gomers $\Delta \varphi$ data on tungsten at the limit $\theta \to 0$

$$ F = \frac{1}{4\pi n_s \alpha} \left(\frac{d\Delta\varphi}{d\theta} \right)_{\theta \to 0} \tag{26} $$

much higher values, between 10^7 and 10^8 V/cm are obtained compared to 1.2×10^6 V/cm given by Beaufils for the gold argon system. But now the electric field is very sensitive to the n_s value, and precise measurements of $\Delta \varphi$ at very low θ are difficult.

(ii) An argument often quoted is the statement of Gomer that the electric field should decay in a very short distance, between 0.5 and 1.5 Å and should thus be unable to polarize the large xenon atom. But it seems not impossible that only the outer electrons should be in this case perturbed by the field, giving rise to a less strong dipole. The lateral repulsive interaction may remain very strong, because the negative charge of the dipole being more or less in the metal surface, we have now an interaction according to a z^{-n} law, n being smaller than 3.

(iii) In the simple calculations based on the electric field conception, α has been taken constant. Actually it is doubtful that in such strong fields this approximation is valid.

(iv) The very nature of this electric field has not been clearly explained until now.

The CTNB theory has all the power of quantum mechanics, and it certainly represents a fair description of the phenomenon in cases of strong physisorption or perhaps weak chemisorption, characterized by surface potential changes. The different behaviour of alkali noble and transition metals may be very well understood due to their very different bandstructure. But in the case of rare gas adsorption the electronic levels of the gas atoms lie very deep, and we need more detailed information about the low lying empty levels in the metal near the surface, for the different crystallographic planes. The only drawback, and it is an important one, is the impossibility of making at present, reasonable calculations.

For instance the resonance integral remains in fact an adjustable parameter. Actually it seems not a too bold hypothesis that the electric field theory and the CTNB theory should represent respectively classical and a quantum mechanical description of the same physical reality. The former being a classical "force" language of the quantum mechanical resonance resulting in charge transfer with a change in energy. In fact the notion of electric field has been introduced in physics to explain the forces between charged bodies. All interactions contributing to physical adsorption are finally essentially electrostatic, and may thus in principle be described by fields. The Hellman-Feynman theorem should be remembered here, as formulated clearly by Hirschfelder *et al.* (1954):

Forces on the nuclei determined from a quantum mechanical potential energy surface are just exactly what we would expect on the basis of classical electrostatics and a knowledge of the electron probability density. In other words, once the distribution of the electron cloud has been determined from the solution of the Schrödinger equation, the forces on the nuclei may be calculated according to the electrostatic interaction formula . . . We should note that spin and exchange effects play an important role in determining the electron cloud, but, once this electron distribution has been fixed, the forces are principally classical electrostatic forces . . .

Thus direct coulombic interaction, induced polarization forces, dispersion forces, charge transfer forces, resonance forces are all electrostatic interaction in principle. If the electric field at the surface were truly measurable and its decay law known, it would be more practical, and thus preferable, to work with this language, even if this electric field concept should be proven to be somewhat artificial in comparison to everyday macroscopic fields. There remains one difficulty, among others, in this synthesis of the two theories. The work function change $\Delta\varphi$ is generally related by the Helmholtz equation to the coverage θ and the dipole moment μ. The dipole moment μ may be expressed as $\mu = ec$ where e is the electronic charge and c the dipole length. There is no problem in the electric field language, but in the case of charge transfer we may have $\mu = e'c'$, which would yield other values for the repulsion energy, and if c' is large, a stronger repulsion than the r^{-3} law.

III. FUTURE WORK

Actually very little reliable data is available of $\Delta\varphi = f(\theta)$ for rare gas adsorption on metals. Careful work must be done now, on different crystallographic planes and extended to very low coverages. Accurate isotherms, together with $\Delta\varphi$ measurements, will allow determination of the virial coefficients as function of temperature, and comparison with calculations of the integrals for different interaction laws. The broadening and displacement of the electron levels in the adatoms may perhaps be studied by measuring the Stark effect, using the resonance in the vacuum ultraviolet between fundamental and the first excited state of the rare gas.

Other optical techniques, for instance photo-electron energy analysis, may throw perhaps more light upon band structure of the metals at their surface.

As far as we know surface plasmons have not been studied yet in relation to rare gas adsorption. Mavroyannis (1963) obtained an expression for the actractive potential U in physisorption on a metal, treated as dielectric, in which, after some simplifications, appears explicitly the surface plasmon loss $\hbar\omega_s$

$$U \approx - \frac{\hbar\omega_s \cdot \alpha}{z^3} \tag{27}$$

This formula gives higher energies than the classical dispersion energy formulas, and it seems now worthwhile to investigate this approach and to compare it with the CTNB theory

REFERENCES

Armand, G. (1971). *Surface Sci.* To be published.
Beaufils, J. P. and Ranchoux, R. (1968).*C.R. Acad. Sci.* **267**, C, 1089.
Engel, J. and Gomer, R. (1970). *J. Chem. Phys.* **52**, 5572.
Guernigou, J. (April 1971). Thesis, Fac. Sciences Orsay, France
Gundry, P. M. and Tompkins, F. C. (1960). *Trans. Faraday Soc.* **56**, 846.
Hirschfelder, J. O., Curtiss, C. F. and Bird, R. B. (1954). In "Molecular Theory of Gases and Liquids" p. 932. J. Wiley, New York.
Mavroyannis, C., (1963). *Mol. Phys.* **6**, 593.
Mignolet, J. C. P. (1953). *J. Chem. Phys.* **21**, 1298.
Mulliken, R. S. (1952). *J. Amer. Chem. Soc.* **74**, 811.
Rhodin, T. N. Jr., (1950). *J. Amer. Chem. Soc.* **72**, 2691.
Ross, S. and Olivier, J. P. (1967). "On Physical Adsorption". Interscience, New York.
Schram, A. (1967). *Nuovo Cimento, Suppl.* I, **5**, 291.
Steele, W. A. (1967). In "The Solid Gas Interface" (E. A. Flood, Ed.) Vol. 1, p. 307.
Van Oirschot, Th. G. J. and Sachtler, W. M. H. (1970). *Ned. Tijdschr. Vacuum Technik,* **8**, 96.

THERMODYNAMICAL CHARACTERISTICS OF SUBSTANCES ADSORBED ON ZEOLITES

G. V. TSITSISHVILI, T. G. ANDRONIKASHVILI, SH. D. SABELASHVILI and T. A. CHUMBURIDZE

Institute of Physical and Organic Chemistry, Academy of Sciences of the Georgian SSR, Tbilisi, USSR

The knowledge of such important physico-chemical quantities as the heat of adsorption (Q), changes of molar entropies (ΔS) and of free energy (ΔG) in adsorption processes allows to obtain valuable information on the character of adsorption interaction. We have determined the above mentioned thermo-dynamical characteristics, as well as entropies of hydrocarbon gases C_1-C_4 and carbon monoxide in the adsorbed state (S_{ads}) on the Type X zeolites containing Na^+, Li^+K^+, Rb^+, Cs^+, Ag^+, Mg^{2+}, Ca^{2+}, Sr^{2+}, Ba^{2+}, Cd^{2+}. The degree of Na^+ replacement by other cations in zeolites was as large as possible. This quantity depended, in the main, on the diameter of that cation which had replaced a sodium ion and its hydration capacity.

Determination of the heat of adsorption may be carried out in different ways: by direct calorimetric measurements, by the use of the data of adsorption isosteres, by theoretical calculations and by means of gas chromatography. In our studies the heat of adsorption of hydrocarbon gases C_1-C_4 and of carbon monoxide on cation exchange forms of zeolites was determined by the data of gas chromatography and compared with the results obtained by other methods.

First the method of gas chromatography for calculations of adsorption heat was worked out by Greene and Pust (1958). They found a rectilinear dependence between the retention time of some chromatographed substances and the inverse value of the column absolute temperature. One may calculate the heat of adsorption from the slope of the straight line. At present, there are a number of works devoted to the determination of heats of adsorption of gases and vapours on different adsorbents, including zeolites, with the use of gas chromatography (Andronikashvili *et al.*, 1965; Cremer, 1959; Eberly, 1961; Eberly, 1962; Kisilev *et al.*, 1962; Moore and Ward, 1960; Neddenriep, 1968; Ross *et al.*, 1962; Turkel'taub *et al.*, 1961). With this method one obtains, as a rule, rather lower values of adsorption heat than those measured by the calorimetric method, calculated on the basis of isosteres of adsorption. That is very clearly displayed for an adsorbate having large molecules or very active forms of adsorbent. Such a situation is caused by adsorption processes going slower than chromatographic ones, and equilibrium has no time to be completely established (Kisilev and Yashin, 1964).

However, as the results of a number of studies show, the values of the heat of adsorption of low molecular hydrocarbons and of some gases, boiling at low temperatures, calculated on the basis of chromatographic, calorimetric and isosteric measurements, are close to each other (Dubinin *et al.*, 1968; Greene and Pust, 1958; Ross *et al.*, 1962; Tsitsishvili and Andronikashvili, 1970). On zeolites

Table 1. *Values of adsorption heat (kcal/mole) obtained by different techniques.*

Compound	Adsorbent	Q chrom.	Q calorim.	Q isosteric	References
CO_2	silicagel	6.290	6.280-7.950	—	a
C_2H_4	silicagel	6.700	6.050-7.540	—	a
Ar	act. charcoal	2.700	3.760-4.220	—	a
N_2	act. charcoal	2.900	3.830-4.100	2.500-4.500	a
Kr	graph. carb.	3.20	—	3.15	b
Xe	graph. black	3.94	—	4.03	b
CH_4	graph. black	2.90	—	3.13	b
C_2H_6	graph. black	3.78	—	4.0	b
C_2H_6	zeolite LiX*	6.0		5.5	c, d
	zeolite NaX	6.2		6.2	c, e
	zeolite CsX	7.0		6.7	c, f
C_2H_4	zeolite LiX*	8.2		8.9	c, d
	zeolite NaX	9.0		9.2	c, e
	zeolite CsX	7.3		7.7	c, f
C_3H_8	zeolite NaX	8.0-8.1	8.5	—	e, g, h

* Note. The degree of replacement of Li^+ by Na^+ is $\sim 50\%$.
a) Greene and Pust, 1958; b) Ross *et al.*, 1962; c) Bezus *et al.*, 1969; d) Tsitsishvili and Andronikashvili, 1970; e) Tsitsishvili *et al.*, 1966; f) Tsitsishvili *et al.*, 1967; g) Kisilev *et al.*, 1965; h) Dubinin *et al.*, 1968.

such a coincidence of results of determination of the heat values of adsorption (table 1) by different methods is characteristic of hydrocarbons up to C_3 and of other gases boiling at low temperatures (Habgood, 1964; Kisilev *et al.*, 1965). Somewhat lower values of adsorption heat determined by gas chromatography are caused by a small filling of the adsorption space at the use of this method. In other words, under such conditions, there is no interaction among adsorbate molecules and hence the values of heats of adsorption are somewhat lowered as they reflect only the interaction in the system adsorbate-adsorbent (Kisilev *et al.*, 1962).

The positive feature of chromatographic methods of determination of adsorption heat is that it does not require the use of comparatively complex equipment and permits the taking of measurements in a short period of time. In addition, one can determine the adsorption heat of several substances simultaneously at high temperatures of the column, and in a wide temperature range.

To have more accurate determination of adsorption heat by the data of gas chromatography, one should choose the experimental conditions under which one observes linearity of isotherms of adsorption of the studied compounds. It is necessary to choose a range of column heating in which chromatographed substances are characterized by symmetry of peaks of the separation curve. With this aim, one should also decrease the amount of the analysed sample put into the chromatographic column (Figueras, 1968). On the basis of the determination of adsorption heat, it is possible to calculate, using formulas, (Tsitsishvili *et al.*, 1966; Andronikashvili *et al.*, 1971; Eremenko *et al.*, 1967; Khan, 1962) the change of free molar energy (ΔG) and the change of molar entropy (ΔS)

$$\Delta G = -RT\ln v_{\mathrm{v}}, \qquad \Delta S = \frac{Q - \Delta G}{T},$$

where R is the universal gas constant kcal/mole, T is the absolute temperature $^{\circ}$K (temperature of column heating) and v_{v} is the specific retention volume.

The entropy of the substance in the adsorbed state (S_{ads}) is calculated as the difference of entropy in the standard state (from the table data) ("Physicochemical Properties of Single Hydrocarbons", 1947; Vukalovich *et al.*, 1953) and changes of molar entropy (ΔS).

The samples for studies in our experiments were prepared as follows. Granules without clay binder were prepared from zeolite powder with the size 15-30 mesh (0.5-1 mm), which after the corresponding thermal activation (dehydration) were put into the chromatographic column. The thermal activation of zeolites was made in two ways:

(1) Moderate activation, i.e. heating of zeolites at 450°C during 5 hours before chromatographic column loading, with the following heating in the column at 300°C in the flow of the gas-carrier.

(2) Deep activation, i.e. heating of zeolites in a chromatographic column first during five hours at 480°C at the continuous pumping out, then under the same conditions, but in the flow of the gas-carrier. The heat of adsorption of hydrocarbon gases C_1-C_4 and of carbon monoxide was determined in the temperature range 20-280°C. Such a temperature range was chosen for each compound in which the chromatographic peak of the substance was highly symmetric.

Values of the adsorption heat for zeolites with singly charged and doubly charged cations are given in tables 2 and 3. In table 3 (in brackets) there are values of the adsorption heat calculated for samples subjected to deep activation. For zeolites containing cadmium, adsorption heat of carbon monoxide and of ethylene are of rather approximate character, since their corresponding peaks on

Table 2. *The adsorption heat of hydrocarbon gases C_1-C_4, and carbon monoxide, on zeolites of Type X with singly charged cations*

Zeolite	Degree of replacement of Na by other cations in %	kcal/mole							$\Delta Q(\text{CO} - \text{CH}_4)$	$\Delta Q(\text{C}_2\text{H}_4 - \text{C}_2\text{H}_6)$	$\Delta Q(\text{C}_3\text{H}_6 - \text{C}_3\text{H}_8)$
		CH_4	CO	C_2H_6	C_2H_4	C_3H_8	C_3H_6	C_4H_{10}			
Lix	91.0	4.1	9.0	5.7	10.7	7.7	14.1	9.5	+4.9	+5.0	+6.4
NaX	0.0	4.5	6.9	6.2	9.0	8.0	11.1	9.9	+2.4	+2.8	+3.1
		(4.2)*	(5.05)*	(6.0)*	(9.2)*	(8.1)**	(10.9)*	(9.2)*			
KX	83.5	4.5	4.8	5.4	7.0	8.2	9.1	9.9	+0.3	+0.6	+0.9
RbX	54.0	5.0	4.8	6.6	6.8	8.3	8.9	10.5	−0.2	+0.2	+0.6
CsX	53.5	5.2	5.0	7.0	7.3	8.5	9.0	10.8	−0.2	+0.3	+0.5
AgX	90.0	5.8	—	8.3	—	9.9	—	—	—	—	—

* Kislev, A. V., et al. (1965)
** Habgood, H. W. (1964).

Table 3. *The adsorption heat of the hydrocarbon gases C_1-C_4, and carbon monoxide, on zeolites of Type X with bivalent cations*

Zeolite	Degree of replacement of Na by other cations in %	kcal/mole							$\Delta Q(CO - CH_4)$	$\Delta Q(C_2H_4 - C_2H_6)$	$\Delta Q(C_3H_6 - C_3H_8)$
		CH_4	CO	C_2H_6	C_2H_4	C_3H_8	C_3H_6	C_4H_{10}			
NaX	0	4.5	6.9	6.2	9.0	8.0	11.1	9.9	+ 2.4	+ 2.8	+ 3.1
$\overline{\text{Mg}}$NaX	65	4.2	5.9	6.1	9.1	7.8	9.9	9.1	+ 1.7	+ 3.0	+ 2.1
		(4.2)	(6.1)	(6.3)	(9.4)	(8.0)	(9.9)	(9.1)	(+ 1.9)	(+ 3.1)	(+ 2.9)
CaNaX	90	4.1	7.9	6.2	10.9	7.2	13.9	9.2	+ 3.8	+ 4.7	+ 6.7
		(4.8)	(9.3)	(6.9)	(28)	(8.9)			(+ 4.5)	(+ 21.1)	
SrNaX	95	5.5	7.8	7.5	13.0	9.6	14.9	11.3	+ 2.3	+ 5.5	+ 5.3
BaNaX	73	5.2	6.9	7.9	10.9	10.3	12.2	10.6	+ 1.7	+ 3.0	+ 1.9
CdNaX	88	4.7	~10	7.1	~27	8.0		10.3	+ 5.3	+ 19.9	

chromatograms are asymmetric. In the three last columns of the tables 2 and 3 the difference of the adsorption heats are given for the following pair components: carbon monoxide-methane, ethane-ethylene, propane-propylene. All zeolites with singly charged cations were moderately activated. The exception was the sample containing lithium, which owing to rather high energy of lithium cation hydration was activated at a higher temperature.

Determination of the heat of adsorption (Q) of saturated hydrocarbons C_1-C_4 on Type X zeolites, with singly charged cations, has shown an increase from the lithium form to cesium (Table 2). On zeolites containing silver the heat of adsorption for saturated hydrocarbons is higher than those obtained on the cesium form of zeolites.

Heats of adsorption of unsaturated hydrocarbon gases C_2-C_3 and carbon monoxide decrease appreciably in the following sequence: $LiX > NaX > KX > RbX$. The adsorption heat of these compounds is higher on a sample containing cesium than on rubidium zeolites. An increase of adsorption heat of saturated hydrocarbons on zeolites containing heavier cations of univalent metals should, obviously, be associated with an increase of dispersion and polarization interactions, which are known to increase with an increase of polarizability of cations entering the zeolite structure. For substances, molecules of which are characterized by the existence of π-bonds, dipole and quadrupole moments, the adsorption interaction is displayed with great strength on zeolites with concentrated positive charges, and characterized by energy heterogeneity.

An increase of the adsorption heat of unsaturated compounds and carbon monoxide on a sample containing cesium, in comparison with the data obtained on rubidium zeolite should apparently be attributed to a strong increase of the fraction of interactions connected with dispersion and polarization effects.

A rectilinear dependence is observed between the values of adsorption heat of hydrocarbon gases C_1-C_4 and the number of carbon atoms in a molecule both in the case of saturated and unsaturated hydrocarbons on zeolites of Type X (figure 1). It should be noted that the first point on the curve, expressing the dependence of the heat of adsorption for unsaturated hydrocarbons on the number of carbon atoms, corresponds to carbon monoxide. The line expressing this dependence, for unsaturated hydrocarbons and carbon monoxide on zeolites containing cations of lithium and sodium, is higher than the line for saturated hydrocarbons. For samples containing cesium the straight lines giving the dependence almost coincide for saturated and unsaturated hydrocarbons.

The influence of the nature of cations of alkaline metals in zeolites is felt by the values of the adsorption heat of separate compounds. Thus, for instance, on zeolites containing cations of lithium, sodium and potassium the heats of adsorption of carbon monoxide are higher than that of methane. And on zeolites containing potassium the values of heats of adsorption of carbon monoxide and of methane are close by their numerical values. On zeolites containing rubidium and cesium the values of adsorption heat for methane are larger than those for carbon monoxide.

The values of the adsorption heat of saturated hydrocarbons of samples containing Na^+, Mg^{++} and Ca^{++} are almost the same (Table 3). Deeper activation of these samples causes only a slight increase of the values of adsorption heat for saturated hydrocarbons, especially for samples containing magnesium. On zeolites containing strontium and barium there is an appreciable increase of Q. One may assume that such a behaviour, characteristic of zeolites with doubly charged cations, should be attributed to lowered values of dispersion and polarization interactions; taking into account replacement of two singly charged cations by one double charged cation, and an increased

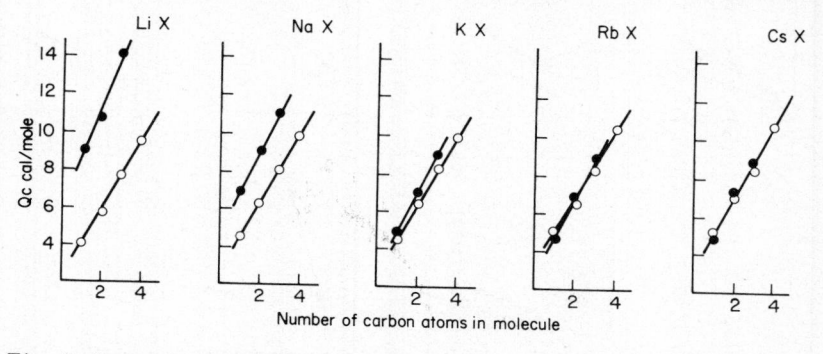

Fig. 1. Dependence of values of the adsorption heat of hydrocarbon gases C_1-C_4, and carbon monoxide, on the number of carbon atoms in a molecule on zeolites Type X containing lithium, sodium, potassium rubidium and cesium. Black circles correspond to carbon monoxide and unsaturated hydrocarbons. White circles for saturated hydrocarbons.

hydrophility of magnesium and calcium containing forms. The opposite picture is observed for strontium and barium replaced zeolites.

The adsorption heat of carbon monoxide and of unsaturated hydrocarbons increases in the sequence $MgX - CaX - SrX$, together with a decrease of hydration energy of cations in the mentioned sequence. Deep activation of zeolites with double charged cations does not cause an appreciable increase of adsorption heat for carbon monoxide, and unsaturated hydrocarbons on samples containing magnesium, but on calcium replaced forms it leads to a considerable increase of the values of adsorption heat for carbon monoxide and especially for ethylene. On BaX zeolites the adsorptions heats for these compounds is, in the main, lower, probably due to partial loss of crystallinity of the sample, but the values of heats of adsorption are still higher than on NaX. Low values of adsorption heat of unsaturated hydrocarbons and carbon monoxide on zeolite containing magnesium are connected with a number of factors. The most important being: high energy of hydration of Mg^{2+}, and the possible migration of these cations from the sites occupied by them into more screened positions under the thermal treatment.

The adsorption heats of saturated hydrocarbons C_1-C_4, on samples containing cadmium, are close to the values obtained on NaX, but the values of heats of adsorption for ethylene and carbon monoxide are strongly increased. This may be because it is connected with complex formation.

On the Type X zeolites with cations of alkaline-earth metals, heats of adsorption of unsaturated hydrocarbons are much higher than those of saturated hydrocarbons with the same number of carbon atoms in a molecule (figure 2).

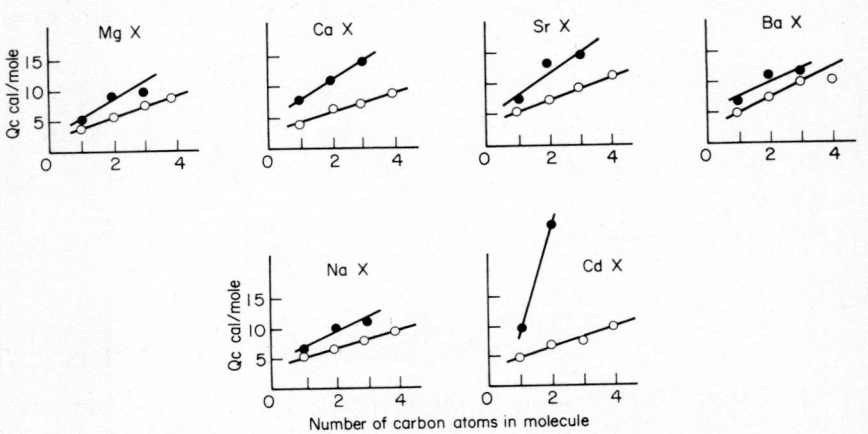

Fig. 2. Dependence of values of the adsorption heat of hydrocarbon gases C_1-C_4, and carbon monoxide, on the number of carbon atoms in a molecule on zeolites Type X containing magnesium, calcium, strontium, barium, cadmium and sodium. Black circles correspond to carbon monoxide and unsaturated hydrocarbon. White circles for saturated hydrocarbons.

Unlike zeolites with single charged cations there is no sharply pronounced decrease of differences in the adsorption heat of systems ethane-ethylene and propane-propylene on this cation-exchange forms with heavier metals. Independently of the nature of double charged cations, heats of adsorption of carbon monoxide are higher than those of methane.

The relative parallelism between the lines characterizing the dependence of heats of adsorption of carbon monoxide, unsaturated hydrocarbons and saturated hydrocarbons on the number of carbon atoms in a molecule, is sharply violated on the sample with replaced cadmium (figure 2). That, apparently, may be explained by a strong specific interaction of ethylene with cadmium cations in zeolites. Changes of molar entropy of adsorption (ΔS) and the values of entropy in the adsorbed state (S_{ads}) of hydrocarbon gases C_1-C_4, and carbon

Table 4a. *Changes of molar free energy (ΔG cal/mole), of molar entropy (ΔS e.u.) and entropy in the absorbed state of hydrocarbon gases C_1-C_4, and of carbon monoxide, on zeolites Type X, containing singly charged cations*

Compound	Temperature, °K	S, entropy e.u.	Samples								
			LiX (91%)			NaX			KX (83.5%)		
			$-\Delta G$	$-\Delta S$	S ads.	$-\Delta G$	$-\Delta S$	S ads.	$-\Delta G$	$-\Delta S$	S ads.
Methane	298	44.5	1729.8	7.9	36.6	1963.3	8.5	36.0	1722.7	9.3	35.2
	350	45.95	977.0	8.9	37.0	1231.0	9.3	36.6	859.3	10.4	35.55
Carbon monoxide	298	47.2	3037.0	20.0	27.2	2430.1	15.0	32.2	1631.6	10.6	36.6
Ethane	350	56.8	2469.0	9.2	47.6	2727.2	9.92	46.9	2653.0	10.7	46.1
	400	58.98	1594.2	10.2	48.7	2030.2	10.42	48.6	1923.8	11.2	47.8
Ethylene	400	55.89	3716.1	17.5	38.4	3325.0	14.2	41.7	2530.6	11.2	44.7
	450	57.5	2643.0	17.9	39.6	2455.2	14.5	43.0	1802.5	11.6	45.9
Propane	400	70.4	3100.0	11.5	58.9	3345.7	11.6	58.8	3321.7	12.2	58.2
	500	75.9	1620.2	12.1	63.8	1900.8	12.2	63.7	1811.0	13.8	62.1
Propylene	500	73.5	3010.3	22.2	51.3	3047.3	16.1	57.4	2451.4	13.3	60.2
n-buthane	500	89.2	2700.0	13.6	75.6	3024.7	13.47	75.7	2991.7	13.8	75.4

Table 4b. *Changes of molar free energy (ΔG cal/mole), of molar entropy (ΔS e.u.) and entropy in the adsorbed state of hydrocarbon gases C_1–C_4, and of carbon monoxide, on zeolites Type X, containing singly charged cations*

Compound	Temperature, °K	S, entropy e.u.	Samples								
			RbX (54%)			Cs (53.5%)			AgX (90.0%)		
			$-\Delta G$	$-\Delta S$	S ads.	$-\Delta G$	$-\Delta S$	S ads.	$-\Delta G$	$-\Delta S$	S ads.
Methane	298	44.5	1904.0	10.4	34.1	1917.9	11.0	33.5	2614.1	10.7	33.8
	350	45.95	1263.1	10.7	35.2	1247.3	11.3	34.6	—	—	—
Carbon monoxide	298	47.2	1729.2	10.3	36.9	1586.1	11.3	35.9	—	—	—
Ethane	350	56.8	2805.0	10.8	46.0	2811.6	12.0	44.8			
	400	58.98	2168.3	11.1	47.9	2193.4	12.0	46.9	3245.0	12.7	46.3
Ethylene	400	55.89	2722.1	10.2	45.7	2607.4	11.7	44.2	3956.0	12.4	44.4
	450	57.5	2043.0	10.6	46.9	1946.6	11.9	45.6	—	—	—
Propane	400	70.4	3506.1	12.0	58.4	3451.5	12.6	57.8	—	—	—
	500	75.9	2272.1	12.1	63.8	2294.3	12.4	63.5	3340.0	13.12	62.8
Propylene	500	73.5	2813.0	12.2	61.3	2724.9	12.6	60.9	—	—	—
n-buthane	500	89.2	—	—	—	3391.6	14.8	74.4	—	—	—

Table 5a. *Changes of molar free energy (ΔG cal/mole), of molar entropy (ΔS e.u.) and entropy in the adsorbed state of hydrocarbon gases C_1-C_4, and of carbon monoxide, on zeolites Type X, containing bivalent cations*

Compound	Temperature, °K	S, entropy e.u.	NaX			MgX (65%)			CaX (90%)		
			$-\Delta G$	$-\Delta S$	S ads.	$-\Delta G$	$-\Delta S$	S ads.	$-\Delta G$	$-\Delta S$	S ads.
Methane	293.15	44.5	1974.0	10.0	34.5	1480.6	9.6	34.9	1827.4	9.9	34.6
	350	45.9	1465.1	9.9	36.1	1019.2	9.3	36.6	1321.8	9.8	36.2
Carbon monoxide	293.15	47.2	2480.9	14.9	32.3	1747.3	14.7	32.5	2921.1	16.9	30.3
Ethane	400	59.0	2329.6	11.5	47.5	1710.8	10.9	48.1	2093.0	11.1	47.9
Ethylene	450	57.5	2886.9	15.2	42.3	1924.6	15.9	41.6	3439.8	16.7	40.8
Propane	400	70.4	3621.8	11.8	58.6	2839.2	12.4	58.0	3312.4	9.7	60.7
Propylene	500	73.5	3594.5	15.9	57.6	2502.5	14.8	58.6	4299.7	20.7	52.8
n-buthane	500	89.2	3594.5	13.2	76.0	2798.25	12.6	76.5	3799.2	12.9	76.2

Table 5b. *Changes of molar free energy (ΔG cal/mole), of molar entropy (ΔS e.u.) and entropy in the adsorbed state of hydrocarbon gases C_1-C_4, and of carbon monoxide, on zeolites Type X, containing bivalent cations*

Compound	Temperature, °K	S, entropy e.u.	SrX (95%)			BaX (73%)			CdX (88%)		
			$-\Delta G$	$-\Delta S$	S ads.	$-\Delta G$	$-\Delta S$	S ads.	$-\Delta G$	$-\Delta S$	S ads.
Methane	293.15	44.5	2134.1	11.38	33.1	2160.8	9.3	35.2	1733.9	10.3	34.2
	350	45.9	1656.2	10.9	35.0	1656.2	9.2	36.7	1242.1	10.0	35.9
Carbon monoxide	293.15	47.2	3067.8	16.2	31.0	2667.7	13.8	33.4	3854.7	21.5	25.7
Ethane	400	59.0	2566.2	13.1	45.9	2857.4	11.1	47.9	2092.0	12.6	46.4
Ethylene	450	57.5	3767.4	20.6	36.9	3480.8	15.7	41.8	6511.1	46.3	11.2
Propane	400	70.4	3840.2	15.2	55.2	4368.0	13.6	56.8	3312.4	11.7	58.7
Propylene	500	73.5	4686.50	20.4	53.1	4368.0	15.6	57.9	—	—	—
n-butane	500	89.2	3685.5	14.5	74.7	4254.3	12.1	77.1	3094.0	14.4	74.8

monoxide both on zeolites with singly charged cations (table 4) and with doubly charged cations (table 5), were calculated from the values of heats of adsorption (Q) and from changes of molar free energies (ΔG). The highest values of entropy changes for unsaturated compounds and carbon monoxide on zeolites with singly charged cations were obtained on lithium samples. From forms containing lithium to rubidium forms the changes of entropy sharply decrease. Values of ΔS on CsX zeolites are rather higher. Changes of entropy for saturated hydrocarbons C_1-C_4 are increased to a considerable extent on AgX and CsX zeolites in comparison with ΔS for lithium zeolites. Thus the following sequence of a decrease in values of changes in the entropies of saturated hydrocarbons can be observed: AgX > CsX > RbX > KX > Nax > LiX (table 4). The mobility of molecules of carbon monoxide and that of unsaturated hydrocarbons is most limited on LiX and of saturated hydrocarbon gases C_1-C_4 on AgX and CsX zeolites. For saturated hydrocarbon gases C_1-C_4 on zeolites with double charged cations there is some slight increase of entropy change in the sequence: NaX—MgX—CaX—SrX (table 5).

On BaX zeolite one observes a sharp decrease of this quantity, probably connected with the sample becoming partially amorphous.

Entropies of saturated hydrocarbon gases C_1-C_4 in the adsorbed state (S_{ads}) experience a decrease in the sequence: MgX—CaX—SrX. Thus, for saturated hydrocarbon gases, a limited freedom of motion on zeolites containing strontium is observed. On zeolites containing barium an increase of S_{ads} of saturated hydrocarbons C_1-C_4 takes place.

Changes of entropy at adsorption (ΔS) of unsaturated hydrocarbons are mostly decreased in the row CaX—SrX—BaX. The highest values of ΔS, in addition to CaX and SrX for unsaturated compounds and carbon monoxide are obtained on zeolites containing cadmium. The freedom of motion of these compounds is limited first of all on zeolites containing cadmium. Thus on CdX zeolite, in particular, the freedom of motion of ethylene is strongly limited in comparison with NaX. This is caused by an intensive interaction, may be with complex formation.

So zeolites with doubly charged cations are characterized by a somewhat different dependence of thermodynamical characteristics of the adsorbed hydrocarbon gases C_1-C_4 and carbon monoxide on cation nature, than in the case of zeolites with singly charged cations. Such a deviation from the regularities observed on zeolites, with singly charged cations must apparently be attributed to the influence of a number of factors and in the case of Mg and Ca zeolites, mainly to high hydrophylity of these ions.

REFERENCES

Andronikashvili, T. G., Tsitsishvili, G. V., Sabelashvili, Sh. D. and Chumburidze, T. A. (1965). "Zeolites: synthesis, properties and application". *Izd. Nauka,* 179.

Andronikashvili, T. G., Tsitsishvili, G. V. and Chumburidze, T. A. (1971). *Bull. Acad. Sci. Georgian SSR,* **61**, 3, 597.

Bezus, A. G., Kisilev, A. V. and Sedlachek, Z. (1969). *Zh. Fiz. Khim.* **43**, 5, 1223.

Cremer, E. (1959). *Z. Anal. Chem.* **170**, 1, 219.

Dubinin, M. M., Isirikian, A. A., Sarakhov, A. N. and Serpinsky, V. V. (1968). *Izv. Akad. Nauk. SSSR, Ser. Khim.* **8**, 1690.

Eberly, P. E. Jr. (1961). *J. Phys. Chem.* **65**, 68.

Eberly, P. E. Jr. (1962). *J. Phys. Chem.* **66**, 812.

Figueras, F. (1968). *Rev. Groupem. Avancem. Methodes Spectrogr.* **4**, 1, 120.

Greene, S. A. and Pust, H. (1958). *J. Phys. Chem.* **62**, 1, 55.

Habgood, H. W. (1964). *Can. J. Chem.* **42**, 2340.

Eremenko, A. M., Korol, A. N., Piontovskaya, M. A. and Neymark, N. E. (1967). *Ukr. Khim. Zh.* **33**, 5, 453.

Khan, M. A. (1962). *Lab. Pract.* **11**, 195.

Kisilev, A. V., Khrapova, E. V. and Shcherbakova, K. D. (1962). *Neftekhimiya,* **2**, 6, 377.

Kisilev, A. V. and Yashin, Ya. N. (1964). *Neftekhimiya,* **4**, 4, 634.

Kisilev, A. V., Chernen'kova, Yu. L. and Yashin, Ya. N. (1965). *Neftekhimiya,* **5**, 4, 589.

Moore, W. R. and Ward, H. R. (1960). *J. Phys. Chem.* **64**, 6, 832.

Neddenriep, B. S. (1968). *J. Colloid Interface Sci.* **28**, 293.

"Physicochemical Properties of Single Hydrocarbons". (1947). *Gostoptekhizdat.* (1960). 343.

Ross, S., Saelens, J. K. and Olivier, J. P. (1962). *J. Phys. Chem.* **66**, 4, 696.

Tsitsishvili, G. V., Andronikashvili, T. G. and Sabelashvili, Sh. D. (1966). *Zh. Fiz. Khim.* **40**, 5, 1128.

Tsitsishvili, G. V., Andronikashvili, T. G., Sabelashvili, Sh. D. and Koridze, Z. N. (1967). *Bull. Acad. of Sci. Georgian SSR,* **46**, 3, 611.

Tsitsishvili, G. V. and Andronikashvili, T. G. (1970). II-Intern. Conf. on Zeolites, Worcester, USA.

Turkel'taub, N. M., Zhukhovitskiy, A. A. and Porshneva, N. V. (1961). *Zh. Prikl. Khim.* **34**, 9, 1946.

Vukalovich, M. P., Kirillin, V. A., Remizov, S. A., Siletskiy, V. A. and Timofeev, V. N. (1953). "Thermodynamical Properties of Gases". p. 152, Mashgiz, Moscow.

INTERACTION OF PAIRS OF ADSORBED HELIUM ATOMS*

W. A. STEELE and E. J. DERDERIAN

Department of Chemistry,
The Pennsylvania State University, Pennsylvania, USA

I. INTRODUCTION

In a preceding paper (Derderian and Steele, in press), a perturbation calculation of the quantum-mechanical second virial coefficient of He^4 and He^3 in the range 10-70°K was developed. In this approach hard-sphere wave functions were chosen as a basis set and the Lennard-Jones 6-12 potential was added to the Hamiltonian as a pertubation. The quantum virial coefficient $B^*(T)$ was given as a sum of terms involving the hard-sphere virial coefficient plus a "classical" perturbation term plus terms in powers of $\hbar^2$ due to the quantum effects. A comparison between our numerical calculations of $B^*(T)$, which include the term proportional to $\hbar^2$, with virial coefficients given by the Wigner-Kirkwood (WK) expansion, *vis-à-vis* the quantum second virial coefficient calculated exactly for a Lennard-Jones gas by Boyd *et al.* (1969), revealed the following: (i) our $B^*(T)$ is superior to $B^*_{WK2}(T)$, the WK second virial coefficient through the term proportional to $\hbar^2$, over the entire temperature range for which it was calculated, (ii) $B^*(T)$ is of comparable accuracy or better than $B^*_{WK6}(T)$, the WK second virial through the term proportional to $\hbar^6$, (iii) our perturbation method can successfully treat exchange contributions, i.e. those due solely to statistics, to the quantum second virial coefficient. In contrast, it is known that the WK expansion is totally unsuitable for treating these exchange effects (Boyd *et al.*, 1969).

In view of the success of this approach, we have applied the same formalism in this paper to calculate the quantum second virial coefficient of a two-dimensional gas. The latter is well approximated experimentally by an adsorbed gas on a smooth uniform surface. We have measured adsorption isotherms of He^4 and He^3 on graphitized carbon black in the range 12-22°K. These data are analysed in terms of the two-dimensional virial isotherm equation to obtain experimental quantum second virial coefficients of He^4 and He^3.

* This work was supported by a grant from the Army Research Office, Durham.

We find that our theoretical, two-dimensional second virial coefficients give good qualitative agreement with these data. Furthermore, it emerges that our perturbation method gives realistic results in the experimental temperature region, where the WK expansion in two dimensions fails badly.

In section II we present and analyse the experimental data. Section III is devoted to an outline of the perturbation method. Our theoretical and experimental results are discussed in section IV.

II. EXPERIMENTAL RESULTS AND ANALYSIS

The apparatus used in this work has been described in detail by Barnes and Steele (1966) and Steele and Aston (1957). Briefly, the adsorbent, in this case a sample of graphitized carbon black weighing 35.44 g with a specific area of ~ 80 m^2/g, is contained in a calorimeter surrounded by an electrically heated shield which is suspended in an evacuated can. The latter is immersed in liquid or solid hydrogen and the temperature of the assembly, consisting of the adsorbent-filled calorimeter plus the adsorbed gas, can be set and maintained to within $0.001°K$ in the range $12\text{-}25°K$. (The temperature was measured with a platinum resistance thermometer.)

Adsorption isotherms of He^4 and He^3 on this surface at coverages up to a monolayer were determined at five temperatures, $12.63°K$, $14.90°K$, $16.89°K$, $19.98°K$ and $22.00°K$. In the analysis of the isotherms, we have assumed that we can regard the adsorbed phase as a dilute, two-dimensional, imperfect gas. The role of the adsorbent then is simply to provide a uniform, external potential field. We can then analyse the experimental isotherms from the standpoint of the two-dimensional virial isotherm, which can be written (Steele, 1967)

$$\frac{N_a}{p} = \frac{Z_s}{kT} + N_a\left(\frac{-2B_{2D}}{A}\right)\frac{Z_s}{kT} + \frac{N_a^2}{2}\left(\frac{2B_{2D}^2}{A^2} - \frac{3C_{2D}}{A^2}\right)\frac{Z_s}{kT} + \text{higher terms} \qquad (1)$$

where N_a is the amount of gas adsorbed on the surface, p is the equilibrium isotherm pressure, Z_s is the configuration integral for a single atom on the surface, A is the specific area of the adsorbent, B_{2D} is the two-dimensional second virial coefficient, and C_{2D} is the two-dimensional third virial coefficient.

N_a and p are, of course, measured experimentally. If one then plots N_a/p vs N_a for a given temperature, the intercept is clearly given by Z_s/kT, and the slope is equal to $(-2B_{2D}/A)Z_s/kT$ from equation (1). B_{2D} is then simply given by

$$B_{2D} = \frac{-A}{2}\frac{\text{slope}}{\text{intercept}} \qquad (2)$$

A typical plot of this nature is given in figure 1.

The virial coefficients for He^4 and He^3, derived in this manner from the experimental data, are presented in Table 1. These values have been reduced by $B_{\text{class-hs}}(T) = \pi\sigma_{LJ}^2/2$, where σ_{LJ} is the parameter in the Lennard-Jones potential and is equal to 2.556 Å.

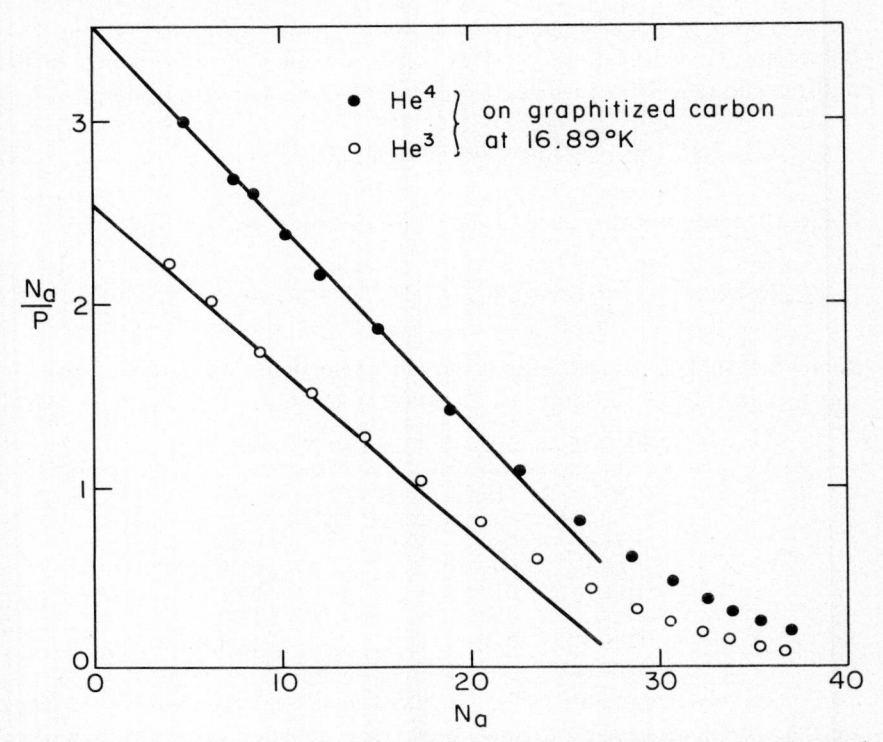

Fig. 1. Typical plots of the isotherm data are shown here. N_a is equal to the total millimoles on the surface of the sample and the pressure is in mm Hg. Slopes and intercepts obtained from these plots are substituted into equation (2) together with the B.E.T. area to obtain two-dimensional second virial coefficients.

III. THE QUANTUM SECOND VIRIAL COEFFICIENT

The perturbation method for the calculation of the quantum-mechanical second virial coefficient has been developed in detail by Derderian and Steele (in press). In this paper we will for the most part simply present the corresponding expressions appropriate to the two-dimensional case.

The heart of our approach is a calculation of the density-independent part of the quantum mechanical pair correlation function, $g(r)$; the integral of this quantity (over the intermolecular separation distance r) gives the second virial coefficient. It is conceptually useful to separate $g(r)$ into two parts

$$g(r) = g_{dir}(r) + g_{exch}(r) \tag{3}$$

where $g_{dir}(r)$ is just the quantum-mechanical pair-correlation function of a Boltzmann gas, whereas $g_{exch}(r)$ embodies the quantum effects due to the statistics and spin of the particles. Explicitly, these are defined as follows

$$g_{dir}(r) = 2\Lambda^2 \int_0^\infty dk \sum_m \psi_{km}^*(r, \phi) e^{-\beta \mathcal{H}_{rel}} \psi_{km}(r, \phi) \tag{4}$$

where ϕ denotes the orientation of the two-dimensional vector $\mathbf{r}$, and

$$g_{exch}(r) = \pm \frac{2\Lambda^2}{(2s+1)} \int_0^\infty dk \sum_m \psi_{km}^*(r, \phi) e^{-\beta \mathcal{H}_{rel}} \psi_{km}(r, \phi + \pi) \tag{5}$$

where $\Lambda = (\hbar^2/2\pi m k T)^{1/2}$ and s is the spin of the individual particle. The $+(-)$ sign in equation (5) corresponds to BE(FD) statistics. The ψ_{km} are spatial

Table 1. *Experimental second virial coefficients*

T	He^4	$B^*(T)$	He^3
12.63	0.643		0.692
14.90	0.692		0.780
16.89	0.789		0.877
19.98	0.887		1.013
22.00	0.994		1.043

hard-sphere wave functions and constitute a complete, orthonormal set. They are exact solutions of the two-dimensional wave equation of relative motion of two particles with a Hamiltonian, $\mathcal{H}_{rel}^0 = (-\hbar^2/2\mu)\nabla^2 + V$, where $\hbar$ is Planck's constant divided by 2π, μ is the reduced mass, and V is the intermolecular potential energy. In the hard-sphere problem, this potential is

$$V_{hs}(r) = +\infty \quad \text{for } r \leqslant \sigma_{hs}$$
$$V_{hs}(r) = 0 \quad \text{for } r > \sigma_{hs}$$

where σ_{hs} is the hard-sphere diameter. In two dimensions, the hard-sphere wave functions are

$$\psi_{km}(r, \phi) = \left(\frac{k}{2\pi}\right)^{1/2} [J_m^2(k\sigma_{hs}) + N_m^2(k\sigma_{hs})]^{-1/2}$$
$$\times \{J_m(kr) N_m(k\sigma_{hs}) - J_m(k\sigma_{hs}) N_m(kr)\} e^{im\phi} \tag{6}$$

where k is the momentum wave number defined by $\mathcal{H}^0_{rel}\,\psi_{km} = (\hbar^2\,k^2/2\mu)\,\psi_{km}$, m is the angular momentum quantum number, and J_m and N_m are integral order Bessel and Neumann functions, respectively. Clearly, the wave functions are identically zero for $r \leqslant \sigma_{hs}$.

Now let us consider the Hamiltonian, $\mathcal{H}_{rel}$, in equations (4), (5), which is defined

$$\mathcal{H}_{rel} = \mathcal{H}^0_{rel} + V_{pert} \tag{7}$$

where V_{pert} is a perturbation potential which we have added to the hard-sphere Hamiltonian, $\mathcal{H}^0_{rel}$. In principle, V_{pert} can be any realistic two-body potential. In this case, we have chosen the Lennard-Jones 6-12 potential to be our perturbation potential.

We are now faced with the task of operating on the wave functions with the operator $\exp(-\beta\mathcal{H}_{rel})$. It is obvious, however, that the hard-sphere wave functions are no longer eigenfunctions of $\mathcal{H}_{rel}$. We have drawn on the work of Friedmann (1962) and expanded the exponential operator in the form

$$e^{-\beta\mathcal{H}_{rel}} = e^{-\beta V_{LJ}}\,T\,e^{-\beta\mathcal{H}^0_{rel}} \tag{8}$$

where T is given by a series expansion in powers of $\hbar^2$:

$$T = T_0 + \hbar^2\,T_1 + \hbar^4\,T_2 + \text{ higher terms} \tag{9}$$

By following the method of solution of equations (8) and (9) that has been outlined by Derderian and Steele (in press), one can show that the operators T_0, T_1, and T_2 have the following explicit form in planar coordinates

$$T_0 = 1 \tag{10}$$

$$\hbar^2\,T_1 = \frac{-\hbar^2\,\beta^2}{2\mu}\left\{ \frac{\partial V_{LJ}}{\partial r}\frac{\partial}{\partial r} + \frac{1}{2r}\frac{\partial}{\partial r}\left(r\frac{\partial V_{LJ}}{\partial r}\right) - \frac{\beta}{3}\left(\frac{\partial V_{LJ}}{\partial r}\right)^2 \right\} \tag{11}$$

and

$$\hbar^4\,T_2 = \frac{-\hbar^4\,\beta^3}{4\mu^2}\left\{ \frac{2}{3}\frac{\partial^2\,V_{LJ}}{\partial r^2}\frac{\partial^2}{\partial r^2} - \frac{\beta}{2}\left(\frac{\partial V_{LJ}}{\partial r}\right)^2\frac{\partial^2}{\partial r^2} + \frac{2}{3r^3}\left(\frac{\partial V_{LJ}}{\partial r}\right)\frac{\partial^2}{\partial\phi^2} \right\} \tag{12}$$

In the present work we have retained only the first two terms in equation (9). Upon substitution of these into equation (4) we obtain

$$g_{dir}(r) = g^{(0)}_{dir}(r) + g^{(1)}_{dir}(r) + \text{ higher terms} \tag{13}$$

where

$$g^{(n)}_{dir}(r) = 2(\hbar^n\,\Lambda)^2 \int_0^\infty dk\, e^{-\beta V_{LJ}} \sum \psi^*_{km}(r,\phi)\,T_n\,e^{-\beta\mathcal{H}^0_{rel}}\,\psi_{km}(r,\phi) \tag{14}$$

It is clear from the form of the operators in equation (9) that the right-hand side of equation (13) is actually an expansion of the pair-correlation function in powers of $\hbar^2$. Analogously, we have

$$g_{exch}(r) = g_{exch}^{(0)}(r) + g_{exch}^{(1)}(r) + \text{higher terms} \tag{15}$$

where

$$g_{exch}^{(n)}(r) = \pm \frac{2(\hbar^n \Lambda)^2}{(2s+1)} \int_0^\infty dk\, e^{-\beta V_{LJ}} \sum \psi_{km}^*(r,\phi) T_n e^{-\beta \mathscr{H}_{rel}^0} \psi_{km}(r,\phi+\pi) \tag{16}$$

We are now in a position to derive the expressions for the quantum second virial coefficient, $B^*(T)$. By definition

$$B(T) = \frac{-1}{2} \int [g(r) - 1]\, d\mathbf{r} \tag{17}$$

We can reduce $B(T)$ by $B_{\text{class-hs}}(T) = \pi \sigma_{LJ}^2/2$ and write equation (17) in a more convenient form

$$B^*(T) = -2(\sigma_{eff})^2 \int_0^\infty [g(R) - 1]\, R\, dR \tag{18}$$

where $\sigma_{eff} = \sigma_{hs}/\sigma_{LJ}$ and $R = r/\sigma_{hs}$. Once again, we find it instructive, as well as useful, to separate the quantum effects.

$$B^*(T) = B_{dir}^*(T) + B_{exch}^*(T) \tag{19}$$

where $B_{dir}^*(T)$ is just the quantum second virial of a Boltzmann gas and embodies quantum effects due to mass only, while $B_{exch}^*(T)$ reflects quantum effects due to the statistics and spin.

It follows quite naturally from the above that one can define

$$B^*(T) = \sum_n [B_{dir}^{(n)*}(T) + B_{exch}^{(n)*}(T)] \tag{20}$$

where

$$B_{dir}^{(0)*} = -2(\sigma_{eff})^2 \int_0^\infty [g_{dir}^{(0)}(R) - 1]\, R\, dR \tag{21}$$

$$B^{(n)*}(T) = -2(\sigma_{eff})^2 \int_0^\infty [g^{(n)}(R)]\, R\, dR \tag{22}$$

Thus equation (20) represents an expansion in powers of $\hbar^2$ of $B_{dir}^*(T)$ and $B_{exch}^*(T)$ and the superscript indicates the power to which $\hbar^2$ is taken.

Let us consider for the moment the term $B_{dir}^{(0)*}(T)$ and the corresponding correlation function, $g_{dir}^{(0)}(R)$. Hard-sphere wave functions are used for the ψ_{km} because they are eigenfunctions of $\mathcal{H}_{rel}^0$. Thus

$$e^{-\beta \mathcal{H}_{rel}^0} \psi_{km} = \psi_{km} e^{-\beta k^2/2\mu} \tag{23}$$

It is obvious from equation (13) that $g_{dir}^{(0)}(R)$ is simply equal to $e^{-\beta V_{LJ}}[g_{dir}(R)]_{hs}$ in this case, where $[g_{dir}(R)]_{hs}$ is just the direct hard-sphere pair correlation function. Therefore, we can further split this term

$$B_{dir}^{(0)*}(T) = [B_{dir}^*(T)]_{hs} + [B_{dir}^{(0)*}(T)]_{att} \tag{24}$$

where we define

$$[B_{dir}^*(T)]_{hs} = -2(\sigma_{eff})^2 \int_0^\infty \{[g_{dir}(R)]_{hs} - 1\} R dR \tag{25}$$

and

$$[B_{dir}^{(0)*}(T)]_{att} = -2(\sigma_{eff})^2 \int_0^\infty (e^{-\beta V_{LJ}} - 1) [g_{dir}(R)]_{hs} R dR \tag{26}$$

Analogous expressions can be written for $B_{exch}^{(0)*}(T)$. After operating with T_1 on V_{LJ}, it can be shown that the second term in the truncated expansion, $B_{dir}^{(1)*}(T)$ is given by

$$B_{dir}^{(1)*}(T) = \frac{48\Lambda^{*2}}{\pi T^{*2}} (\sigma_{eff})^{-10} \int_0^\infty dR[4(\sigma_{eff})^{-12} R^{-25}$$
$$-4(\sigma_{eff})^{-6} R^{-19} + R^{-13}] e^{-\beta V_{LJ}} [g_{dir}(R)]_{hs} \tag{27}$$

with a similar expression for $B_{exch}^{(1)*}(T)$. Here, $\Lambda^* = \Lambda/\sigma_{hs}$ and $T^* = kT/\epsilon$.

We now have explicit expressions for all of the terms we have included in our perturbation expansion. We anticipate, however, that our results will be rather sensitive to the size of σ_{eff}, and indeed, this turns out to be the case. A variational method has been discussed (Derderian and Steele, in press) which allows us to calculate a "best" value for σ_{eff}. This value would give the closest agreement with the exact $B^*(T)$ for a Lennard-Jones gas, if enough terms are included in the perturbation expansion to give good convergence. In this variational method, we have minimized the total free energy of our system as a function of σ_{eff} at a given temperature. More specifically, the working expression in two dimensions is

$$B_{hs}^*(T) + \frac{8}{T^*}(\sigma_{eff})^{-4} \int_0^\infty dR[(\sigma_{eff})^{-6} R^{-11} - R^{-5}][g(R)]_{hs} \geqslant B_{real}^*(T)$$

(28)

where $B_{real}^*(T)$ is the quantum second virial coefficient of the real system. Without explicitly knowing $B_{real}^*(T)$, (which is independent of σ_{eff}), we can minimize the *lhs* as a function of σ_{eff} at a given temperature. We have found that the "best" σ_{eff} values for He^4 and He^3 are 0.790 and 0.780, respectively, with almost no variation as a function of temperature in the range 10-70°K. It is interesting to note that these values are identical to those obtained in the three-dimensional case. This seems reasonable, since σ_{eff} is independent of the dimensionality in the classical limit. It should also be pointed out that the calculations show that exchange terms are negligible in the range 10-70°K, so that the only differences in the virial coefficients of He^4 and He^3 are due to mass effects.

Consequently, the smaller value found for σ_{eff} in the case of He^3 reflects its more quantum nature, since the classical values of σ_{eff} are considerably higher than those for either He^3 or He^4.

IV. DISCUSSION OF THE THEORETICAL AND EXPERIMENTAL RESULTS

We have used the perturbation method presented in detail by Derderian and Steele (in press) and outlined in the preceding section, to calculate the two-dimensional quantum second virial coefficient, $B^*(T)$, of He^4 and He^3 in the range 10-70°K. In this paper the perturbation expansion is taken only through $B^{(1)*}(T)$, the term proportional to $\hbar^2$. We will compare these theoretical values for $B^*(T)$ with our experimental second virial coefficients and with the two-dimensional quantum virial coefficients calculated by Sams (1965) and Klein (1966) using a Wigner-Kirkwood expansion.

However, we would like to discuss first our calculated values for $B_{exch}^*(T)$. These are given in Tables 2 and 3 for He^4 and He^3, respectively. We notice immediately the striking fashion in which $B_{exch}^*(T)$ is dependent on the temperature. For all but the lowest temperatures the exchange second virial coefficient is negligibly small. This remarkable behavior is identical to that of the three-dimensional $B_{exch}^*(T)$ which has been described (Derderian and Steele, in press). Furthermore, exact calculations of $B^*_{exch}(T)$ with a Lennard-Jones potential in three dimensions show the same rapid disappearance of the exchange virial with rising temperatures (Boyd et al., 1969). There is every reason to believe, therefore, that the exact two-dimensional $B^*_{exch}(T)$ will exhibit the same behavior. It appears then that the present perturbation method gives the correct behavior, at least qualitatively, of the two-dimensional exchange virial coefficient.

Table 2. *Exchange second virial coefficient of* He^4 *with* $\sigma_{eff} = 0.790$.

T	$[B^*_{exch}(T)]_{hs}$	$[B^{(0)*}_{exch}(T)]_{att}$	$B^{(1)*}_{exch}(T)$	$B^*_{exch}(T)$
3.07	-4.59×10^{-4}	-3.29×10^{-3}	2.23×10^{-2}	1.86×10^{-2}
4.67	-3.44×10^{-5}	-1.19×10^{-4}	4.36×10^{-4}	2.83×10^{-4}
9.54	-1.25×10^{-7}	-2.43×10^{-7}	2.11×10^{-7}	-1.57×10^{-7}
12.98	0	0	0	0
18.69	0	0	0	0

Table 3. *Exchange second virial coefficient of* He^3 *with* $\sigma_{eff} = 0.780$.

T	$[B^*_{exch}(T)]_{hs}$	$[B^{(0)*}_{exch}(T)]_{att}$	$B^{(1)*}_{exch}(T)$	$B^*_{exch}(T)$
4.19	2.24×10^{-4}	7.37×10^{-4}	-3.71×10^{-3}	-2.75×10^{-3}
6.39	1.68×10^{-5}	3.09×10^{-5}	-1.01×10^{-4}	-5.33×10^{-5}
13.04	6.08×10^{-8}	8.02×10^{-8}	-6.98×10^{-8}	7.12×10^{-8}
17.75	0	0	0	0
25.56	0	0	0	0

The same cannot be said about the WK expansion. It has been customary to include in the latter, in both two and three dimensions, an exchange term which corresponds to the exchange virial coefficient for a perfect gas. The reason for this seems to be to insure the proper limiting behavior of the WK expansion when applied to a perfect gas. It is known, however, that a WK expansion is totally incapable of dealing realistically with exchange effects in three dimensions. The same behavior seems to occur in two dimensions. Consequently, the two-dimensional WK calculations of Sams (1965) and Klein (1966), which include the perfect gas exchange virial coefficient, are in error in this respect. This exchange term, which is even larger in two than in three dimensions, remains finite even as high as 85°K.

We now turn our attention to the complete virial coefficient, $B^*(T)$, as given by our perturbation method. $B^*_{hs}(T)$, $B^{(0)*}_{att}(T)$, $B^{(1)*}(T)$ and $B^*(T)$ are given in

Table 4 for He^4 with σ_{eff} = 0.790 and in Table 5 for He^3 with σ_{eff} = 0.780. We recall that $B^*_{exch}(T)$ is negligible in this temperature range and therefore, the difference in $B^*(T)$ for He^4 and He^3 simply reflects the difference in their masses.

In figure 2 we plot the following quantities for He^4 as a function of temperature: (i) our experimental values for the second virial coefficient, (ii) our theoretical values for $B^*(T)$ with σ_{eff} = 0.790, (iii) our theoretical values for

Table 4. *Second virial coefficient of* He^4 *with* σ_{eff} = 0.790.

T	$B^*_{hs}(T)$	$B^{(0)^*}_{att}(T)$	$B^{(1)^*}(T)$	$B^*(T)$
9.54	1.367	−1.275	0.593	0.685
12.98	1.246	−0.928	0.324	0.642
14.18	1.215	−0.847	0.278	0.646
14.90	1.199	−0.805	0.256	0.650
15.88	1.178	−0.754	0.231	0.655
16.45	1.168	−0.727	0.218	0.649
16.95	1.158	−0.704	0.208	0.662
17.62	1.147	−0.676	0.196	0.667
18.69	1.130	−0.636	0.179	0.673
21.47	1.092	−0.550	0.145	0.687
22.92	1.075	−0.513	0.132	0.694
24.41	1.060	−0.479	0.120	0.701
25.87	1.046	−0.449	0.110	0.707
27.79	1.029	−0.415	0.099	0.713
29.20	1.018	−0.393	0.092	0.717
31.52	1.002	−0.360	0.083	0.725
33.22	0.992	−0.339	0.077	0.730
35.55	0.978	−0.314	0.069	0.733
38.14	0.965	−0.289	0.063	0.739
51.91	0.912	−0.197	0.040	0.755
74.75	0.858	−0.119	0.024	0.763

$B^*(T)$ with σ_{eff} = 0.700, (iv) $B^*_{WK2}(T)$, the WK virial coefficient through the term proportional to $\hbar^2$, (v) $B^*_{WK4}(T)$ the WK virial coefficient through the term proportional to $\hbar^4$, (vi) $B^*_{class}(T)$, the classical second virial coefficient for a Lennard-Jones gas. It should be noted that $B^*_{WK2}(T)$ and $B^*_{WK4}(T)$ contain only Boltzmann terms; i.e., we have discarded the perfect gas exchange virial coefficient. Figure 3 gives analogous plots for He^3.

First of all, let us compare in figure 2 our $B^*(T)$ with σ_{eff} = 0.790 and σ_{eff} = 0.700 with $B^*_{WK2}(T)$ and $B^*_{WK4}(T)$. It appears that both $B^*(T)$ with σ_{eff} = 0.790 and σ_{eff} = 0.700 are considerably more realistic than $B^*_{WK2}(T)$ over the entire temperature range. Also, $B^*_{WK4}(T)$ seems to give poor results below 70°K.

Indeed, the magnitude of the term proportional to $\hbar^4$ in this region indicates very poor convergence of the WK expansion. It is interesting to note that, as in three dimensions, $B^*(T)$ with $\sigma_{\text{eff}} = 0.700$ appears to be better than $B^*(T)$ with the "best" value of σ_{eff}. The reason is, of course, the same as in the three-dimensional case; namely, the σ_{eff} value which gives the best results for a particular calculation is a function of the number of terms included in the expansion, unless, of course, enough terms are taken to attain convergence, in

Table 5. *Second virial coefficient of* He^3 *with* $\sigma_{\text{eff}} = 0.780$.

T	$B_{hs}^*(T)$	$B_{\text{att}}^{(0)*}(T)$	$B^{(1)*}(T)$	$B^*(T)$
13.04	1.332	−0.867	0.376	0.841
17.75	1.214	−0.640	0.226	0.800
19.40	1.185	−0.586	0.198	0.797
20.38	1.169	−0.557	0.184	0.796
21.71	1.149	−0.522	0.168	0.795
22.49	1.138	−0.504	0.160	0.794
23.18	1.129	−0.488	0.153	0.794
24.09	1.118	−0.469	0.145	0.794
25.56	1.101	−0.441	0.134	0.794
29.36	1.064	−0.380	0.110	0.794
31.35	1.048	−0.354	0.101	0.795
33.38	1.033	−0.330	0.093	0.796
35.38	1.020	−0.309	0.086	0.797
38.01	1.004	−0.286	0.078	0.796
39.94	0.993	−0.269	0.072	0.796
43.11	0.977	−0.246	0.065	0.796
45.44	0.967	−0.231	0.061	0.797
48.63	0.954	−0.213	0.055	0.796
52.16	0.940	−0.195	0.050	0.795
71.00	0.889	−0.130	0.033	0.792

which case $\sigma_{\text{eff}} = 0.790$ would be by definition the "best" value. Figure 3 for He^3 shows the same features and the same observations are valid there also.

It would be instructive to include in figures 2 and 3 exact values of the two-dimensional virial coefficient for a Lennard-Jones potential but, as far as we know, these have not been calculated. Nevertheless, we can make use of the situation in three dimensions, where exact results are available, to make a general comment on the ability of perturbation methods to approximate the exact $B^*(T)$. We have found (Derderian and Steele, in press) that our perturbation method was generally superior to the WK expansion, not only in its successful treatment of the exchange effects, but also in approximating the exact $B^*(T)$ over the range 10-70°K. Furthermore, our perturbation expansion appeared to converge less slowly than the WK expansion. In two dimensions, as is apparent from figures 2 and 3, our perturbation method appears to fare better than the

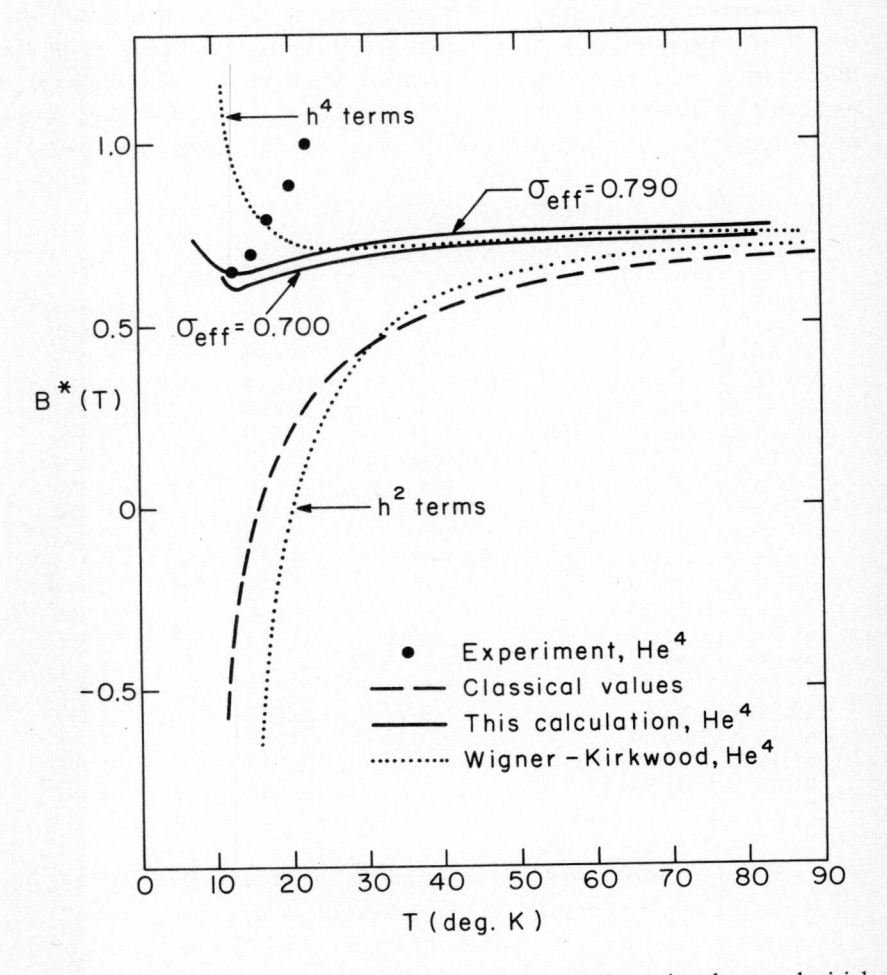

Fig. 2. Various calculations of the reduced two-dimensional second virial coefficient $B^*(T)$ are plotted here versus temperature and compared with the experimental values for He^4 on a graphitized carbon black surface.

WK expansion. Here the question of convergence is accentuated especially in the case of He^3. In general, the convergence of these expansions appears to be poorer in two than in three dimensions. The two-dimensional WK expansion is considerably worse than in three-dimensions; the large difference between

$B^*_{WK2}(T)$ and $B^*_{WK4}(T)$ is due to the size of the term proportional to $\hbar^4$ which simply overwhelms the rest of the terms in $B^*_{WK4}(T)$.

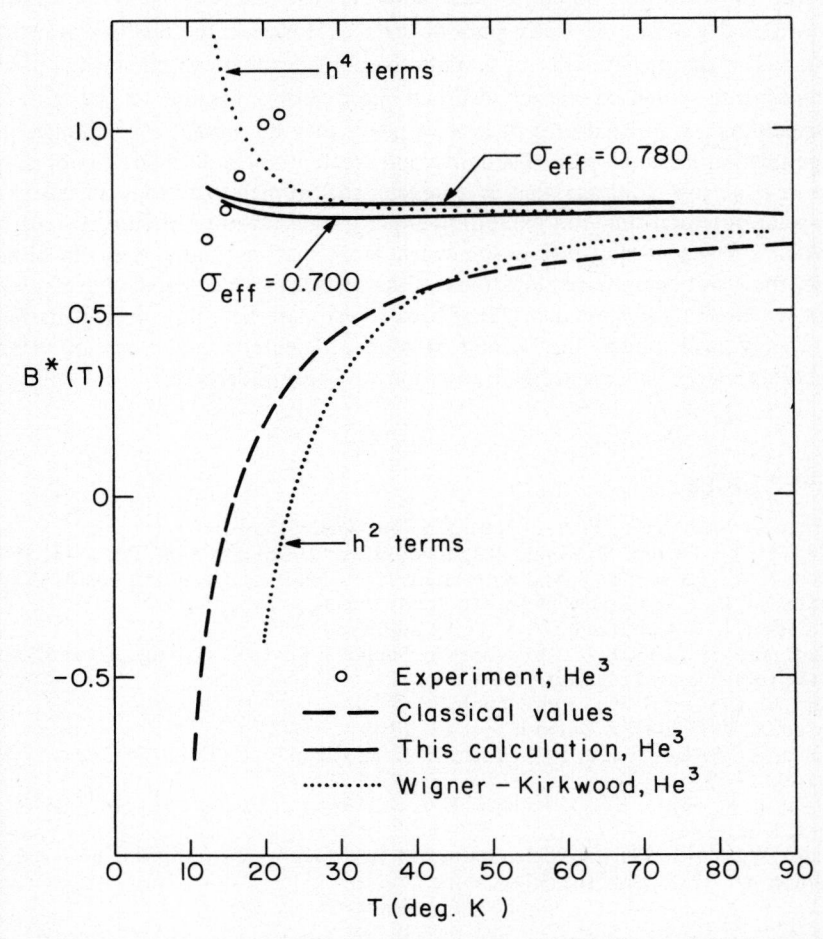

Fig. 3. These curves are the same as in figure 2, but this figure gives the results for He^3 rather than He^4. In both cases, the reducing factor is $\pi\sigma^2_{hs}/2$, the classical hard sphere virial, and the potential parameters used for the He-He Lennard-Jones curve are those given by Boyd *et al.* (1969).

When the theoretical values are compared with the experimental virial coefficients, some interesting points emerge. Clearly, for both He^4 and He^3, the experimental virial coefficient is approximated qualitatively by the theoretical values. The experimental values are, however, somewhat larger and have a stronger dependence upon temperature. There are several possible reasons for this behavior. We recall that our perturbation method, and specifically, the form of the perturbation potential, are designed for the treatment of a two-dimensional gas on a perfectly smooth surface. However, the real experimental surface is made up primarily of graphite basal planes that exhibit small periodic variations in potential as an adsorbed atom moves parallel to the surface. Furthermore, a slight degree of heterogeneity may be present even though the experimental isosteric heats (Derderian and Steele, to be published) do not show the rise at low coverage that is characteristic of heterogeneous surfaces. A gas-solid potential function exhibiting either of these features should give rise to a virial coefficient that differs somewhat from that calculated here. In either case, there will be preferred locations on the surface (with a degree of preference that is temperature-dependent). It is hard to estimate here how these variations in energy will affect the second virial coefficients; calculations of $B^*(T)$ including energy barriers to free translation are being undertaken.

REFERENCES

Barnes, M. and Steele, W. A. (1966). *J. Chem. Phys.* **45**, 461.
Boyd, M. E., Larsen, S. Y. and Kilpatrick, J. E. (1966). *J. Chem. Phys.* **45**, 499.
Boyd, M. E., Larsen, S. Y. and Kilpatrick, J. E. (1969). *J. Chem. Phys.* **50**, 4034.
Derderian, E. J. and Steele, W. A. (In press). *J. Chem. Phys.*
Derderian, E. J. and Steele, W. A. To be published.
Friedmann, H. (1962). *In* "Advances in Chemical Physics" (I. Prigogine, ed.) vol IV. Interscience Publishers, New York.
Klein, M. L. (1966). *Mol. Phys.* **10**, 87.
Larsen, S. Y. (1968). *J. Chem. Phys.* **48**, 1701.
Larsen, S. Y., Kilpatrick, J. E., Lieb, E. K. and Jordan, H. F. (1965). *Phys. Rev.* **140A**, 129.
Larsen, S. Y., Witte, K. and Kilpatrick, J. E. (1966). *J. Chem. Phys.* **44**, 213.
Sams, J. R. (1965). *Mol. Phys.* **9**, 17.
Steele, W. A. (1967). *In* "The Solid Gas Interface" (E. A. Flood, ed.) vol 1, chap. 10. M. Dekker Inc., New York.
Steele, W. A. and Aston, J. G. (1957). *J. Amer. Chem. Soc.* **79**, 2393.

QUANTUM STATES AND HEAT CAPACITY OF HELIUM ADSORBED ON GRAPHITE

D. E. HAGEN, A. D. NOVACO and F. J. MILFORD

Battelle Memorial Institute, Columbus, Ohio, USA

I. INTRODUCTION

Current experimental interest in physically adsorbed sub-monolayer helium films has stimulated theoretical studies of the single adsorbed helium atom. Substrates which seem very suitable for helium film studies are the graphite-like substrates: graphitized carbon black and exfoliated graphite. These substrates, besides having a large surface to volume ratio, possess a high degree of homogeneity (Ross and Olivier, 1964). However, before it is possible to understand the physics of these films, it is necessary to understand the nature of the adsorption of a single helium atom. A logical first step must then be the investigation of the single particle states of a helium atom adsorbed on a graphite solid.

Calculations for helium adsorbed on a variety of substrates have shown either marked localization of the adatom or strong band effects. Theoretical results for helium adsorbed on the (100) face of rare gas solids show the adatom is well localized within the area of the adsorption cell (Ricca *et al.,* 1969; Novaco and Milford, 1970). The corresponding band structure shows the lowest band to have a width some two orders of magnitude smaller than the first band gap. Theoretical results for helium adsorbed on argon-plated copper show the adatom to be mobile, but indicate strong band effects in the $1°K$ to $6°K$ temperature range (Milford and Novaco, 1971). Experimental results for helium adsorbed on exfoliated graphite (Bretz and Dash, 1971) show the adatom to be very mobile and indicate that the periodic nature of the surface has little effect upon low-coverage films at $4°K$. This degree of mobility of the helium atom can be inferred from rough estimates of the potential barrier between adsorption sites and the zero point energy of localization at a site. However, no definite statement about band effects can be made without a detailed quantum calculation. The problem is then to solve the single particle Schrödinger equation for a helium atom adsorbed on an exposed basal plane of a graphite solid. This is done by extending techniques developed elsewhere (Milford and Novaco, 1971).

II. POTENTIAL ENERGY

The substrate is assumed to be a semi-infinite graphite solid having an undistorted basal-plane surface. Defects and thermal vibrations of the solid are not considered, so that each carbon atom is fixed at its lattice site. Since surface stretching and distortion are neglected, the geometry of the surface basal plane is identical to that of the interior ones. The substrate is therefore an idealized one.

A graphite basal plane is an open hexagonal lattice having, as shown in figure 1(a), a carbon atom at each hexagon vertex. This hexagonal carbon ring forms an

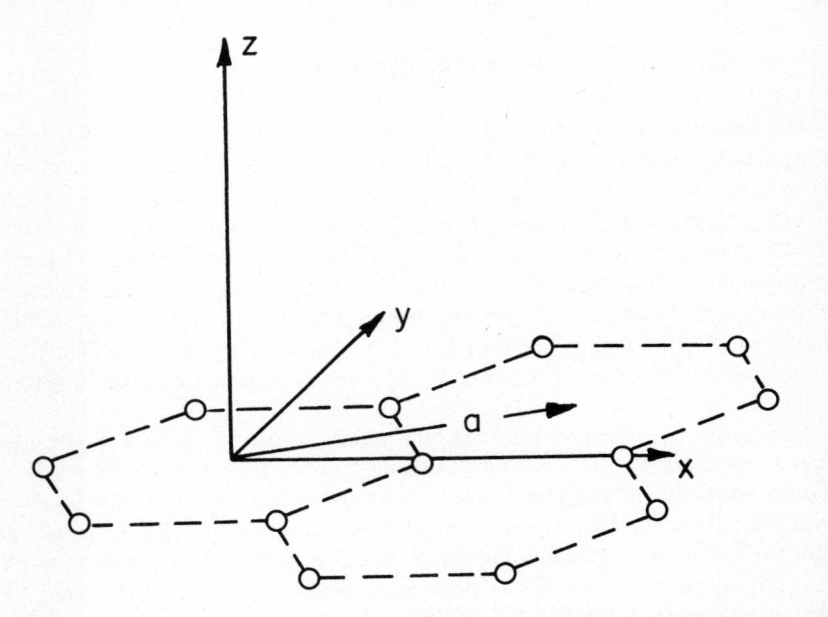

Fig. 1. (a) Orientation of the x-y-z coordinate system on the graphite basal plane. Each circle represents a surface carbon atom. a denotes the distance between the centers of neighboring hexagons.

adsorption cell. The carbon-carbon nearest neighbor distance is 1.415×10^{-8} cm. The graphite solid consists of a staggered parallel arrangement of such planes with an interplanar spacing of 3.5×10^{-8} cm. This staggering is such that the center of an adsorption cell in one plane can be projected onto a carbon atom in the next plane. The two such possible arrangements are conventionally denoted by ABCABC and ABAB. The ABAB arrangement is the one considered here, but the potential energy of a helium atom adsorbed on the ABCABC type substrate is virtually identical to that of the ABAB type. The coordinate system used is illustrated in figure 1(a). It is oriented such that the x and y axes are in the surface basal plane with the positive z-axis pointing out of the solid. The origin is located at the center of an adsorption cell.

The interaction of a rare-gas atom with a graphite substrate can be written as a sum over all carbon atoms of a two-body gas-carbon potential (Crowell and Steele, 1961). This two-body potential is taken to be of the Lennard-Jones (12-6) type. Values for the Lennard-Jones parameters ϵ_0 and ρ_0 can be obtained by conventional interpolation methods. The values so obtained for the helium-graphite system are $\epsilon_0 = 17.0°K$ and $\rho_0 = 3.34 \times 10^{-8}$ cm. If $V(\mathbf{r})$ is the

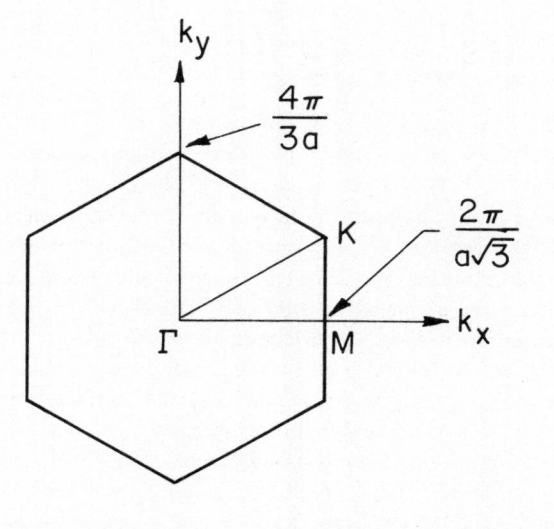

Fig. 1. (b) The Brillouin zone corresponding to the lattice of adsorption sites.

potential energy of a helium atom with position vector $\mathbf{r}$, and ρ_i is the distance between the helium atom and the ith carbon atom, then

$$V(\mathbf{r}) = \sum_i \epsilon_0 \left[\left(\frac{\rho_0}{\rho_i} \right)^{12} - 2 \left(\frac{\rho_0}{\rho_i} \right)^6 \right] \tag{1}$$

where the sum extends over all carbon atoms. Although summing (1) to convergence offers no problem to a computer, it is convenient to speed convergence by an appropriate integral approximation to all terms not directly summed. Typically this is done by directly summing over all carbon atoms with a ρ_i less than some chosen ρ_m, and treating the rest of the solid as a continuum.

The value of ρ_m is chosen so as to obtain the desired degree of accuracy. Upon performing the appropriate integrals, (1) becomes

$$
V(\mathbf{r}) = \sum_{\rho_i < \rho_m} \epsilon_0 \left[\left(\frac{\rho_0}{\rho_i} \right)^{12} - 2 \left(\frac{\rho_0}{\rho_i} \right)^6 \right]
$$
$$
+ \frac{8\pi \epsilon_0 \rho_0^3}{\sqrt{3} \; da^2} \left[\left(\frac{\rho_0}{\rho_m} \right)^9 \left(\frac{1}{9} - \frac{1}{10} \frac{z}{\rho_m} \right) - 2 \left(\frac{\rho_0}{\rho_m} \right)^3 \left(\frac{1}{3} - \frac{1}{4} \frac{z}{\rho_m} \right) \right] \tag{2a}
$$

for $z < \rho_m$, and

$$
V(\mathbf{r}) = \frac{8\pi \epsilon_0 \rho_0^3}{\sqrt{3} \; da^2} \left[\frac{1}{90} \left(\frac{\rho_0}{z} \right)^9 - \frac{1}{6} \left(\frac{\rho_0}{z} \right)^3 \right] \tag{2b}
$$

for $z \geqslant \rho_m$. In (2a) and (2b), the parameter a is the distance between two nearest adsorption cell centers, and d is the interplanar spacing.

The behavior of $V(\mathbf{r})$ can be displayed by plotting V as a function of z for fixed values of x and y. In figure 2 this is done for three specific locations in the basal plane. These locations are: (1) the center of an adsorption cell, (2) midway between two carbon atoms, and (3) a carbon atom site. The potential varies little around the perimeter of an adsorption cell, but deepens somewhat toward its center. The potential well in the direction perpendicular to the surface is very deep and narrow, so excitations in the z-direction are energetically well separated. Periodic variations in $V(\mathbf{r})$ along the surface are small, causing excitations along the surface to occur in wide bands.

The potential given by (2a) is nonseparable. That is to say, it cannot be written as a function of z plus a function of x and y. However, this characteristic is not as exaggerated in this case as it is for other substrates (Novaco and Milford, 1970). The result is some slight mixing of motion parallel to the plane with motion perpendicular to the plane.

III. QUANTUM STATES

The single particle states of either a He^3 atom or a He^4 atom in the potential field given by (1) are found by solving the Schrödinger equation

$$
\left[-\frac{\hbar^2}{2m} \nabla^2 + V(\mathbf{r}) \right] \psi(\mathbf{r}) = E\psi(\mathbf{r}) \tag{3}
$$

where m is the mass of the helium atom. The technique used in solving (3) is tailored to the two notable characteristics of $V(\mathbf{r})$, its two-dimensional periodicity and its asymmetric shape as a function of z. The periodicity suggests

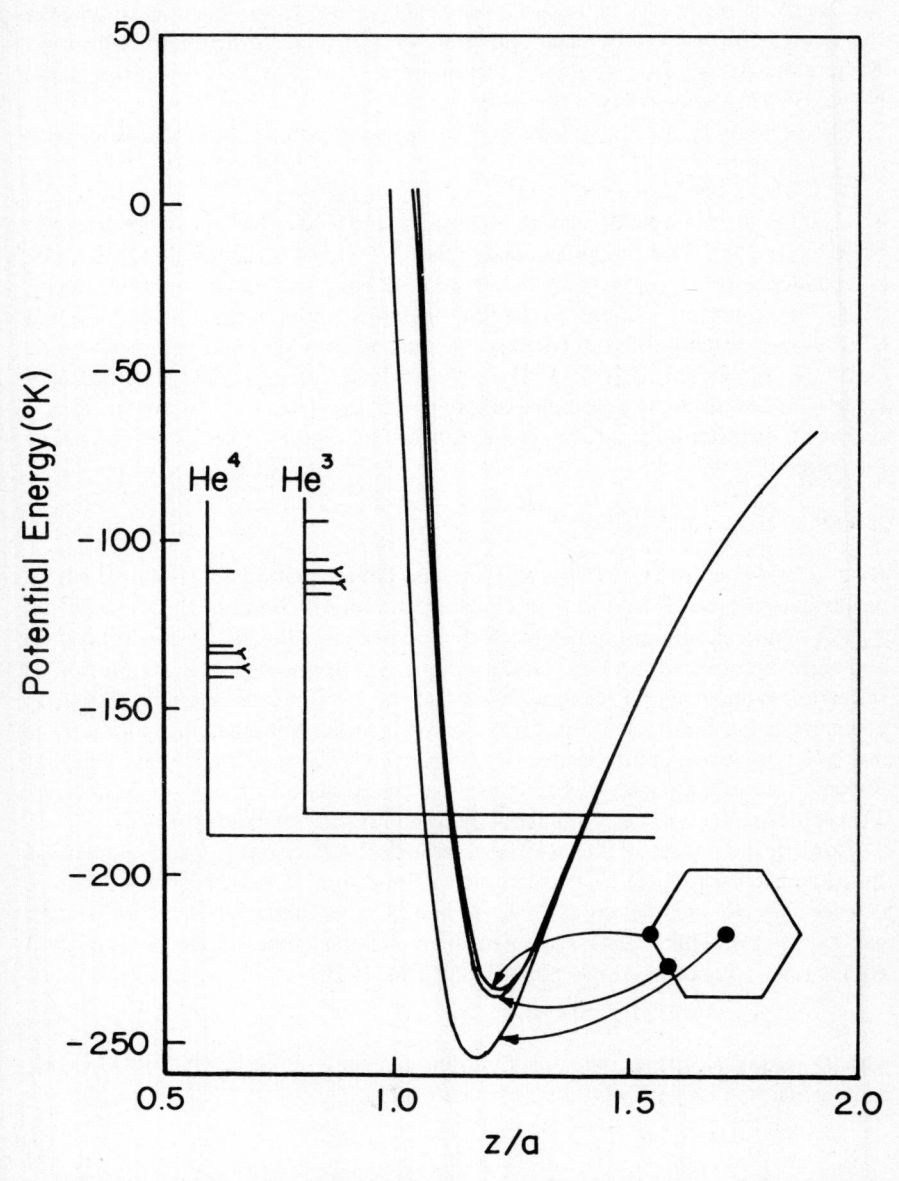

Fig. 2. Potential profiles showing $V(\mathbf{r})$ as a function of z for three fixed points in the x-y plane. Also shown are energy level diagrams for the lowest eight Γ ($\mathbf{k} = 0$) states of He^4 and He^3.

the use of band theory methods (Jones, 1960), while the z-dependence suggests the expansion of $\psi(\mathbf{r})$ as a linear combination of orthonormal functions having a discrete spectrum. The combined technique is described fully by Milford and Novaco (1971), and is only outlined here.

The periodicity of $V(\mathbf{r})$ allows $\psi(\mathbf{r})$ to be decomposed into Bloch waves with

$$\psi(\mathbf{r}) = e^{i\mathbf{k}\cdot\mathbf{r}} U_{\mathbf{k}}(\mathbf{r}) \tag{4}$$

The vector $\mathbf{k}$ is perpendicular to the z-axis and $U_{\mathbf{k}}(\mathbf{r})$ has the same periodic behavior as $V(\mathbf{r})$. The energy associated with $\psi(\mathbf{r})$ in (4) is denoted by $E(\mathbf{k})$. The z-dependence of $U_{\mathbf{k}}(\mathbf{r})$ is expanded as a linear combination of functions $M_j(z)$. These functions are chosen to be the eigenfunctions of the one-dimensional Schrödinger equation for a particle of mass m in a potential given by some conveniently chosen $V_m(z)$. The x-y dependence of $U_{\mathbf{k}}(\mathbf{r})$ is expanded as a Fourier series using the complex exponential exp $(i\mathbf{K}\cdot\mathbf{r})$. The vector $\mathbf{K}$ is a vector of the reciprocal lattice. The expansion for $U_{\mathbf{k}}(\mathbf{r})$ is a sum over j and $\mathbf{K}$ of the form

$$U_{\mathbf{k}}(\mathbf{r}) = \sum_j \sum_{\mathbf{K}} \xi_{\mathbf{K}}^{\,j}(\mathbf{k}) M_j(z) e^{i\mathbf{K}\cdot\mathbf{r}} \tag{5}$$

By use of (4) and (5), (3) becomes an eigenvalue equation for a matrix whose eigenvalues are the $E(\mathbf{k})$ and whose eigenvectors have components given by the $\xi_{\mathbf{K}}^{\,j}(\mathbf{k})$. Once all the components of this matrix are calculated, the eigenvalues and eigenvectors are found by standard numerical techniques. The assumption of reflection symmetry of $V(\mathbf{r})$ about the $x = 0$ and $y = 0$ planes results in all matrix elements being real rather than complex quantities. Although the symmetry is not exact for the graphite lattice (it is, however, for a single basal plane), in that region of coordinate space where the wave functions of interest are important, $V(\mathbf{r})$ exhibits this symmetry numerically to about six significant figures.

Numerical solution of the matrix equations for $E(\mathbf{k})$ and $\xi_{\mathbf{K}}^{\,j}(\mathbf{k})$ requires that the expansion for $U_{\mathbf{k}}(\mathbf{r})$ be truncated to a finite sum. If $V_m(z)$ is chosen well, it is necessary to include only $j = 1$, 2, and 3 (the fundamental and first two excited $M_j(z)$). This means $V_m(z)$ must mimic the curves in figure 2. A good choice for $V_m(z)$ is the Morse potential (Morse, 1929)

$$V_m(z) = De^{-\beta(z-z_0)}[e^{-\beta(z-z_0)} - 2] \tag{8}$$

with D, β, and z_0 being judiciously chosen constants. In order that the $M_j(z)$ be easily generated numerically, the parameter

$$\alpha_0 = \frac{2\sqrt{2mD}}{\cdot\hbar\beta} - 1 \tag{9}$$

is constrained to be an integer. Typical values of the Morse potential parameters appropriate for the helium-graphite system are $D = 243.3°K$, $z_0 = 3.03 \times 10^{-8}$ cm, and $\alpha_0 = 8$ (He^3) or 10 (He^4).

The sum in (5) over $\mathbf{K}$ is truncated in such a way as to insure the finite sum has the same point symmetry as the original expansion. This is accomplished by including a reciprocal lattice vector ($\mathbf{K} = 0$) and its first three shells of neighbors. The final expansion for $U_\mathbf{K}(\mathbf{r})$ has 57 (3 x 19) terms.

The ground-state energy of a He^3 atom is $-181.4°$K, while the corresponding energy for He^4 is $-188.8°$K. The small difference between these energies ($\sim 4\%$) is indicative of the fact that the major contribution to the zero point energy comes from localization in the z-direction. There is little localization of the adatom in the plane. The zero point energies (73.5°K for He^3 and 66.1°K for He^4) are much larger than the 17.2°K potential barrier to site-to-site tunnelling, indicating high mobility of the adatoms. Energy-level diagrams for He^3 and He^4 illustrating the lowest eight $k = 0$ states are shown in figure 2.

Examination of probability distributions is a good way of investigating the nature of the adsorption states. Two such distributions are $P_1(x, y)$, the probability of locating the adatom at some location in the plane irrespective of its z coordinate; and $P_2(z)$, the probability of locating the adatom at some height z above the plane irrespective of its x-y position. These two distributions are given by

$$P_1(x, y) = \int_{-\infty}^{\infty} dz \, \psi^*(\mathbf{r}) \, \psi(\mathbf{r}) \tag{10a}$$

and

$$P_2(z) = \int_A dxdy \, \psi^*(\mathbf{r}) \, \psi(\mathbf{r}) \tag{10b}$$

A three-dimensional perspective view of the surface formed by plotting $P_1(x, y)$ as a function of x and y for both the ground state and the first excited $k = 0$ state of He^4 is shown in figure 3. The probability of locating the adatom is greatest at the center of an adsorption cell (adsorption site), but the ground state is extremely delocalized since P_1 throughout the adsorption cell is never less than 47% of its maximum value. The He^3 P_1 functions are very similar to the He^4 ones, showing just slightly increased delocalization. Examination of the P_1 and P_2 functions for various $k = 0$ states of both He^3 and He^4 show that the ground state and the lowest six excited $k = 0$ states are basically excitations in the plane with z behavior being that of the fundamental mode. The first z-excitation in the $k = 0$ states occurs in the seventh excited state, having an energy some 80-85°K above that of the ground state. Figure 4 shows the P_2 function for the ground state, first excited state, and seventh excited state (all $k = 0$ states) of He^4. The results for He^3 are very similar, showing only a slight spreading of the distributions.

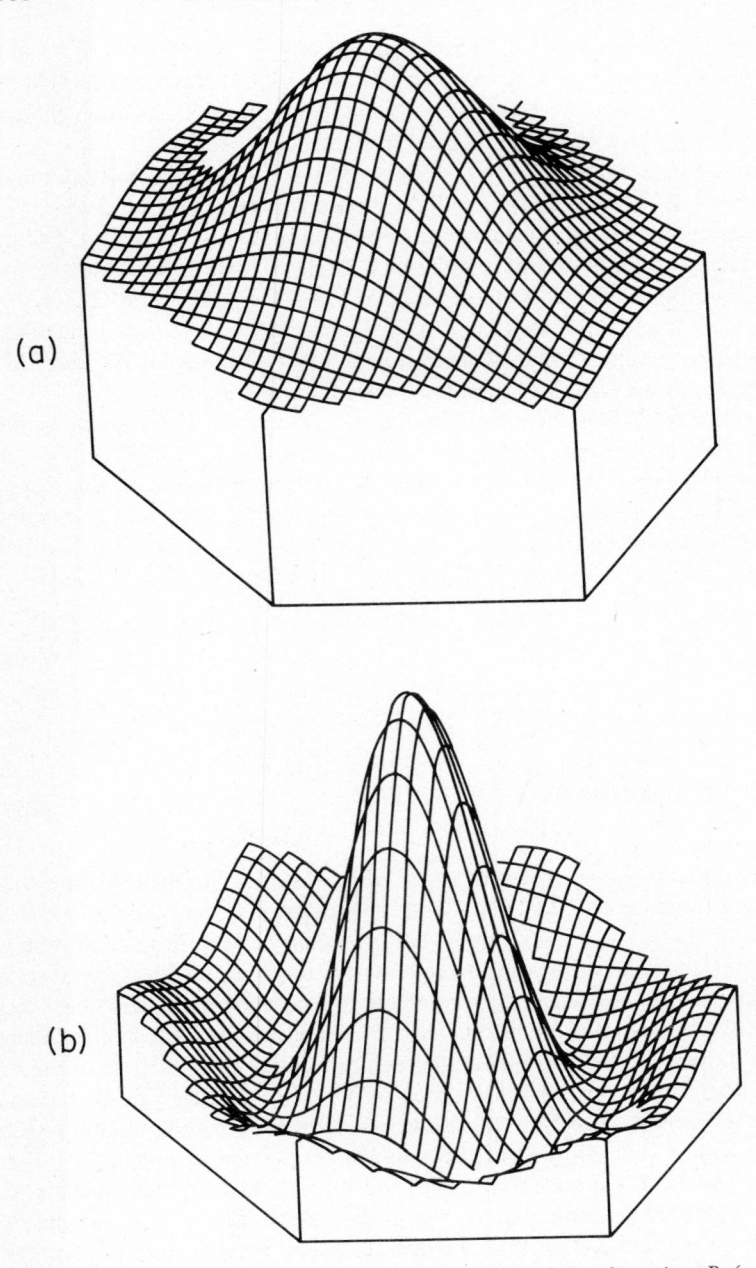

(a)

(b)

Fig. 3. A three-dimensional perspective plot of the function $P_1(x, y)$ for (a) the ground state and (b) the first excited state of He^4 (both **k** = 0 states).

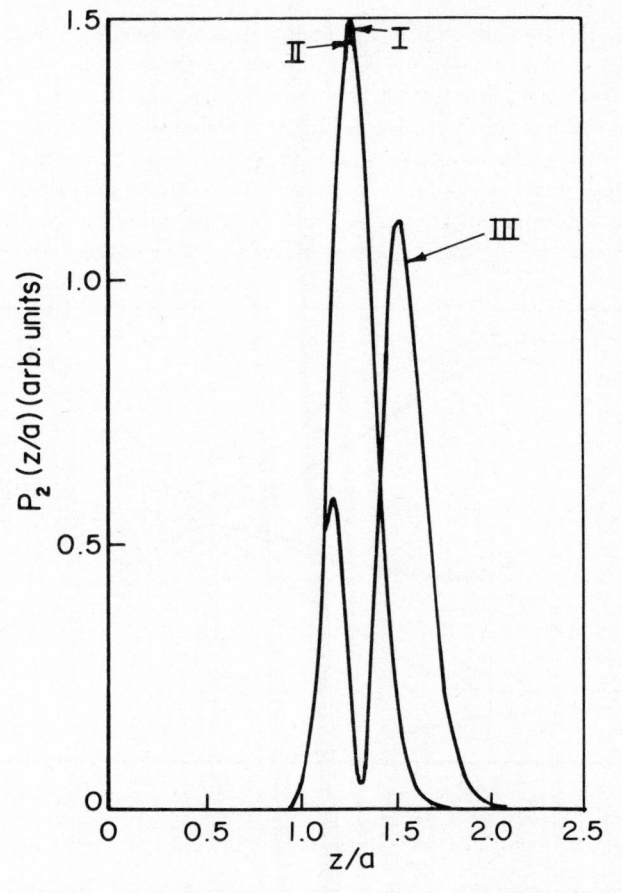

Fig. 4. The function $P_2(z)$ for the (I) ground state, (II) first excited state, and (III) seventh excited state (all $\mathbf{k} = 0$ states) of He^4.

IV. BAND STRUCTURE AND SPECIFIC HEAT

The complete description of the adatom quantum states requires that (3) be solved for values of $\mathbf{k}$ throughout the reduced Brillouin zone. The lattice of adsorption sites is a two-dimensional simple hexagonal lattice with lattice constant $a = 2.45 \times 10^{-8}$ cm. The corresponding Brillouin zone is a hexagon with an edge length of $4\pi/3a$, and is shown in figure 1(b). The points of high symmetry are, in the conventional manner, labeled Γ, M, and K. The energy eigenvalues are most conveniently displayed using the standard band theory approach of plotting $E(k)$ vs k in the high symmetry directions. Figure 5 shows the He^4 band structure along the $KM\Gamma K$ contour. Only the lowest seven bands are plotted. This band structure is characterized by very wide band widths and very narrow band gaps, indicating a highly mobile adatom. In fact, this band structure is sufficiently similar to that of the two-dimensional empty lattice that the differences may accurately be described as small perturbations. These differences are exemplified as seen in figure 5 by the small gap between the first and second band and the slight splitting of the first six excited Γ states. Both the

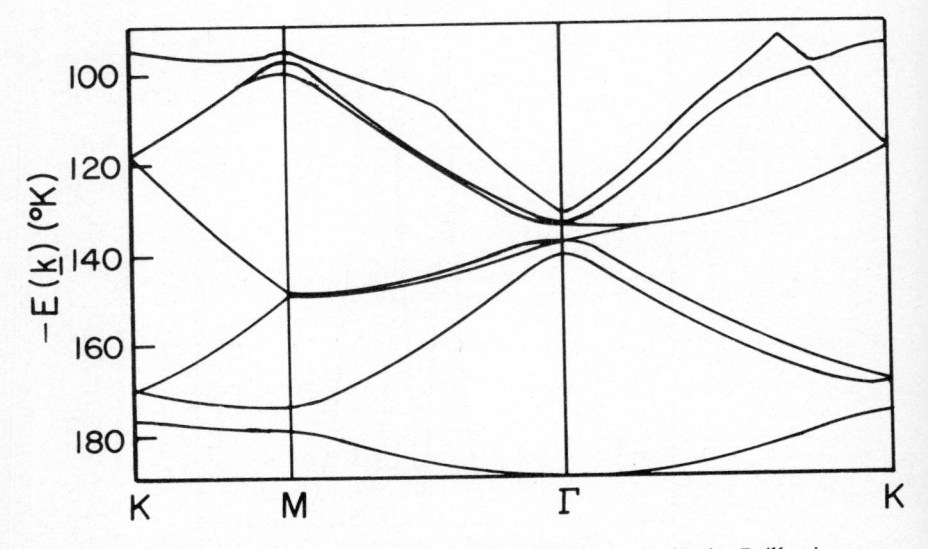

Fig. 5. Energy bands for He^4 along the $KM\Gamma K$ contour in the Brillouin zone, showing the lowest seven bands.

gap and the splitting are non-existent in the empty lattice band structure. The eighth band of He^4 differs significantly from that of the empty lattice. The reason is this He^4 band involves a z-excitation. However, this band lies some 80°K above the lowest band, and does not influence low temperature behavior.

The thermodynamic functions for the helium film are readily calculable if helium-helium interactions are ignored. The model is then a system of non-interacting Bosons or Fermions with an energy spectrum given by the eigenvalues $E(k)$. The treatment, using the grand canonical ensemble, is entirely conventional (Huang, 1963) and is only outlined here. The internal energy is

$$U = \sum_{j\,\mathbf{k}} E(j,\mathbf{k})\,n(E) \tag{11}$$

where $E(j, \mathbf{k})$ is the jth energy level corresponding to wave vector $\mathbf{k}$ in the reduced Brillouin zone. The function $n(E)$ is just the grand canonical distribution function

$$n(E) = (2s + 1)/[e^{(E-\mu)/k_B T} \pm 1) \tag{12}$$

where s is the spin of the adatom and the plus sign is used for Fermions while the minus sign is used for Bosons. The chemical potential μ is determined by

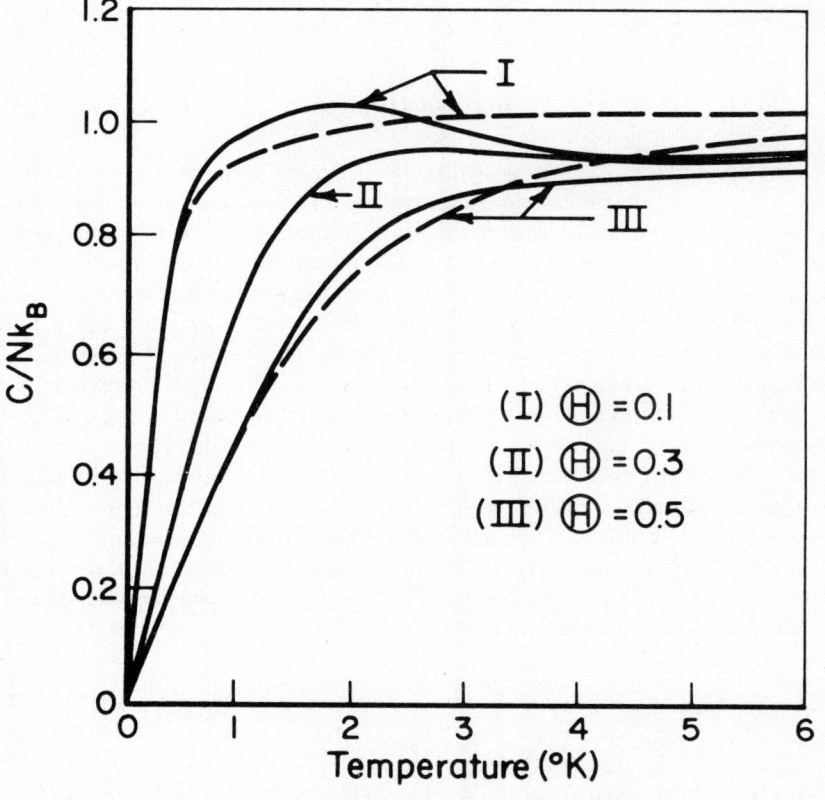

Fig. 6. Specific heat per atom as a function of temperature for He[4] films of coverages $\Theta = 0.1$, 0.3, and 0.5. The dashed lines show the specific heat for an ideal gas at $\Theta = 0.1$ and 0.5.

setting the sum of $n(E)$ equal to the number of adatoms. The specific heat is obtained by differentiating (11) with respect to temperature T, which leads to a set of equations displayed elsewhere (Milford and Novaco, 1971).

The calculated heat capacity for He[4] is shown in figure 6 for several coverages, Θ. The coverages are specified as fractions of a monolayer, while a monolayer is defined as one adatom per unit cell. This would undoubtedly not exactly correspond to the experimental coverages, since it would place the helium atoms (at $\Theta = 1$) well within each other's repulsive core. However, this is still a convenient parameterization for the investigation of thermal behavior at low coverages. The temperature dependence of the specific heat is very similar to that of the two-dimensional free Bose gas. For low coverages, there is a slight peak which is a result of the small band gap, but this disappears at higher coverages. The ideal two-dimensional gas specific heats are shown for the highest and lowest coverages.

V. CONCLUSION

The helium atom exhibits a high mobility on the graphite surface. Both the nature of the ground-state wavefunction and the structure of the lower bands support this statement. The $P_1(x, y)$ distribution shows the ground-state wavefunction to be very delocalized. The concept of an adsorption site is, in fact, only meaningful in terms of the small maximum in the ground state $P_1(x, y)$ function at the center of an adsorption cell. The structure of the lower bands is very nearly that of the two-dimensional free particle, showing periodic variations in the surface potential to be small compared to the zero-point energy.

The high excitation energy ($\sim 80°K$) for the first z-excitation shows that the adsorption states are of a highly two-dimensional character. The $P_2(z)$ distribution is nearly the same for each of the lower bands, indicating little mixing of x-y motion with z-motion. In final summation, the most important conclusions to be drawn from this study are: (1) for all but the most detailed calculations, helium adsorbed on a graphite substrate behaves at low temperatures as a two-dimensional system, and (2) the effects of the periodic nature of the graphite surface are very small.

REFERENCES

Bretz, M. and Dash, J. G. (1971). *Phys. Rev. Lett.* **26**, 963.
Crowell, A. D. and Steele, R. B. (1961). *J. Chem. Phys.* **34**, 1347.
Huang, K. (1963). "Statistical Mechanics". Wiley, New York.
Jones, H. (1960). "The Theory of Brillouin Zones and Electronic States in Crystals". North-Holland, Amsterdam.
Milford, F. J. and Novaco, A. D. (1971). *Phys. Rev.* **A4**, 1136.
Morse, P. M. (1929). *Phys. Rev.* **34**, 57.
Novaco, A. D. and Milford, F. J. (1970). *J. Low Temp. Phys.* **3**, 307.
Ricca, F., Pisani, C. and Garrone, E. (1969). *J. Chem. Phys.* **51**, 4079. Errata in *J. Chem. Phys.* (1970) **53**, 2546.
Ross, S. and Olivier, J. P. (1964). "On Physical Adsorption". Interscience, New York.

LOCALIZED AND DELOCALIZED STATES OF ADSORBED RARE-GAS ATOMS

F. RICCA, C. PISANI and E. GARRONE

University of Turin, Italy

I. INTRODUCTION

Increasing attention is being devoted to the physical adsorption of rare-gas atoms on crystal solids, from the experimental and theoretical point of view.

Recent theoretical works have defined the potential energy of interaction of the adsorbed atoms (Ricca and Pisani, 1967; Ricca, 1967, Neustadter and Bacigalupi, 1967) and given evidence of the band structure of adsorption energy, as introduced by two-dimensional periodicity at the solid surface.

As to the latter, the adsorption of He has been studied on the (100) face of solid Xe, Kr and Ar, on the (110) face of Xe, on close-packed Ar monolayers and on graphite (Ricca *et al.,* 1969; Novaco and Milford, 1970; Ricca and Garrone, 1970; Hagen *et al.,* 1971). Two different approaches have been adopted in these studies, which respectively recall the "tight binding" and "plane wave" method usually employed in solid-state physics. Obviously these two approximations have two different regions of optimal application, and further methods should probably be employed in intermediate regions, as the OPW method for instance.

The band-energy approach should, in principle, make it possible to compare some theoretical views on physical adsorption with the most refined experimental results that have been recently obtained for the specific heats of He[3] and He[4] on various solid substrates at very low temperatures (Goodstein *et al.,* 1965; Dash, 1968; McCormick *et al.,* 1968; Stewart and Dash, 1970; Wallace and Goodstein, 1970). Yet it could well appear that the single-particle approximation usually employed is inadequate for a satisfactory description of the real state of atoms adsorbed in sub-monolayers, which are not extremely rarefied.

However, this cannot be considered as the sole result that quantum mechanical treatments can produce in studying simple physisorption models. Indeed quantum mechanics is needed adequately to clarify the meaning and the importance of surface heterogenities in physisorption, since the potential energies at the adsorption sites are certainly not exhaustive. A local approach

111

appears to be essential in most cases: of course it favours the use of tight-binding approximation, but this in turn requires preliminary evaluation of the degree of delocalization of adsorbed states.

This work is devoted to considering, from this point of view, the quantum states of single rare-gas atoms (He^3, He^4, Ne, Ar) adsorbed on Xe crystals in the most typical cases of the (100) and (111) face. Lighter rare gases have been chosen as adsorbate, and the heaviest one as adsorbent, in order to reduce to a minimum the relaxation effects that adsorption may introduce in the solid structure. The model adopted for the solid assumes a semi-infinite perfect crystal, having at the surface an unrelaxed structure identical to the bulk (Xe has a f.c.c. structure with fundamental parameter $a^0 = 6.24$ Å).

Adsorption has been considered as a single particle problem, concerning the motion of a single adsorbed atom in the field due to the solid atoms at rest in their equilibrium position. Potential energy has been evaluated in the pairwise additive approximation using the Lennard-Jones "6-12" potential

$$\epsilon = \epsilon^0 \left[\left(\frac{r^0}{r} \right)^{12} - 2 \left(\frac{r^0}{r} \right)^6 \right] \tag{1}$$

The parameters employed in Lennard-Jones potentials for the interaction of different adsorbed atoms with a single Xe atom are given in table 1. Obviously they are the same for He^3 and He^4.

Table 1. *Parameters in the Lennard-Jones potentials for different pairs of rare-gas atoms (Rica and Pisani, 1967)*

Atom pair	$\epsilon^0 (10^{-16}$ erg)	r^0 (Å)
Xe–He	66.24	3.72
Xe–Ne	123.70	3.82
Xe–Ar	226.53	4.20

It is well known that the adsorption sites for rare-gas atoms on the (100) face of a Xe crystal are of a unique type, at the centre of a square of surface atoms. This gives a simple square lattice of adsorption sites.

As far as the (111) face is concerned, it has been pointed out (Ricca and Pisani, 1967) that this represents, in principle, the simplest example of regular heterogeneity, since adjacent sites have not quite the same potential energies of adsorption. However, since the differences are negligible in practice, all the sites at the (111) face will be considered identical in this work, at the centre of each equilateral triangle of surface atoms. This gives a graphite-like structure for the adsorption-site lattice to be considered.

On both faces the identical adsorption sites are interconnected through saddle points, while the less-favoured adsorption positions are above the surface atoms of the solid. Table 2 gives the potential energies of adsorption at these three particular positions on the solid surface.

Table 2. *Potential energies of adsorption for various rare-gas atoms adsorbed on the (100) and (111) face of a* Xe *crystal(*)*

Adsorbed atom	(100)			(111)		
	Site	Saddle	Atom	Site	Saddle	Atom
He	−339.9	−209.3	−135.2	−270.5	−233.6	−147.2
Ne	−641.2	−403.8	−268.4	−518.4	−452.9	−291.2
Ar	−1251.1	−854.9	−607.9	−1071.8	−970.8	−676·9

* All the energies are in units of 10^{-16} erg.

II. PERIODIC SOLUTIONS

Direct and reciprocal lattices for both faces are given in figure 1 (a) and (b), where the description of the reciprocal lattice has been limited to the first Brillouin zone. In both the adsorption-site lattices the edge of the primitive unit cell is given by $a = a^0/\sqrt{2}$, where a^0 is the edge of the conventional unit cell in the underlying Xe crystal. For the (100) face the unit cell contains only one adsorption site, while for the (111) face each cell contains two identical sites which have been labelled A and B (Lomer, 1955).

An approximate solution for the motion of the adsorbed atom in the doubly-periodic potential field at the surface of a crystal may be found through the application of the variational method to a suitable set of Bloch functions, which can be constructed from the wave functions of local oscillators centred at the adsorption site.

A Bloch function for the wave vector **k** will have the form

$$\Psi_j^{\mathbf{k}}(\mathbf{r}) = N^{-1/2} \sum_\nu \psi_j(\mathbf{r} - \mathbf{R}_\nu) \exp(i\mathbf{k}\cdot\mathbf{R}_\nu) \tag{2}$$

where ν characterizes the cell in the direct lattice and j gives the quantum state to which the local function ψ_j corresponds. The vector **r** gives the position of the centre of the adsorbed atom and the $\mathbf{R}_\nu$'s are the translation vectors for the periodic field, parallel to the adsorbing surface.

In the case of the (100) face, the local function may be given by the eigenfunction of a single oscillator centred at the adsorption site. In the case of the (111) face, however, the local function will be represented by a linear combination of the eigenfunctions of two oscillators centred at the sites A and

B, respectively. Because sites A and B are identical, the constant coefficients in such a combination will entirely depend on the symmetry properties of the space group for the adsorption-site lattice.

By considering the local oscillators (described in cylindrical polar coordinates) as resulting from a two-dimensional harmonic oscillator parallel to the

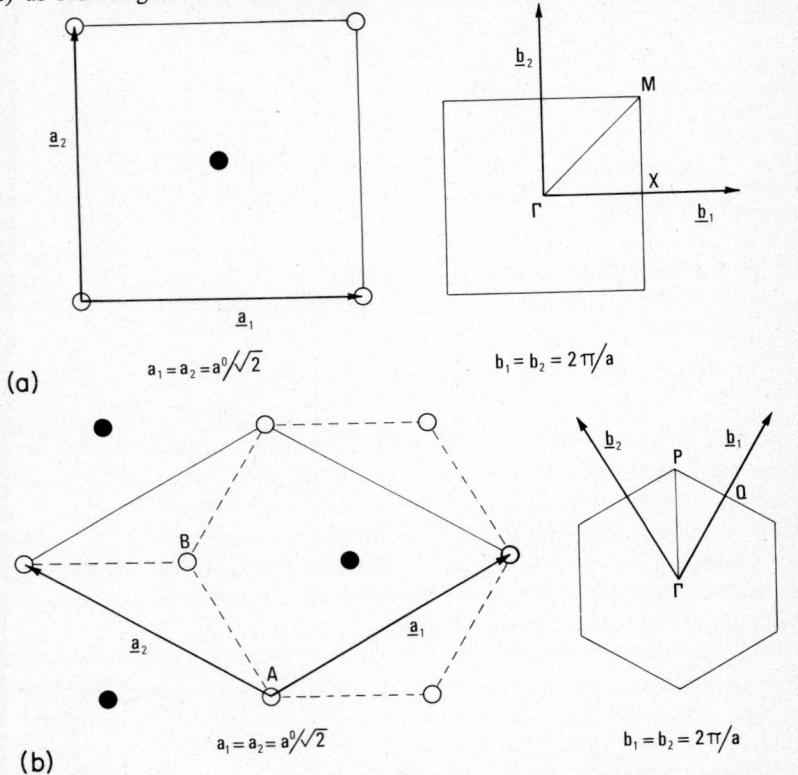

(a) $a_1 = a_2 = a^0/\sqrt{2}$ $b_1 = b_2 = 2\pi/a$

(b) $a_1 = a_2 = a^0/\sqrt{2}$ $b_1 = b_2 = 2\pi/a$

Fig. 1. (a). Elementary unit cell and first Brillouin zone for the simple square lattice of the adsorption sites at the (100) face. (b). Elementary unit cell and first Brillouin zone for the graphite-like lattice of the adsorption sites at the (111) face. ● = surface atom, ○ = adsorption site.

surface and a Morse oscillator in a direction normal to this surface, the Bloch functions for the (100) and (111) face respectively become

$$\Psi^{\mathbf{k}}_{n,\,m,\,p}(\mathbf{r}) = N^{-1/2} \cdot M_p(z) \sum_v H_{n,\,m}(\rho_v, \phi_v) \exp(i\mathbf{k} \cdot \mathbf{R}_v) \tag{3}$$

$$\Psi^{\mathbf{k}}_{n,\,m,\,p}(\mathbf{r}) = N^{-1/2} \cdot M_p(z) \sum_v [c_A H_{n,\,m}(\rho_{vA}, \phi_{vA})$$
$$+ c_B H_{n,\,m}(\rho_{vB}, \phi_{vB})] \exp(i\mathbf{k} \cdot \mathbf{R}_v) \tag{4}$$

The m, n, p quantum numbers correspond to the j-th quantum state of the previous notation. The function $H_{n,m}$ is given by

$$H_{n,m}(\rho,\phi) = R_{n,m}(\rho)(2\pi)^{-1/2} e^{im\phi} \tag{5}$$

with

$$R_{n,m}(\rho) = \left[2 \frac{n! \, \alpha^{(|m|+1)}}{(n+|m|)!} \right]^{1/2} e^{-(\alpha\rho^2)/2} \rho^{|m|} L_n^{|m|}(\alpha\rho^2) \tag{6}$$

where $L_n^{|m|}$ is the Laguerre polynomial.

The Morse oscillator is expected to work well in adsorption problems, since the Morse potential

$$V(z) = D[e^{-2\beta(z-z^0)} - 2e^{-\beta(z-z^0)}] \tag{7}$$

satisfactorily fits the asymmetric potential at the surface of the solid.

The form of the corresponding energy eigenfunction for bound states is

$$M_p(z) = \left[\frac{1}{\beta} \sum_{i=0}^{p} \frac{\Gamma(2s_p+i)}{i!} \right]^{-1/2} e^{-\xi/2} \xi^{2s_p} L_p^{2s_p}(\xi) \tag{8}$$

where $L_p^{2s_p}$ is the generalized Laguerre polynomial (Tricomi, 1955), with

$$\begin{cases} s_p = \dfrac{\sqrt{2mD}}{\beta\hbar} - (p+\tfrac{1}{2}) \\[2mm] \xi = \dfrac{2\sqrt{2mD}}{\beta\hbar} e^{-\beta(z-z^0)} \end{cases} \tag{9}$$

Harmonic and Morse oscillators have been defined for different gases so as to minimize the total energy of each adsorbed atom, when treated as a three-dimensional oscillator localized at a single adsorption site, in its fundamental state. This has been done by the variational method, by considering α (in the harmonic oscillator) and both β and z^0 (in the Morse oscillator) as variational parameters. The third parameter D in the latter case has been assumed to be given by the value of the potential $V(r)$ at $z = z^0$. The values found for these parameters for the different rare-gas atoms adsorbed on the (100) and (111) face are given in table 3. In this same table, the number of bound states for the various Morse oscillators is also given.

Applying the variational method to linear combinations of Bloch sums, which belong to suitable irreducible representations, a matrix equation can be obtained in the form

$$(S^{-1}H)A = AE \tag{10}$$

where H and S are the hamiltonian and overlap matrices, A is the matrix of the coefficients in the linear combination, and E is the diagonal matrix of the energy eigenvalues.

The matrix elements S_{jl} and H_{jl} refer to Bloch functions which correspond to the same wave vector k, and to the sets n_j, m_j, p_j and n_l, m_l, p_l of quantum numbers for the local oscillators. Both S_{jl} and H_{jl} may be expanded in terms corresponding to the successive shells of neighbours in the direct lattice, around

Table 3. *Parameters for harmonic and Morse oscillators, corresponding to various rare-gas atoms adsorbed on the (100) and the (111) face of a* Xe *crystal*

Crystal face	Adsorbed atom	$\alpha(\text{Å}^{-2})$	$\beta(\text{Å}^{-1})$	z_0 (Å)	$D(10^{-16}$ erg)	n^*
(100)	He^3	0.411	0.295	2.334	311.1	5
	He^4	0.497	0.298	2.714	318.4	6
	Ne	1.769	0.287	2.187	640.5	21
	Ar	2.778	0.272	2.762	1254.7	45
(111)	He^3	0.247	0.375	2.938	253.4	4
	He^4	0.280	0.349	2.938	253.4	5
	Ne	1.072	0.340	2.884	517.3	16
	Ar	1.908	0.333	3.285	1068.4	33

* n = number of bound states in the Morse oscillator.

a central cell ($\nu = 0$). From equations (3) and (4) it appears that their calculation reduces to evaluating sums of integrals of this kind

$$\begin{cases} (S_{jl}^{AB})_{0,\nu} = \delta_{p_j p_l} \, c_A^* c_B \int (H_{n_j m_j}^A)_0 (H_{n_l m_l}^B)_\nu \, d\sigma \\ (H_{jl}^{AB})_{0,\nu} = c_A^* c_B \int [M_{p_j}(H_{n_j m_j}^A)_0] \hat{H} [M_{p_l}(H_{n_l m_l}^B)_\nu] \, d\tau \end{cases} \quad (11)$$

where $(S_{jl}^{AB})_{0,\nu}$ indicates the overlap integral between the j-th eigenfunction for the oscillator centred in the A site of the 0-th central cell, and the l-th eigenfunction for the oscillator centred in the B site of the ν-th cell. Similar meaning may be attributed to the indexing of other terms. These expressions can be easily modified to apply to the case where both functions are centred on sites of the same kind, and to the simpler case of the (100) face.

Integrals of the form $(S_{jl}^{AB})_{0,\nu}$ may be evaluated analytically, while the other integrals may conveniently be put in the form

$$(H_{jl}^{AB})_{0,\nu} = E_l (S_{jl}^{AB})_{0,\nu}$$
$$+ c_A^* c_B \int [M_{p_j}(H_{n_j m_j}^A)_0](V - V_\nu^B)[M_{p_l}(H_{n_l m_l}^B)_\nu] \, d\tau \quad (12)$$

where V is the potential energy of the adsorbed atom at each point over the surface, and $V_\nu{}^B$ is the potential function for the oscillator centred in the site B of the ν-th cell. The bulk of the calculation work consists of the numerical evaluation of this second integral.

Those Bloch functions that are bases for irreducible representations of the two space groups, corresponding to the adsorption site lattices on the (100) and (111) face, can easily be obtained from equations (3) and (4).

In the case of the (100) face, assuming that the lower and upper limit for the lowest energy band corresponds to points Γ and M in the first Brillouin zone respectively, the bases for the Γ_1 and M_1 irreducible representations, excluding the normalizing factor, are

$$\begin{cases} \Psi_{n,4m,p}^{\Gamma_1} = M_p(z) \sum_\nu R_{n,4m}(\rho_\nu) \cos(4m\phi_\nu) \\ \Psi_{n,4m,p}^{M_1} = M_p(z) \sum_\nu (-1)^{\nu_1+\nu_2} R_{n,4m}(\rho_\nu)\cos(4m\phi_\nu) \end{cases} \tag{13}$$

where $m = 0, 1, 2, \ldots$ and ν_1, ν_2 are defined by $\mathbf{R}_\nu = \nu_1 \mathbf{a}_1 + \nu_2 \mathbf{a}_2$.

For the (111) face, both the lower and upper limit for the lowest energy band are found at point Γ and the bases for the corresponding Γ_1 and Γ_4 irreducible representations, excluding the normalizing factor, are

$$\begin{cases} \Psi'_{n,3m,p}^{\Gamma_1} = M_p(z) \sum_\nu [R_{n,3m}(\rho_{\nu A}) \cos(3m\phi_{\nu A}) \\ \qquad\qquad + R_{n,3m}(\rho_{\nu B}) \cos(3m\phi_B)] \\ \Psi'_{n,3m,p}^{\Gamma_4} = M_p(z) \sum_\nu [R_{n,3m}(\rho_{\nu A}) \cos(3m\phi_{\nu A}) \\ \qquad\qquad - R_{n,3n}(\rho_{\nu B}) \cos(3m\phi_B)] \end{cases} \tag{14}$$

Standard sets of 36 Bloch functions have been taken into account for the various adsorbed atoms and the various irreducible representations here considered. They result from the combination of the first four Morse eigenstates with the first nine two-dimensional harmonic eigenstates belonging to each irreducible representation. From table 3 it appears that four is the least number of bound states shown by the different Morse oscillators considered here, corresponding to the case of the He^3 atom adsorbed on the (111) face. As far as the number of harmonic eigenstates is concerned, figure 2 shows that it provides a satisfactory convergence in the energy eigenvalues, with the only exception of the He atoms adsorbed on the (111) face, which is discussed below.

For truncating the neighbour expansion, attention was first paid to the S matrix, whose terms were computed precisely by analytically evaluating overlap integrals up to a distance, where their contribution was entirely negligible: in the case of He atoms on the (111) face this required considering neighbours up to the eighth ones. The relative importance of successive neighbour shells, in the S

matrix elements, was then used as a standard in truncating the neighbour expansion of the H matrix elements. These expansions were cut at terms whose contribution to the S elements was lower than contribution from the nearest neighbours by at least four orders of magnitude.

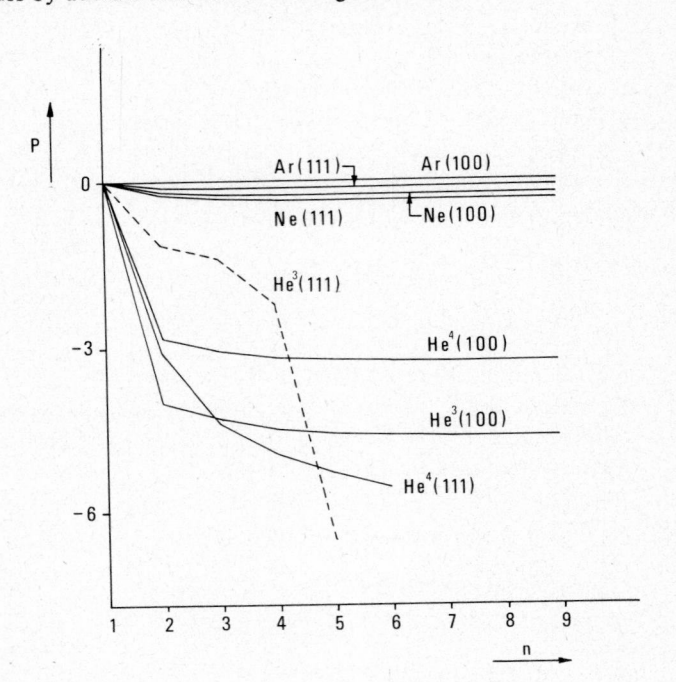

Fig. 2. Percentage gain in energy P for the fundamental Γ_1 state, as a function of the number n of harmonic eigenstates included in the variational function.

$$P = [(E_1 - E_n)/E_1] \times 100$$

III. CHARACTERIZATION OF THE DEGREE OF LOCALIZATION THROUGH THE PROPERTIES OF THE OVERLAP MATRIX

From equation (10) it appears that the overlap matrix plays a critical role in the solution of our problem. Three different situations may arise, which correspond to three different aspects of such a matrix.

The first case is found when the local functions pertaining to different cells of the adsorption site lattice do not overlap: i.e. when the adsorbed atom is in a strictly localized state. In this case the neighbour expansion of S_{jl} and H_{jl} will reduce to the first term, which gives $H = H^0$ (the hamiltonian matrix for the local, non-periodic, problem), but also $S_{jl} = \delta_{jl}$, i.e. $S = S^{-1} = I$. In such conditions equation (10) becomes

$$H^0 A = AE \tag{15}$$

which may be solved by simply diagonalizing the local hamiltonian matrix. In this case the periodic treatment vanifies and single energy levels are found for different adsorbed states.

On the other hand, when the state of the adsorbed atom is largely delocalized, the higher terms of the local functions do overlap considerably. In this case, as we know from the theory of solids, our hope of getting close approximations by building up a sufficiently large set of basis functions, is frustrated. At any given point the Bloch sums are composed of contributions

Table 4. *Classification of the examined adsorption systems according to S matrix properties, for a set of 36 Bloch functions belonging to the Γ_1 irreducible representation*

	He^3	He^4	Ne	Ar
(100)	NS	NS	I	I
(111)	O	O	NS	I

from local functions centred on a large number of cells, overlapping and combining so as to produce a sort of smearing of the probability density. An increasing interdependence is introduced in this way between different Bloch sums, since these poorly structured Bloch functions are obviously far from orthogonal. Thus the rows in the S matrix become mutually proportional and the matrix itself approaches singularity (Parmenter, 1952). It then becomes impossible to construct the $S^{-1} H$ matrix, which has to be diagonalized, for solving the problem.

This may be artificially prevented by reducing the number of terms in the local function, which of course produces a loss of accuracy in describing the system. Furthermore, it is then necessary to determine the Bloch sums very accurately, since the solution of the matrix equation depends upon the minute deviations from proportionality of the rows in the S matrix. Indeed, the trend to the singularity of the S matrix reflects the poorness of the local approximation and calls for a treatment different from the Bloch method: for instance the OPW method.

In those cases where the S matrix is neither identical nor singular, the application of the Bloch method illustrated above gives the best results.

The adsorption systems that have been considered may be classified in three different groups owing to the features of the corresponding S matrices built up from the standard set of 36 Bloch functions belonging to the Γ_1 irreducible representation which corresponds to the fundamental state.

This classification is given in table 4, where the symbols have the following meaning: I = identical matrix, local solution; O = singular matrix, roughly approximated periodic solution; NS = non-singular matrix, accurate periodic solution.

As seen previously, the cases O require the number of Bloch functions to be reduced until the singularity disappears. Non-singular matrices for He^4 and He^3 adsorbed on the (111) face are obtained using the first 24 and 16 terms respectively, instead of the standard 36. Though a more adequate method should be adopted for these cases, yet the results which may be obtained in this way are able to give the essential features of the system and the order of magnitude for energies and band-widths.

IV. RESULTS AND DISCUSSION

Energies and band-widths which have been obtained for the fundamental state and the lowest energy band of different rare-gas atoms adsorbed on the (100) and the (111) face of a Xe crystal are given in table 5.

Table 5. *Lowest energy level* E_0 *and band-width* $(\Delta E)_0$ *for the fundamental state of rare-gas atoms adsorbed on the (100) and the (111) face of a* Xe *crystal*[*]

Adsorbed atom	E_0	$(\Delta E)_0$	E_0	$(\Delta E)_0$
He^3	-192.8	0.02	-165	13
He^4	-209.6	0.01	-175	9
Ne	-558.5	$-$	-454.1	0.005
Ar	-1181.4	$-$	-1008.1	$-$
		(100)		(111)

[*] All the energies are in units of 10^{-16} erg.

He^3 and He^4 show a very limited mobility on the (100) face, but such a large one on the (111) face that the adopted method of approximation, based on the use of local functions, has revealed itself to be inadequate. Ne and Ar are strictly localized, except for the case of Ne on the (111) face, where a small band-width has been calculated.

The probability density distributions for the different atoms on the two faces of the Xe crystal can be illustrated using isodensity surfaces, similar to those which are normally employed to describe electron orbitals in atoms and

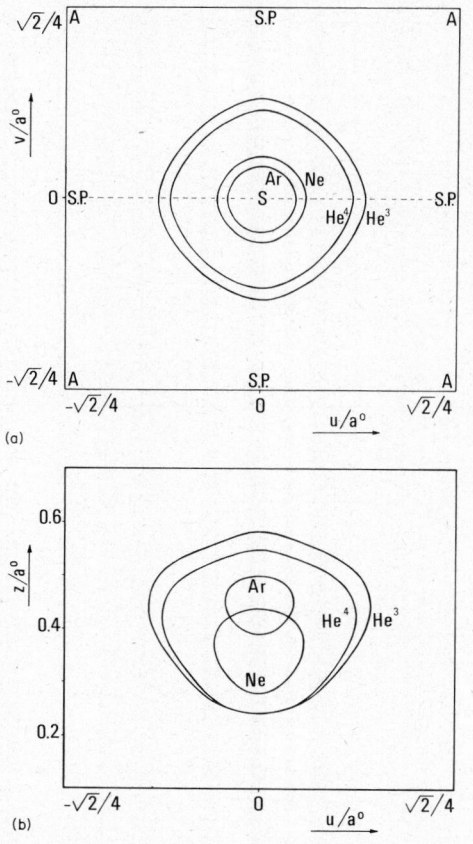

Fig. 3. Isodensity surfaces including 95% of the total probability for the fundamental state of various rare-gas atoms adsorbed on the (100) face of a Xe crystal. (a) Horizontal sections through the average z ordinate for each gas. (b) Vertical sections on a plane joining two saddle points on opposite sides of the adsorption site. (The z coordinate is measured from the plane through the centre of surface atoms). A = surface atom, S = adsorption site, S.P. =saddle point, $-----$ trace of the vertical plane.

molecules. The isodensity surface including 95% of the total probability can be usefully employed to this end.

In figure 3 horizontal and vertical sections of this surface are shown for the fundamental state of the atoms Ar, Ne, He^4 and He^3 on the (100) face. Due to the very thin energy band which characterizes the adsorption even of the partially delocalized He^3 and He^4, differences between the "bonding" Γ_1 and "anti-bonding" M_1 states, are not detectable in curves of this kind. All the atoms

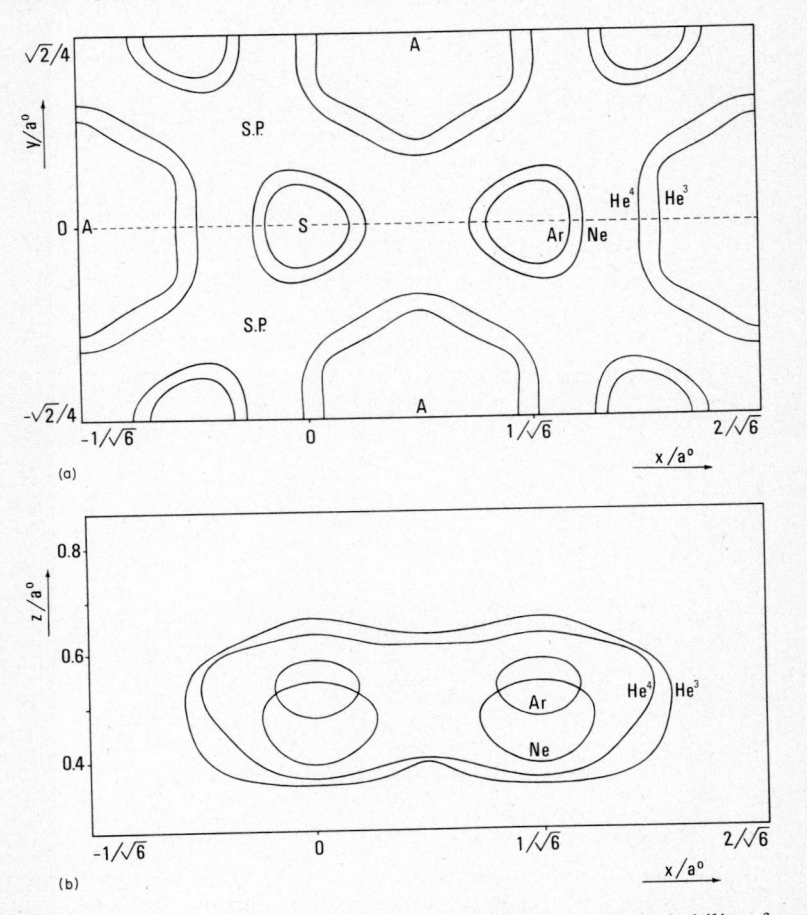

Fig. 4. Isodensity surfaces including 95% of the total probability for the fundamental Γ_1 state of various rare-gas atoms adsorbed on the (111) face of a Xe crystal. (a) Horizontal sections through the average z ordinate for each gas. (b) Vertical sections on a plane joining two adjacent adsorption sites. Symbols as in figure 3.

appear to be well centred near the adsorption site, at the middle of the unit cell formed by the surface atoms of the solid. The different height which is shown by the centre of gravity of Ne and Ar distributions reveals the influence of their atomic radius on the "penetration" in the surface structure of the solid.

Figure 4 reproduces similar curves for the fundamental state of the same atoms on the (111) face. Two adjacent sites of adsorption are considered in the figure. In this case a qualitative difference exists between the localized atoms of

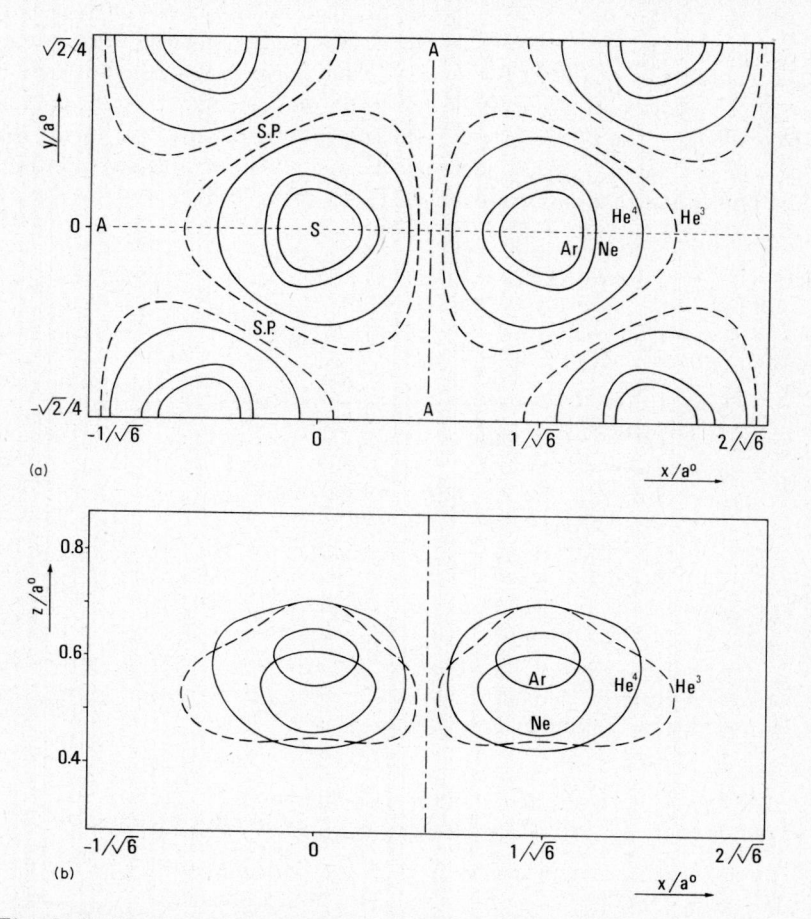

(a)

(b)

Fig. 5. Isodensity surfaces including 95% of the total probability for the Γ_4 state of various rare-gas atoms adsorbed on the (111) face of a Xe crystal. Γ_4 state corresponds to the top of the lowest energy band. (a) and (b) sections as in Fig. 4; symbols as in Fig. 3. $-----$ trace of the vertical plane of figure (b); $-\cdot-\cdot-$ trace of the nodal plane.

Ar and Ne and the delocalized atoms of He[4] and He[3]. The "orbitals" for the He atoms are not inscribed in the cell, but extend from site to site over all the crystal surface. Differences between the "bonding" Γ_1 and "anti-bonding" Γ_4 states are now introduced by the appreciable extent of the lowest energy band for He^4 and He^3. These differences are illustrated in figure 5, that however must be considered as essentially indicative, since the approximation is certainly a poor one, especially for He^3.

A further insight on the different behaviour of the various adsorbed atoms on the two crystal faces here considered can be furnished by table 6, which refers to the fundamental Π_1 states. The average value $\bar{z}$ of the z ordinate above the adsorption site is compared in this table to the value z^0_{site} of the ordinate corresponding to the minimum potential energy over the site, and to the value z^0_{saddle} of the ordinate corresponding to the minimum potential energy over the saddle points. Also the residual energies T_0 are given in table 6, where they are

Table 6. *Average ordinates and residual energies for the fundamental state of rare-gas atoms adsorbed on the (100) and (111) face of a Xe crystal.* [*]

Crystal face	Adsorbed atom	$\bar{z} - z^0_{site}$ (Å)	$\bar{z} - z^0_{saddle}$ (Å)	T_0 (10^{-16} erg)	$T_0 - \Delta V$ (10^{-16} erg)
(100)	He^3	+0.576	−0.726	146.5	+15.9
	He^4	+0.488	−0.502	130.3	− 0.3
	Ne	+0.136	−0.786	82.7	−154.7
	Ar	+0.060	−0.677	69.7	−326.5
(111)	He^3	+0.371	+0.047	105	+68.1
	He^4	+0.431	+0.106	95.4	+58.5
	Ne	+0.111	−0.109	64.3	−1.2
	Ar	+0.051	−0.119	63.7	−37.3

* Data referring to He^3 on the (111) face are purely indicative here, owing to the poorness of the Bloch approximation.

compared to the potential barriers to surface migration ΔV.

After considering table 6 the following comments may be advanced:

The asymmetric nature of the field in the normal direction to the crystal surface always produces a positive value for the difference $\bar{z} - z^0_{site}$, which is largely enhanced for the very mobile helium atoms.

The average ordinates of the adsorbed atoms are generally lower than the ordinates which correspond to the easiest passage from site to site through the saddle point. A significant inversion, however, is found for He^4 and He^3 over the (111) face, drawing attention to the essentially delocalized nature of the adsorption in these cases.

Finally, comparisons between table 6 and table 5 enable us to discuss the mobility of the adsorbed species. In fact, large band-widths appear when the residual energy is much higher than the potential energy barrier to migration, while on the other hand, low residual energies correspond to single levels in the energy spectrum. Only for those cases in which residual energy is near to the potential barrier, are small bandwidths detectable even below this barrier and tunnel effects can occur. In short, at a first approximation the mobility of adsorbed atoms may be looked at from a semi-classical point of view.

It is apparent from all these data that the two faces (100) and (111) of solid Xe show extremely different adsorption properties. Not only is the total energy for the fundamental state appreciably different, but the mobility of the adsorbed atoms shows a tremendous increase when passing from (100) to (111). This offers a particularly striking example of regular heterogeneity, which is not due to imperfections, but to the different arrangements of surface atoms, which are specific for different crystal faces.

Such a large heterogeneity makes it necessary to be careful when assuming approximate models for ill-defined adsorbing structures, as in the case of solids plated with pre-adsorbed rare-gas atoms. But it could, perhaps, give an acceptable physical basis to the patch-structure models that are required to justify the heat-capacity behaviour experimentally found for physisorbed He.

REFERENCES

Dash, J. G. (1968). *J. Chem. Phys.* **48**, 2820.
Goodstein, D. L., Dash, J. G. and McCormick, W. D. (1965). *Phys. Rev. Lett.* **15**, 447.
Hagen, D. E., Novaco, A. D. and Milford, F. J. (1971). This volume, I, 8, p. 99.
Lomer, W. M. (1955). *Proc. Royal Soc. Ser. A* **227**, 330.
McCormick, W. D., Goodstein, D. L. and Dash, J. G. (1968). *Phys. Rev.* **168**, 249.
Neustadter, H. E. and Bacigalupi, R. J. (1967). *Surface Sci.* **6**, 246.
Novaco, A. D. and Milford, F. J. (1970). *J. Low Temp. Phys.* **3**, 307.
Parmenter, R. H. (1952). *Phys. Rev.* **86**, 552.
Ricca, F. (1967). *Nuovo Cimento, Suppl.* **5**, 339.
Ricca, F. and Pisani, C. (1967). *Ric. Sci.* **37**, 536, 547.
Ricca, F. Pisani, C. and Garrone, E. (1969). *J. Chem. Phys.* **51**, 4079.
Ricca, F. and Garrone, E. (1970). *Trans. Faraday Soc.* **66**, 959.
Stewart, G. A. and Dash, J. G. (1970). *Phys. Rev. A* **2**, 918.
Tricomi, F. (1955). *In* "Vorlesungen über Orthogonalreihen", p. 212. Springer Verlag, Berlin.
Wallace, J. L. and Goodstein, D. L. (1970). *J. Low Temp. Phys.* **3**, 283.

ADSORPTION OF ^{3}HE AND ^{4}HE ON VARIOUS SUBSTRATES BELOW 30°K*

J. G. DAUNT and E. LERNER

Stevens Institute of Technology, Hoboken, New Jersey, USA

I. INTRODUCTION

This paper gives a brief preliminary report on some recent measurements of adsorption isotherms of ^{3}He, ^{4}He and neon deposited on copper and on argon-coated copper. It also presents some previous (Daunt and Rosen, 1970a) data on the adsorption of ^{3}He and ^{4}He on synthetic zeolite for comparison with the recent data. From the data the isosteric heats of adsorption are calculated as a function of coverage, and information regarding the monolayer coverage deduced.

II. THE EXPERIMENTAL ARRANGEMENTS

The measurements were made using apparatus which has been described elsewhere (Daunt and Rosen, 1970a). The range of pressures covered was from about 0.25 mmHg to 75 mmHg and the temperature range from 4.2°K to 26°K. The filling tube going from room temperature down to the cold specimen chamber had an inner diameter 2.16 mm. The thermomolecular pressure corrections necessitated by this tube, which connected the specimen at the low temperature to the pressure gauge at room temperature, were not greater than 3% for pressures of 0.5 mmHg and above. All necessary corrections for void volumes, gas non-ideality, etc. were carried out in the same manner as described previously (Daunt and Rosen, 1970a).

The copper surface for these adsorption studies was made in the form of a sponge by sintering pressed copper powder ("Druid Copper", Grade MD60 made by Alcan Metal Powders Inc.) of average particle size 2 x 2 x 0.5 microns in a hydrogen furnace at 650°C for 0.5 hours. This construction assured temperature

* Work supported by a contract with the Department of Defense (Themis Program), the Office of Naval Research and by a Grant from the National Science Foundation.

homogeneity over the whole adsorbing surface. The specimen used in the measurements comprised three such copper sponges in the form of discs, 5.00 cm dia. and 0.68 cm thick each, with a total mass of 170 g. The filling factor was 47.7%.

III. THE SURFACE AREA OF THE COPPER SPONGE

The surface area, Σ, of the specimen was obtained from measurements of adsorption isotherms of N_2 and argon taken at $77.3°K$. The isotherms proved very reproducible and showed no hysteresis. They were typical Type II isotherms (Young and Crowell, 1962) and the value of the monolayer coverage, V_m was estimated by the "Point B" method (Young and Crowell, 1962). The results are given in table 1.

Table 1. *The monolayer coverage,* V_m, *and the surface area* Σ *as deduced from* N_2 *and argon isotherms on a copper sponge at* $77.3°K$ *(see text). The molecular areas,* σ, *are taken from Young and Crowell (1962)*

Substance	V_m cm^3(STP)/g	σ Å^2	Σ m^2/g
N_2	0.11	16.2 (liquid)	0.48
N_2	0.11	13.8 (solid)	0.41
Ar	0.11	13.8 (liquid)	0.41
Ar	0.11	12.8 (solid)	0.38

The surface area Σ in m^2/g was calculated from V_m, using the formula (Young and Crowell, 1962)

$$\Sigma = 0.269\ V_m \sigma$$

where σ is the molecular area of the adsorbed gas in Å^2 and V_m is in cm^3(STP)/g. As has been discussed many times, and reviewed, for example in a previous paper (J. G. Daunt and C. Z. Rosen, 1970a), there is a question as to whether the σ for N_2 should be taken to be that for the liquid state (16.2 Å^2) or that for the solid state (13.8 Å^2). In table 1 we show the values of V_m and Σ for N_2 and Argon obtained from our $77.3°K$ isotherms, where Σ is deduced from σ values both for the liquid and the solid state. It will be seen that Σ deduced from the N_2 (solid) isotherm is about 8% lower than that obtained from the argon (solid) isotherm. A similar discrepancy between the N_2 and argon data for evaluation of Σ was observed previously (Daunt and Rosen, 1970a) in our synthetic zeolite adsorption measurements. The cause of the discrepancy is not clear, but may be associated with some uncertainty in the value of σ for argon.

We have chosen σ for solid N_2 as appropriate for calculation of Σ, giving a result of $\Sigma = 0.41$ m²/g for our copper sponge. This value has been used in all data presented in this paper.

IV. THE MEASURED ISOTHERMS

Figures 1 and 2 show the adsorption isotherms of ⁴He on the clean bare copper sponge and on a monolayer of argon deposited on the same sponge at the following temperatures: 6.18°K, 7.90°K, 9.65°K, 11.60°K, 13.50°K, 15.08°K

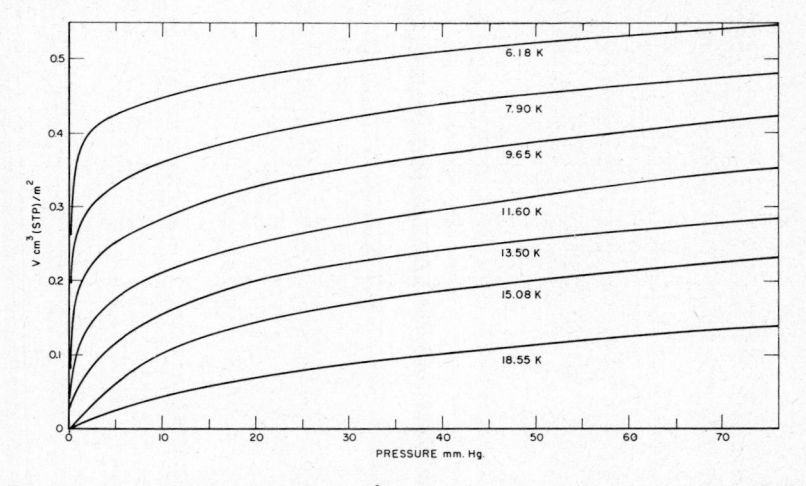

Fig. 1. Adsorption isotherms for ⁴He on bare Cu at temperatures as marked.

and 18.55°K. In depositing the argon, an amount of argon gas corresponding to the measured amount required at 77.3°K to form a monolayer was admitted to the sample cell at room temperature, the cell was then cooled slowly over a period of 19 hours from room temperature to liquid helium temperature. This permitted adequate time for the argon to diffuse throughout the copper sponge and to deposit evenly on it. By comparing figures 1 and 2, it will be noted that at any given temperature and pressure, the amount of ⁴He adsorbed on the argon covered copper sponge is less than that adsorbed on the bare sponge.

The data for ³He adsorbed on bare copper at the same selected temperatures are shown in figure 3. By comparison with figure 1 for ⁴He on bare copper, it will be seen that the ³He curves show smaller adsorption at any given temperature and pressure. However at the highest temperatures and pressures reported these differences are very small and within experimental error. This result is in qualitative agreement with the comparison of adsorption of ³He and ⁴He on synthetic zeolite 13X as shown in figures 4 and 5 and as reported previously (Daunt and Rosen, 1970a).

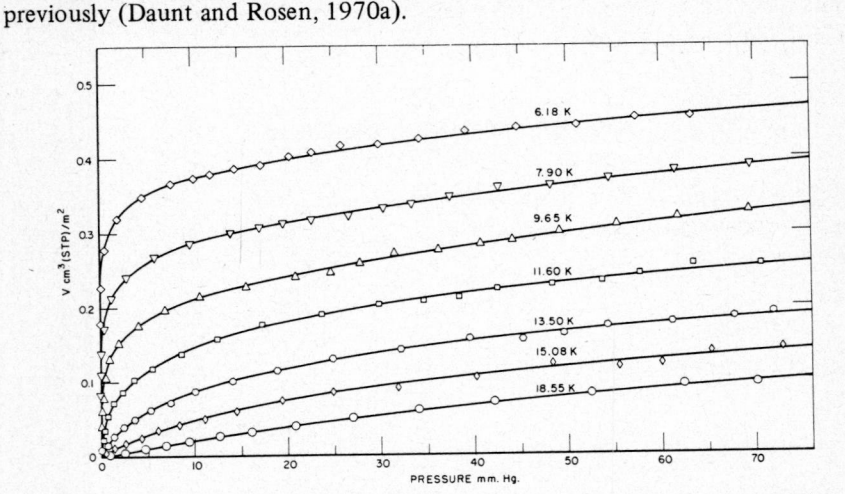

Fig. 2. Adsorption isotherms for ⁴He on monolayer of argon on Cu at temperatures as marked.

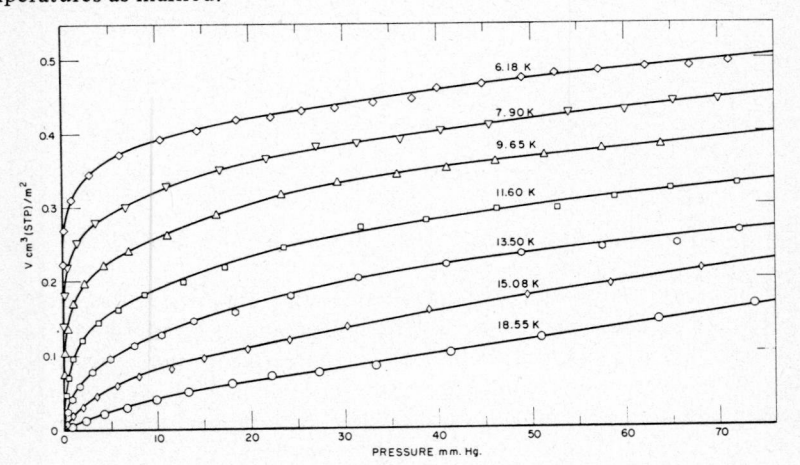

Fig. 3. Adsorption isotherms for ³He on bare Cu at temperatures as marked.

Comparison of figures 2 and 3 shows that the introduction of an argon monolayer on the copper lowers the adsorption of ⁴He, even below that of ³He on bare copper.

In order to make a qualitative and quantitative comparison of the results using the copper sponge with earlier ones using synthetic zeolite, we show in figures 4 and 5, on the same scale in $cm^3(STP)/m^2$, the adsorption isotherms for ⁴He and ³He on synthetic zeolite (Linde Molecular Sieve, Type 13X) as measured in our laboratory previously (Daunt and Rosen, 1970a) at the

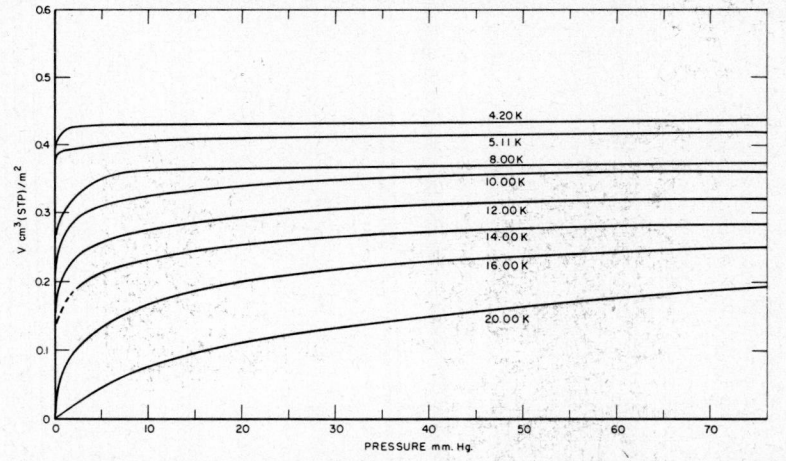

Fig. 4. Adsorption isotherms of ⁴He on synthetic zeolite 13X at temperatures as marked.

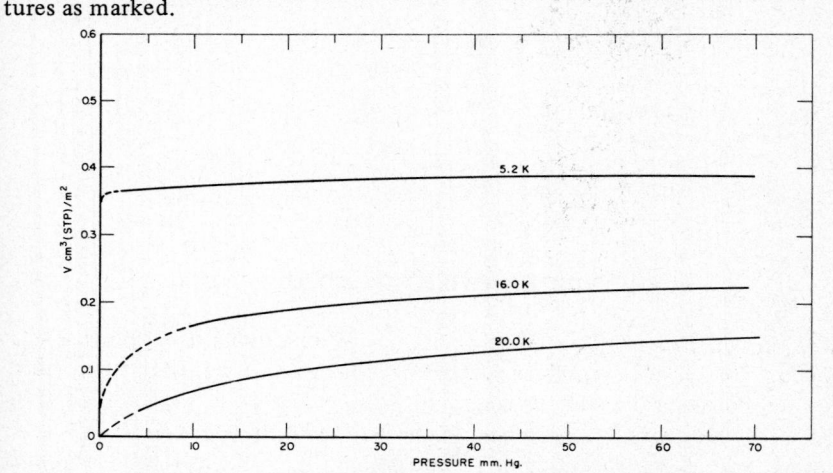

Fig. 5. Adsorption isotherms of ³He on synthetic zeolite 13X at temperatures as marked.

following temperatures: 4.2°K, 5.11°K, 8.0°K, 10.0°K, 12.0°K, 14.0°K, 16.0°K and 20.0°K. The character of these adsorption curves seems to differ from those found with the copper sponge; they appear much "flatter" in the higher pressure ranges. This may be associated with much smaller sizes of the voids in the zeolite.

Figure 6 shows the measured neon adsorption isotherms on the bare copper sponge and on the copper sponge coated with a monolayer of argon at 22.6°K and at 25.75°K. Here the much lower adsorption on the argon monolayer as compared with the adsorption on the bare copper is clearly evident.

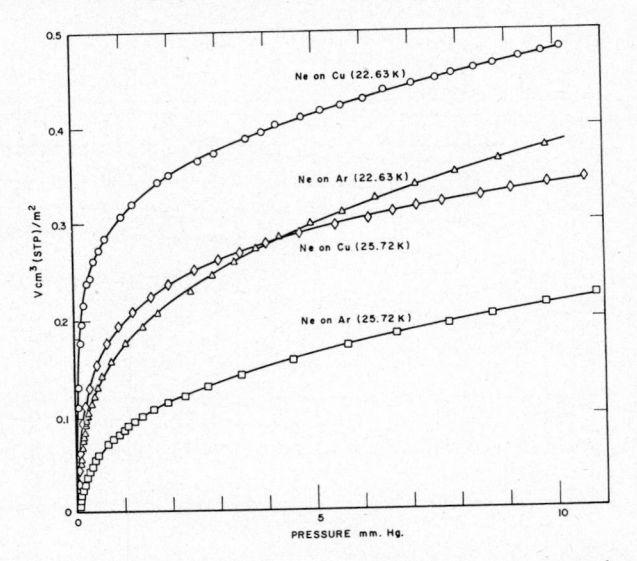

Fig. 6. Adsorption isotherms for neon on bare Cu and on monolayer of argon on Cu at temperatures as marked.

V. THE ISOSTERIC HEAT OF ADSORPTION

The data of figures 1, 2, 3, 4, and 6 have been used to construct plots of $\ln p$ versus $1/T$ for various constant coverages (V constant). It was found that these plots were linear over a wide range of temperature. The slope, $|\partial \ln p/\partial(1/T)|V$, is equal to Q_{st}/R, where Q_{st} is the isosteric heat of adsorption. The results of these computations are shown in figures 7 and 8 which give Q_{st}/R in degrees Kelvin as a function of the coverage, V in $cm^3(STP)/m^2$.

Figure 7 shows the results for neon on bare copper and on copper covered with a monolayer of argon; for ^4He on the bare copper and, for comparison, for ^4He on synthetic zeolite (Linde Molecular Sieve, 13X). The latter results were recomputed from data presented in a previous paper (Daunt and Rosen, 1970a).

It will be seen that Q_{st} is significantly lower at all coverages for neon on a monolayer of argon deposited on copper than for neon on bare copper. At a

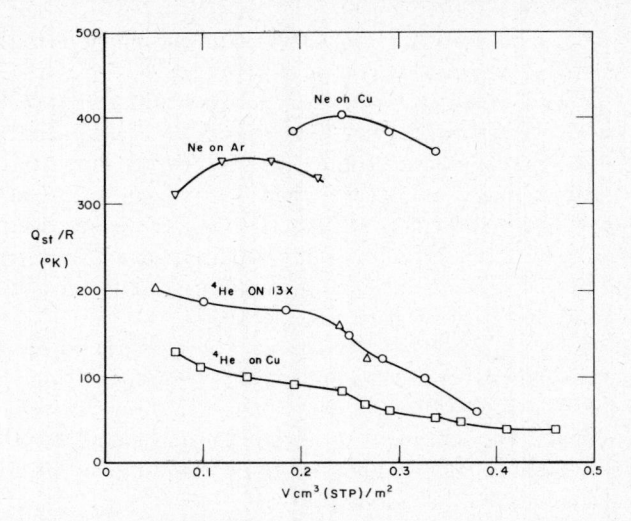

Fig. 7. The isosteric heat of adsorption, Q_{st}/R, as a function of coverage V/Σ for: Neon on bare Cu and on monolayer of argon on Cu; ⁴He on synthetic zeolite 13X and ⁴He on bare Cu.

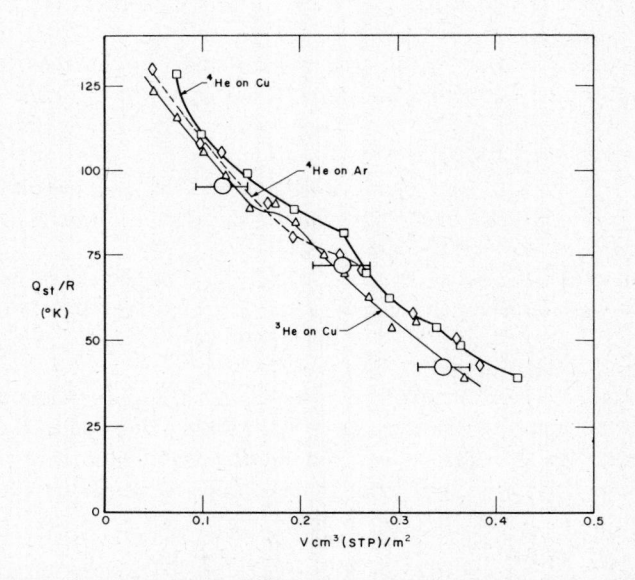

Fig. 8. The isosteric heat of adsorption, Q_{st}/R, as a function of coverage V/Σ for ⁴He on bare Cu and on monolayer of argon on Cu; —○— (Wallace and Goodstein, 1970) and ³He on bare Cu.

coverage of 0.24 cm^3(STP)/m^2, Q_{st}/R for neon on copper is 400°K and for neon on a monolayer of argon on copper Q_{st}/R is 320°K.

Figure 7 shows that the Q_{st} values for ^{4}He are significantly lower than those for neon. At a coverage of 0.24 cm^3(STP)/m^2 Q_{st}/R is 83°K for ^{4}He on bare copper and 162°K on synthetic zeolite 13X. As is clearly shown in figure 7, Q_{st} for ^{4}He on bare copper is much lower than that on zeolite 13X at any chosen coverage below about 0.4 cm^3(STP)/m^2. The Q_{st} values for neon are not as secure as those for ^{4}He, because of the relatively much smaller number of isotherms of neon from which calculations are made. This relative imprecision particularly refers to the coverage dependence of Q_{st}.

Figure 8 shows Q_{st}/R versus coverage for ^{3}He and ^{4}He on the bare copper and for ^{4}He on a monolayer of argon on copper. It also shows three points taken from extensive data by Wallace and Goodstein (1970) for ^{4}He on a monolayer of argon on copper. The agreement between our results and those of Wallace and Goodstein is fair and probably within the errors introduced in the normalization of Wallace and Goodstein's data to our own.*

It appears that Q_{st} at any chosen coverage for ^{4}He on the bare copper is higher than that for ^{3}He also on the bare copper. In comparing ^{4}He on bare copper and on the argon monolayer, it appears that below a coverage of about 0.25 cm^3(STP)/m^2, Q_{st} on argon is smaller than on the bare copper, however above the stated coverage there appears to be little difference between the observed Q_{st} values.

Although the precision in the Q_{st} values is much poorer than those in the observed adsorption isotherms, a feature which makes it difficult to interpret closely the variation of Q_{st} with coverage, two main features appear in figures 7 and 8 for the helium data, namely

(1) There appears to be a change in slope, or perhaps an inflexion at about $V = 0.24$ for ^{4}He both on bare copper and on the argon monolayer and at about 0.17 cm^3(STP)/m^2 for ^{3}He on bare copper.

(2) Both above and below these coverages, Q_{st} is dependent on coverage and shows little sign of becoming independent of coverage even at the coverages observed.

We interpret these two features as follows:

(1) The inflexions are interpreted as indicating the approximate coverage at which the first layer of adsorbed gas is completed. This yields V_m values as follows: For ^{4}He on bare copper *and* on the argon monolayer on copper $V_m \simeq 0.24$ cm^3(STP)/m^2, For ^{3}He on bare copper 0.17 cm^3(STP)/m^2. These

---

* To normalize Wallace and Goodstein's (1970) data to our own, we have taken our ^{4}He on the argon monolayer to have a monolayer capacity, V_m, of 0.24 cm^3(STP)/m^2 as explained in the text above. No change has been made in Wallace and Goodstein's data, nor can a comparison of our V_m value with their data be made, since they did not specify the surface area of their adsorbent.

values are to be compared with some previous values, such those of McCormick *et al.* (1968) who quoted $V_m \simeq 0.33$ for ⁴He and 0.29 for ³He on bare copper and those of Daunt and Rosen (1970a) who quoted V_m to be probably less than 0.29 cm³(STP)/m² for ⁴He on synthetic zeolite and 0.26 cm³(STP)/m² for ³He on the same adsorbent (Daunt and Rosen, 1970b).

Table 2. *Data of isoteric heat* (Q_{st}) *and estimated monolayer coverage* (V_m).

	Q_{st}/R (K) At monolayer capacity	V_m cm³(STP)/m²
He³ on bare copper	88	0.17
He⁴ on bare copper	83	0.24
He⁴ on Ar monolayer on Cu	76	0.24

	Q_{st}/R (K) At $V = 0.24$ cm³(STP)/m²
Neon on bare copper	400
Neon on Ar monolayer on Cu	320

The observation that V_m for ⁴He appears to be approximately the same for adsorption on bare copper as on the argon monolayer, taken with the fact that the Q_{st} values do not differ significantly for the two substrates, presumably indicates that the surface area for adsorption is not different for the two substrates.

(2) The way Q_{st} continues to increase as the coverage is reduced below the monolayer coverage for the helium on all substrates used is interpreted as indicating inhomogeneity of adsorbing sites.

We sum up our data on Q_{st} and V_m values in table 2, which gives our evaluation of the quantities on the various substrates at monolayer coverage.

REFERENCES

Daunt, J. G. and Rosen, C. Z. (1970a). *J. Low Temp. Phys.* **3**, 89.
Daunt, J. G. and Rosen, C. Z. (1970b). *Proc. Symp. Thermo. Phys. Prop.* 5th, p. 319.
McCormick *et al.* (1968). *Phys. Rev.* **168**, 249.
Wallace, J. L. and Goodstein, D. L. (1970). *J. Low Temp. Phys.* **3**, 283.
Young, D. M. and Crowell, A. D. (1962). *In* "Physical Adsorption of Gases". Butterworth, London.

AN EXPERIMENTAL INVESTIGATION OF THE PHYSISORPTION OF HELIUM ON ARGON, KRYPTON AND XENON

T. J. LEE and L. GOWLAND

Royal Observatory, Edinburgh; U.K.A.E.A. Culham Laboratory, Abingdon, England

I. INTRODUCTION

As workers in the field of laboratory astrophysics at low temperatures we have a deep interest in adsorption phenomena of light gas atoms and molecules. About 1% of interstellar matter is in the form of solid dust grains with temperatures probably between 3°K and 30°K, most of the rest is in the form of hydrogen atoms which, under favourable conditions can be associated into molecules by interactions with the grains. In very dense clouds these molecules may even be frozen onto grains to form solid hydrogen mantles. Helium gas is also present with number densities about 10% of those of the hydrogen. To calculate rates of processes such as molecule formation as a function of temperature information on sticking probability, surface mobility and adsorption energies is required. Gas molecules of low mass raise special problems, e.g. they can have zero point energies comparable with adsorption energies. Theoretical models must therefore include calculations of both the potential and kinetic energy of the adsorbed atom. Such calculations have been carried out for the adsorption energy of helium on rare-gas solids by Ross and Steele (1961) by Ricca *et al.* (1969) and by Ricca and Garrone (1970). The present experiments were performed to test these calculations, as we are interested in the development of the theory of physisorption to include the cases of atomic and molecular hydrogen on substrates such as graphite, silicates and ice.

II. EXPERIMENTAL APPARATUS

The apparatus used in these experiments, the main features of which are shown in figure 1, is based on the model F cryopump developed by Chubb (1970). Two modifications have been made in order to reduce the thermal radiation flux incident on the condensing surface:

(1) The gas inlet tube has been replaced by one of smaller diameter.

(2) Radiation traps have been fitted into the connections to the gas density measuring intruments.

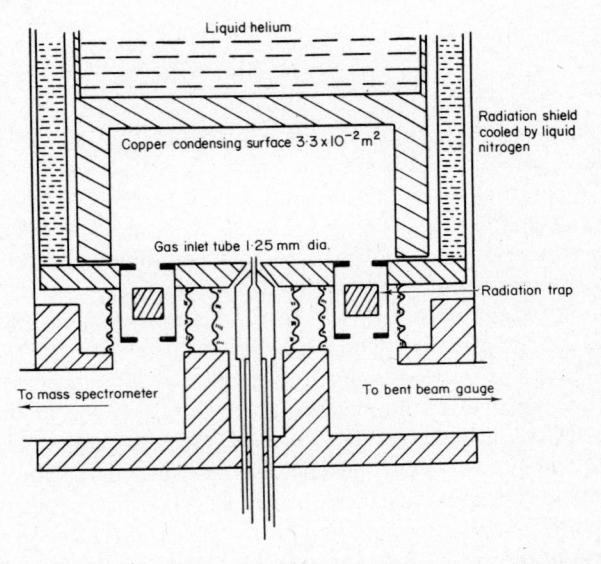

Fig. 1. Model G. Cryopump—Enlargement showing the radiation traps and gas inlet tube.

These traps are cylindrical copper structures, the internal surfaces of which are coated with glass as an absorber. The modified cryopump is referred to as the model G cryopump; the radiation load to the cooled condensing surface is less than 3×10^{-1} Watts m^{-2}. Gas densities are measured with an A.E.I. MS10 mass spectrometer and with a "bent beam" ionisation gauge, similar to that described by Helmer and Hayward (1966). The condensing surface is of copper and has a total area of 3.3×10^{-2} m^2. This surface is cooled by liquid helium and its temperature can be varied over the range $1.6°K$ to $5.3°K$ by pumping or pressurising the helium. The temperature of the helium reservoir is monitored by a Germanium resistance thermometer or derived from the vapour pressure in the cryostat. The calibrated gas handling system and gauge calibration techniques have been described by Chubb and Gowland (1967) and the data acquisition system by Chubb *et al.* (1968).

III. MEASUREMENTS

Conventional techniques were used to evacuate the system to below 10^{-9} Torr, the radiation shield was cooled with liquid N_2 and the cryostat filled with liquid helium. The temperature control system was set to maintain the cryostat

Table 1. *Physical parameters during the growth of substrate and adsorbed layers*

Substrate Gas	Argon	Krypton	Xenon
Temperature of growth in °K	3.53	2.44	2.85
Thickness x 10^{20} molecules m^{-2}	215	242	13.5
Temperature for helium deposition	3.25	3.2	3.4
Sticking coefficient	0.47	0.41	0.38
Helium densities x 10^{20} molecules m^{-2}	0.0144	0.244	0.0183
	0.0288	0.492	0.0448
	0.0666	0.109	0.115
	0.112		0.157
	0.162		

at a constant temperature and the substrate gas to be studied was deposited until the desired thickness was achieved. Thicknesses of several times 10^{20} molecules m^{-2} were used, as experience has shown that for condensed gas layers of thickness greater than $\sim 2 \times 10^{20}$ molecules m^{-2} the adsorption-desorption properties are independent of thickness. Helium gas at room temperature was then injected in pulses 15 seconds long at a flow rate of $\sim 10^{16}$ molecules m^{-2} s^{-1}. After 2 or 3×10^{18} molecules m^{-2} of helium inflow gas injection was interrupted and the temperature of the cryostat was raised to above 5°K and then lowered to about 3.5°K, data records were taken at intervals of 0.1°K or less throughout this procedure. This cycle was sometimes repeated before further helium deposition took place. Gas injection was recommenced for a similar duration and one or two temperature cycles performed. Table 1 lists the substrate gas, its thickness and temperature of growth, the temperature at which helium was deposited and the surface densities of helium for which temperature cycles were performed.

Also shown in table 1 are average sticking coefficients of helium on each substrate. These were measured in the manner described by Chubb (1970). Average values are shown since actual values varied little over the surface density range investigated. For krypton and xenon the sticking coefficients at the highest coverages were about 10% greater than those at the lowest. For argon no variation was apparent.

IV. VARIATIONS OF GAS DENSITY WITH TEMPERATURE

In all cases the variation of helium density as the cryostat temperature was cycled followed a similar pattern; illustrated in figure 2 on a logarithm of helium density against reciprocal temperature plot. As the temperature was increased at a uniform rate from that at which the helium was deposited, point A, the density rises as shown; at B the temperature was held constant and the gas density fell to about half its highest value in about 20 seconds; on decreasing the temperature the logarithm of the density falls linearly, with reciprocal temperature, until D where it deviated along the path D to E. In cases where a second cycle was performed the data did not retrace either of its former paths during the increase in temperature, but, after an initial rise in density, followed a nearly linear relation until the temperature again reached the peak value. On decreasing the temperature the variation was nearly that followed during the first decrease in temperature. Under conditions of equilibrium the gas density n, in a gauge at uniform temperature is related to R, the rate of desorption of atoms from the surface by the equation

$$R = \tfrac{1}{4} n \gamma \bar{c} \qquad \qquad (1)$$

where $\bar{c}$ is the mean molecular velocity at the temperature of the gauge and γ the sticking coefficient. If the desorption of helium from the rare-gas substrates is governed by first-order kinetics then

$$R = N\nu \exp\left(-E/kT\right) \qquad \qquad (2)$$

where, N is the surface density of helium, E is the adsorption energy and the other symbols have their usual meaning, then the form of the curve in figure 2 can be discussed in terms of variations of N and T. For constant N the logarithm of the gas density varies linearly with $1/T$ and values of the adsorption energy may be calculated from the slope, such regions exist in the experimental data, that in figure 2 is depicted by the full line C-D. The evidence that N is substantially constant for this region is that firstly the line can be retraced by increasing and decreasing the temperature within certain limits, and secondly no variations in density with time at constant temperature are observed, whereas changes in density with time occur in the other regions (shown by the broken

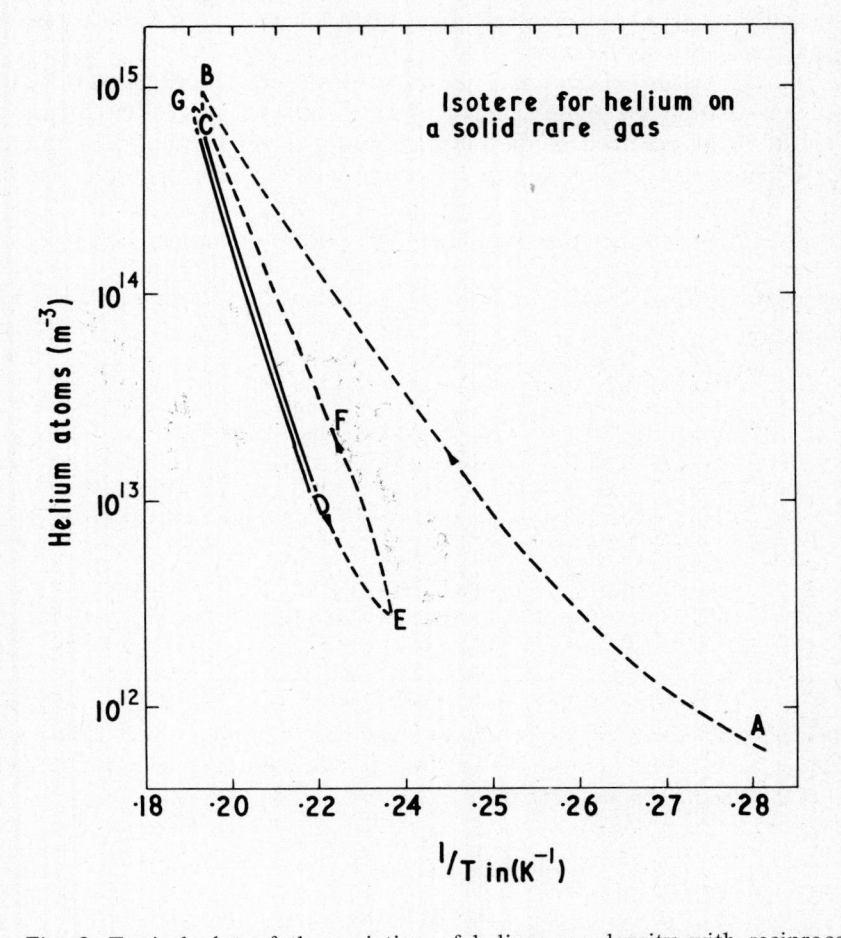

Fig. 2. Typical plot of the variation of helium gas density with reciprocal temperatures for helium adsorbed on a rare-gas solid.

lines). Such changes, like the decrease in density with time at B are thought to be caused by rearrangement of the adsorbed helium with time, either by diffusion into and out of the bulk or by a configuration change on the surface. A qualitative explanation of the experimentally observed behaviour shown in figure 2 can be constructed in terms of these ideas.

V. ADSORPTION ENERGIES

Figures 3, 4 and 5 show those portions of the experimental data from which adsorption energies are obtained. Here logarithm desorption rate is plotted against reciprocal temperature.

The adsorption energy and pre-exponential term in equation (2) are determined from the slope and intercept. A large fraction of the helium is thought to be adsorbed in the substrate during these determinations and an appropriate value of the average percentage concentration of helium can be

Table 2. *Experimental values of constants in the desorption rate equation*

Substrate	He%	E eV	E Joules	E J mol^{-1}	$N\nu$ molecules m^{-2} s^{-1}
A	0.01	0.0176	2.82×10^{-21}	1700	1.6×10^{32}
A	0.02	0.0133	2.13×10^{-21}	1285	3.2×10^{29}
A	0.04	0.0133	2.13×10^{-21}	1285	3.6×10^{30}
A	0.06	0.0126	2.01×10^{-21}	1220	2.3×10^{30}
A	0.08	0.0125	1.99×10^{-21}	1200	4.8×10^{29}
Kr	0.01	0.0135	2.16×10^{-21}	1300	2.2×10^{29}
Kr	0.02	0.0141	2.25×10^{-21}	1355	3.8×10^{30}
Kr	0.05	0.0126	2.01×10^{-21}	1210	1.6×10^{30}
Xe	0.064	0.0173	2.77×10^{-21}	1670	8.4×10^{31}
Xe	0.3	0.0161	2.57×10^{-21}	1550	8.0×10^{31}
Xe	0.85	0.0136	2.17×10^{-21}	1310	2.6×10^{30}
Xe	1.2	0.0137	2.19×10^{-21}	1320	6.0×10^{30}

obtained from the numbers of substrate and helium atoms injected. These are listed in table 2 together with the substrate gas, the adsorption energy in various units and the pre-exponential, $N\nu$.

DISCUSSION

The results in table 2 show a decrease in adsorption energy with quantity injected. The values for argon and xenon at the lowest coverage probably about 1% of a monolayer are high, and almost certainly pertain to adsorption at defects which must be abundant in rare-gas solids condensed at about 3°K. Adsorption energies measured from quantities of helium injected between 2×10^{18} molecules m^{-3} and 10^{-19} molecules m^{-3} are considered to be adsorption energies of helium atoms on the regular portions of the surfaces. There is no

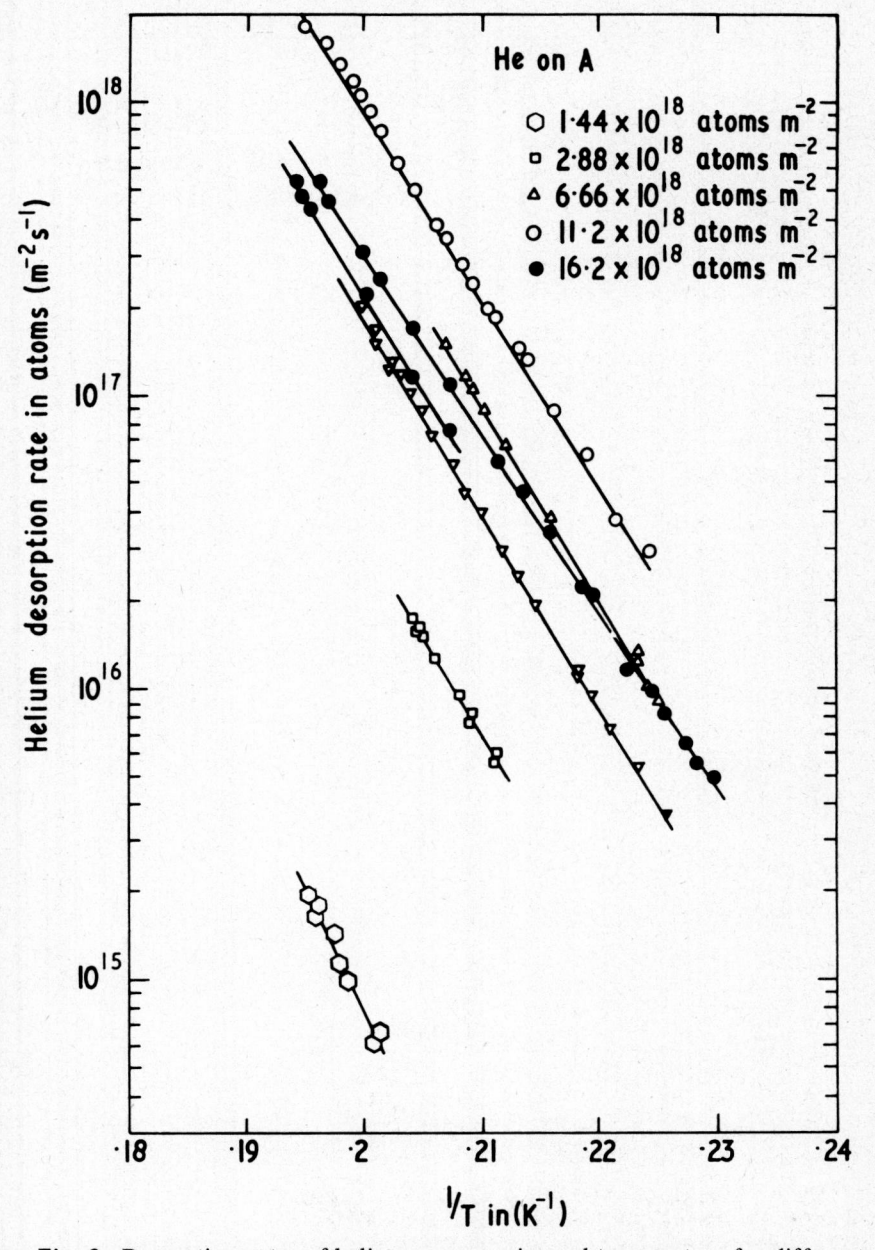

Fig. 3. Desorption rates of helium versus reciprocal temperature for different quantities of helium adsorbed on solid argon.

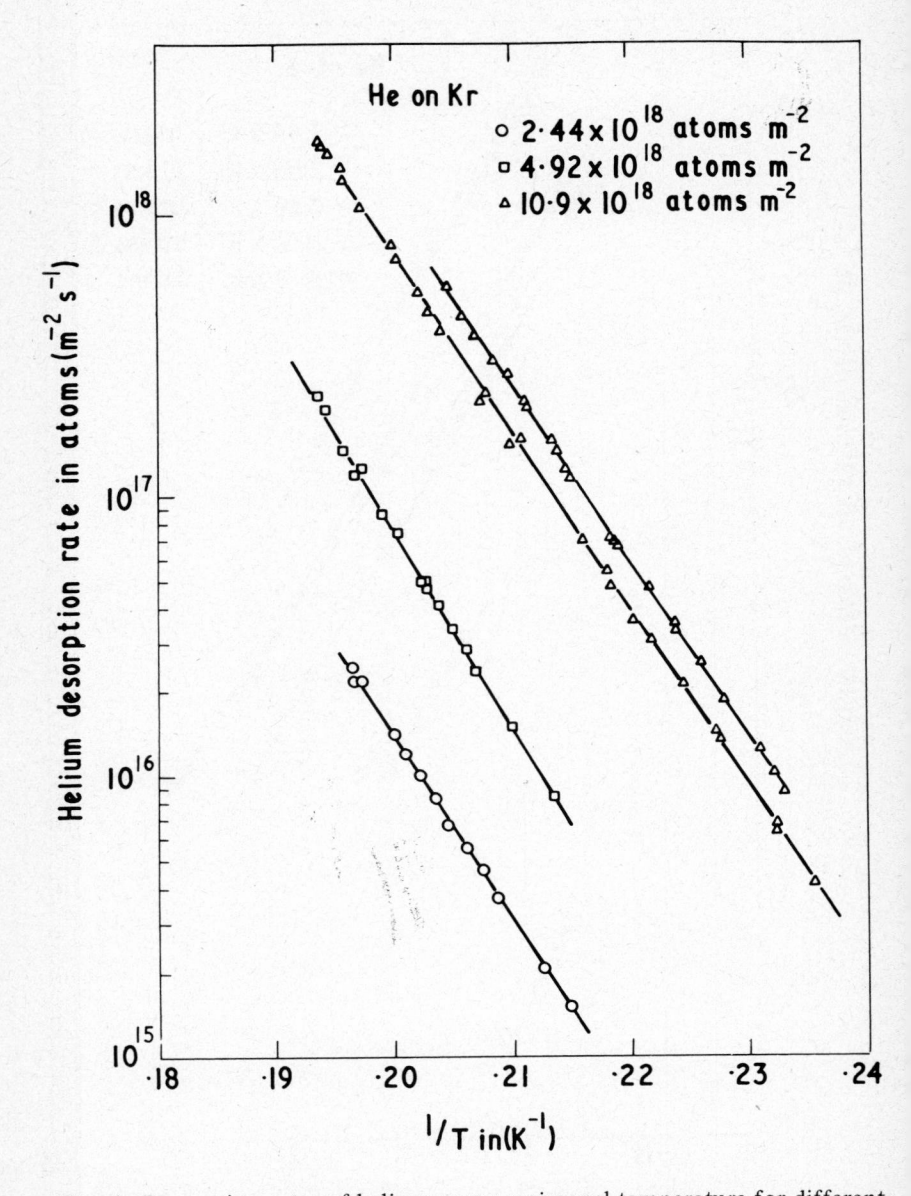

Fig. 4. Desorption rates of helium versus reciprocal temperature for different quantities of helium adsorbed on solid krypton.

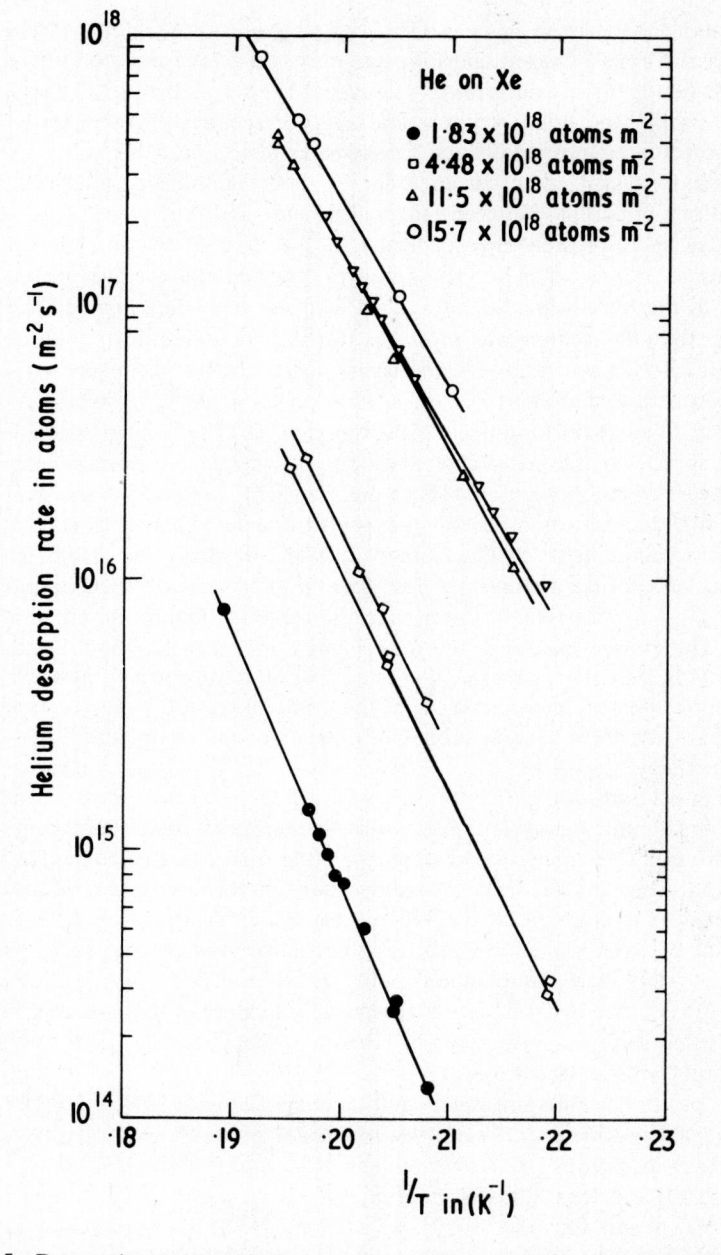

Fig. 5. Desorption rates of helium versus reciprocal temperature for different quantities of helium adsorbed on solid xenon.

significant difference in energies when helium quantities are doubled within this range in the cases of argon and krypton, only one value for xenon was obtained Average values for the adsorption energy are 1285 J mol^{-1}, 1325 J mol^{-1} and 1550 J mol^{-1} for argon, krypton and xenon respectively. At higher coverages the adsorption energy is lower. If helium does diffuse into the substrate as our experiments suggest, then the presence of these helium atoms will reduce the interaction potential between the adsorbate and substrate. The greatest reduction in adsorption energy occurs in the case of xenon, here since the substrate is thinner than for the argon or krypton the average percentage of helium in the bulk is greater for a given quantity of helium injected. Again it is stressed that the helium was probably distributed non-uniformly through the substrate and concentrations in the layers nearer the surface might be different.

Experimental studies of helium adsorption on argon have been reported by Ross and Steele (1961) and by McCormick *et al.* (1968), high-area substrates, copper sponge in the latter case and TiO_2 powder in the former, coated with monolayers of argon were used. Ross and Steele (1961) derived a value of 920 J mol^{-1} for the adsorption energy of helium from the exponential isotherms and by a calorimeter method. The isotherms of McCormick *et al.* (1968) indicated that the adsorption energy was less than that measured by Ross and Steele (1961). This is considerably lower than the value for helium on argon reported above. The atomic spacing in the monolayer thick substrates could be different from that in bulk specimens, leading to a different adsorption potential. Further, higher pressures, and hence coverages, than those used in the present experiment might have given rise to some mixing of the helium and argon and thus a reduced binding energy.

The most interesting feature of physisorbed light gas atoms is that their zero-point kinetic energy is as large as several tenths of the potential energy and in some cases this energy might be greater than the potential barrier to surface migration. Thus any theoretical considerations of the adsorbed state must make a satisfactory evaluation of the kinetic energy. Calculations of the interaction potential between helium and rare-gas solids have been performed by summing the Van der Waals interactions over infinite crystal lattices; the three-dimensional Schrödinger equation for the helium atom in the potential field was solved using various approximations (Ross and Steele, 1961; Ricca *et al.*, 1969; Ricca and Garrone, 1970).

Energies of the fundamental bound state are found to be 792 J mol^{-1} for He on Ar (100) by Ross and Steele (1961); the later and more extensive calculations of Ricca *et al.* (1969, 1970) give values 1043 J mol^{-1} and 1267 J mol^{-1} for He on Kr (100) and Xe (100) (Ricca *et al*, 1969) and 1519 J mol^{-1} for He on Xe (110) (Ricca and Garrone, 1970), 1550 J mol^{-1}. Our experimental value for helium on xenon is very close to the last case, theoretical adsorption energies on A (110) and Kr (110) would be extremely interesting. Should these agree with our experimental values (and the trends shown in the results obtained so far indicate that this is probable) there would be a good case for performing experiments in which both surface structure and adsorption energy may be determined. Such experiments could use a well-collimated molecular beam to probe the surface of a specimen, condensed on a rotatable substrate. Examination of the shape of the reflected beam as a function of the angular position of the substrate would give information on the order and symmetry of its surface structure. Though our experiments have the serious limitation that the structure of our surfaces is unknown, the quantitative and qualitative agreement which exists between our results and theoretical predictions is very satisfying, and there are grounds for believing that good models now exist for at least one class of surface phenomena.

ACKNOWLEDGEMENTS

One of the authors is a Senior Research Fellow on the staff of the Royal Observatory, Edinburgh, which finances the project; facilities are provided by U.K.A.E.A., Culham Laboratory. The astronomical aspects of the program are under the direction of Dr. V. C. Reddish.

REFERENCES

Chubb, J. N. (1965). U.K.A.E.A. Research Group report CLM-R54.
Chubb, J. N. and Gowland, L. (1967). *Vacuum*, **17**, 449.
Chubb, J. N., Gowland, L., Pollard, I. E. and Sinton, E. K. (1968). *Proc. 4th International Vacuum Congress*, pp. 414-418.
Chubb, J. N. (1970). *Vacuum*, **20**, 477.
Helmer, J. C. and Hayward, W. H. (1966). *Rev. Sci. Instrum.* **37**, 1652.
McCormick, W. D., Goodstein, D. C. and Dash, J. G. (1968). *Phys. Rev.* **168**, 249.
Ricca, F., Pisani, C. and Garrone, E. (1969). *J. Chem. Phys.* **51**, 4079.
Ricca, F. and Garrone, E. (1970). *Trans. Faraday Soc.* **166**, 959.
Ross, M. and Steele, W. A. (1961). *J. Chem. Phys.* **35**, 862.

II. PARTICLE BEAMS IN GAS-SOLID INTERACTION STUDIES

ATOMIC AND MOLECULAR SCATTERING, DIFFRACTION AND TRAPPING ON TUNGSTEN AND PLATINUM SURFACES

W. H. WEINBERG* and R. P. MERRILL

Department of Chemical Engineering,
University of California,
Berkeley, California, USA

I. INTRODUCTION

The study of the scattering of collimated molecular and atomic beams from solid surfaces is a technique which can, in principle, be used to study a number of the important phenomena associated with the fundamentals of the gas-solid interaction. In addition, the scattering of light gases (e.g. helium, hydrogen, and deuterium) can profitably elucidate several important features of the structure of the surfaces themselves, and in some cases the structure of absorbates present on the surface (Smith and Merrill, 1969). To achieve these objectives careful attention must be given to the preparation and characterization of the surfaces themselves (Merrill and Smith, 1970) and to ways of limiting the contamination of the surfaces by system background gases. Though in some cases it has been possible to achieve well defined surfaces by specialized techniques such as continuous deposition (Smith *et al.,* 1969) or by carefully exploiting the thermochemistry of certain surfaces (Yamamoto and Stickney, 1970), it seems clear that experiments must be performed at ultra-high vacuum (10^{-10} Torr and below) to provide the experimental flexibility needed for full exploitation of the method. And if some alternative characterization of the surface (e.g. by LEED, Auger spectroscopy, flash desorption, work function measurements etc.) can be made *in situ* the reliability of the conclusions reached is enhanced considerably. Relatively few scattering results have approached such ideal experimental conditions (Smith *et al.,* 1969; Yamamoto and Stickney, 1970; O'Keefe and French, 1969; Smith and Merrill, 1970; Stoll *et al.,* 1971; West and Somorjai, 1971). In much of the scattering data from metals where the state of the surface is highly questionable, or at best uncertain, has been reported over the years.

* Permanent address: Department of Chemical Engineering, California Institute of Technology, Pasadena, California.

In this laboratory ultra-high vacuum system (base pressure 5 × 10⁻¹¹ Torr) has been constructed which provides for molecular and atomic beam scattering with detection in the d.c. mode. The beam is produced by effusion from a simple Knudsen cell and single stage collimation to a divergence of 2°. Details of the apparatus are given elsewhere (Smith and Merrill, 1969).

This paper's purpose is to identify, from scattering data already reported (Smith and Merrill, 1970; Stoll et al., 1971; Weinberg and Merrill, 1970) and

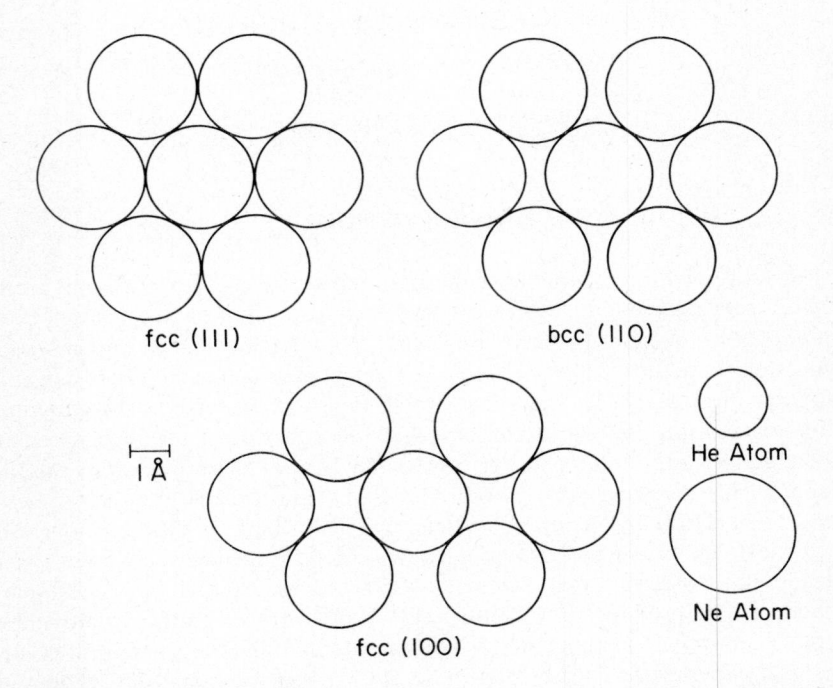

Fig. 1. Geometric arrangement of atoms in selected low index faces of fcc and bcc crystals.

data soon to be reported in detail elsewhere (Weinberg and Merrill, 1972a, 1972b), what may be some rather general modes of interaction for scattering from metals and to examine the sort of structural information which is potentially available from diffractive and non-diffractive scattering of helium and deuterium.

Scattering from three different metal surfaces, platinum (111), platinum (100), and tungsten (110) will be discussed. The surface atoms of all three are co-planar and the arrangement of the atoms in each surface is shown in figure 1.

Platinum has the fcc structure and tungsten the bcc structure. The atomic radii of the metals are nearly identical, as indicated in the figure. Also shown to scale are the atomic sizes of helium and neon atoms. These three surfaces represent a regular progression of the smoothest possible surface packing, fcc (111), to the more open structure of fcc (100), with bcc (110) being intermediate in atomic surface roughness. Furthermore, the differences in surface roughness among these surfaces is appreciable on the scale of the size of a helium atom and, rather smaller, on the scale of the size of a neon atom.

II. PHENOMENOLOGICAL SCATTERING REGIMES

Three phenomenological regimes of scattering can be identified from the scattering observed from these surfaces and the scattering previously reported

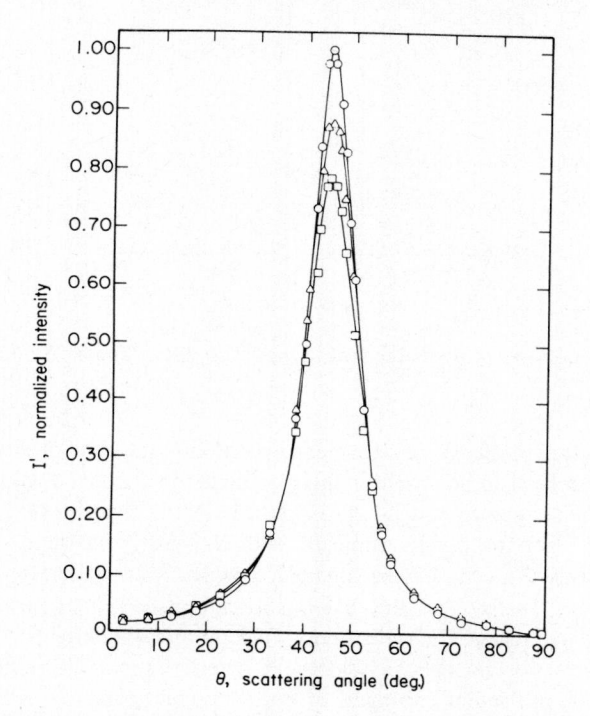

Fig. 2. Helium scattering from W(110) surface at $T_g = 295°$K and $\theta_i = 45°$. ○ $T_S = 375°$K, △ $T_S = 575°$K, □ $T_S = 775°$K. Direct beam intensity is 5250 pa and the maximum reflected intensity at $T_S = 375°$K is 820 pa.

from other fcc (111) surfaces (Smith *et al.*, 1969). A mode termed quasi-elastic, is observed for helium scattering from all surfaces. In this mode it appears that the interactions are governed primarily by elastic collisions with the repulsive portion of the potential energy of interaction. An example of this mode is shown in figure 2 where helium scattering from a tungsten (110) surface is depicted. The beam at room temperature impinges at an incident angle of 45° at surface temperatures of 100°C, 300°C, and 500°C. The beam intensities are normalized with respect to the peak maximum of the 100°C surface which is 16% of the incident beam flux. All three distributions are centered about the specular angle (45°) and changes in the beam temperature do not result in measurable shifts away from the specular (Stoll *et al.*, 1971).

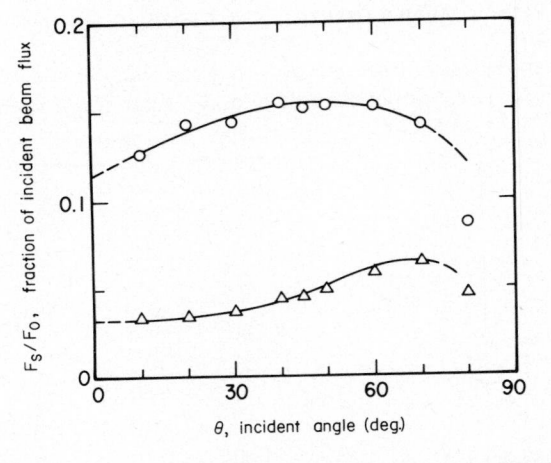

Fig. 3. Dependence of specular scattering on incident angle; $T_g = T_S = 25°C$, ○ He, △ D_2.

In figure 3 the intensity of the peak maximum (always observed at the specular angle) is plotted against the angle of incidence, where the larger angles represent the more grazing angles of incidence. In the figure both helium and deuterium are shown to rise in intensity with increasing incidence angle. The decrease in intensity at the highest incidence angles results because the crystal intersects only a portion of the beam at near grazing incidence. Figure 4 illustrates why an increase in the intensity with increasing angles of incidence would be expected for quasi-elastic scattering. A microscopically rough surface is shown with beam molecules impinging at various impact parameters. This results in a variety of local angles of incidence and hence dispersion in the scattered beam, with respect to the direction of the incident beam, even though the local scattering is highly elastic. As the angle of incidence increases, a portion of the surface becomes "shadowed" so that the variation in the local angles of incidence will be less and hence this geometric dispersion is also less. As the surface temperature is increased the effective surface roughness is increased also and this thermal roughening of the surface is probaly responsible for the attenuation in the scattering as a function of surface temperature (see figure 2). It is also apparent from these geometrical arguments that the rougher surfaces on a microscopic scale (see figure 1) should scatter more diffusely. The results depicted in figures 2 and 3 are also characteristic of the data reported on other fcc (111) surfaces (Smith *et al.*, 1969; Palmer *et al.*, 1969), and the data of Yamamoto and Stickney (1970) on tungsten (110).

The scattering of helium from several fcc (111) surfaces is seen in figure 5.

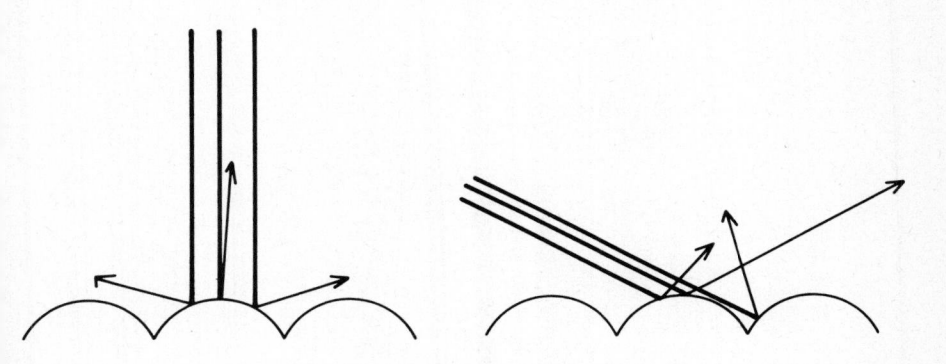

Fig. 4. Broadening of scattering distribution by microscopically rough surface.

The data on gold, silver, and nickel (Smith *et al.*, 1969) was taken at an incidence angle of 50° but has been shifted five degrees, for comparison with the platinum data taken at 45°. Since the scattering is narrower for the more grazing angles of incidence, the actual scattering from gold, silver, and nickel, (Smith *et al.*, 1969) should be slightly broader than that shown in figure 5, but not enough to obscure the trends. The width of the scattering does not correlate at all with the gas to surface mass ratio as expected from the hard cube theory (Logan and Stickney, 1966) but rather with the "lattice stiffness parameter" of the soft cube theory (Logan and Keck, 1968) and/or the mean square displacement of the surface atoms. Since the helium scattering does not exhibit the shifts away from the specular angle predicted by the cube theories it may be concluded that thermal roughening is responsible for the correlation shown in figure 5. This is certainly consistent with the increase in specularity with increasing incidence angle and with the idea that elastic scattering from the repulsive portion of the potential is the dominant mode of interaction in this quasi-elastic regime.

Helium scattering data from tungsten (110) are also plotted in figure 5. Based on its expected mean square displacement the scattering from tungsten should be somewhat narrower than that from nickel, yet the data shows its width to be intermediate, between platinum and nickel. It seems likely that this apparent discrepancy occurs because the tungsten surface is microscopically rougher than the fcc (111) surface (see figure 1). The other characteristics of quasi-elastic

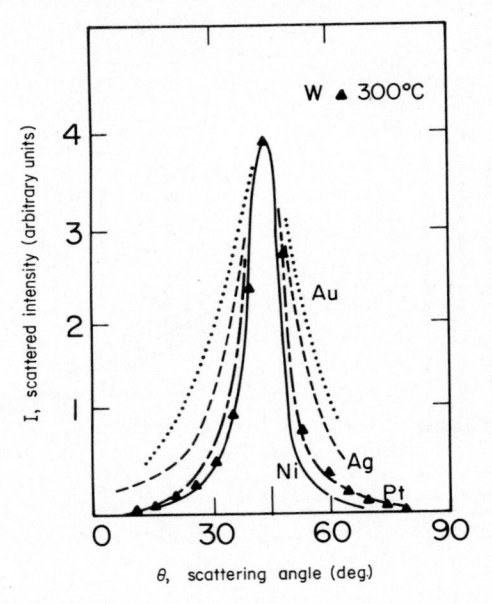

Fig. 5. Room temperature He scattering from W(110) at 575°K, Au(111) at 575°K, Ag(111) at 570°K, Ni(111) at 700°K and Pt(111) at 575°K.

scattering are met with tungsten (110) as well as with the fcc (111). A more detailed comparison of helium scattering from these metals is given elsewhere (Smith and Merrill, 1970; Weinberg and Merrill, 1972a)

Helium is the only inert gas which exhibits the characteristics of this quasi-elastic scattering. In figure 6 the scattering of a 700°C krypton beam from platinum (111) is shown. The intensities of the scattered beam are much lower than for the helium scattering, the peak maxima are no longer observed at the specular angle, and appreciable intensities are observed at scattering angles of 0° and below. This back-scattering is indicative of gas being trapped initially and re-emitted diffusely (Stoll *et al.*, 1971). Note that scattering maxima below the specular angle and maxima above the specular angle are observed, indicating inelastic interactions which can cause a net transfer of momentum either from the surface to the gas or vice versa; depending on the ratio of the gas temperature to the solid temperature. The peak maxima attenuate some with surface temperature but not so drastically as the helium scattering, especially from platinum (Stoll *et al.*, 1971).

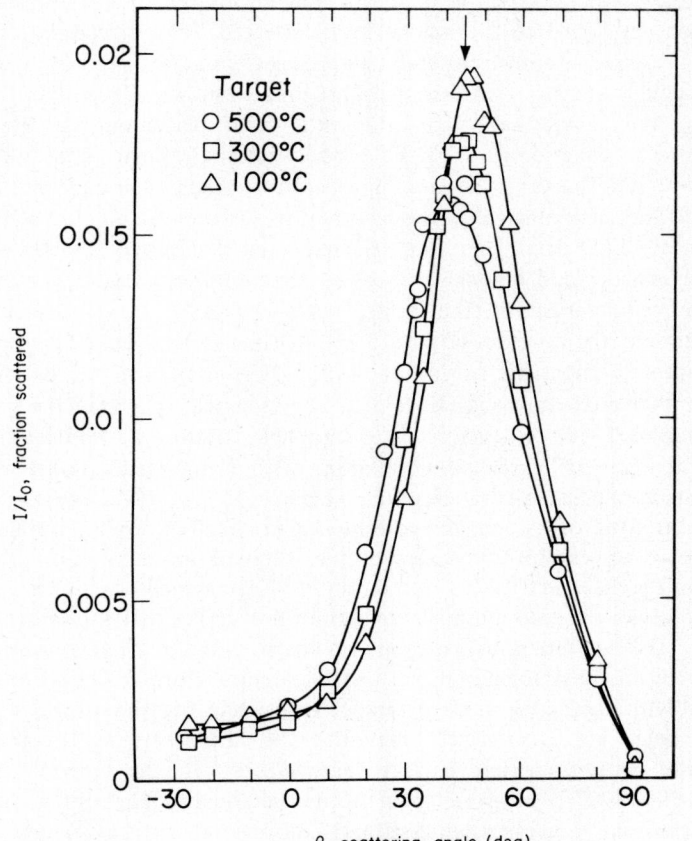

Fig. 6. Krypton scattering from Pt(111) 700°C beam.

The effect of incident angle is shown in figure 7 for argon scattering from platinum (111). Here the results are opposite to those of helium scattering. The peak maxima decrease, with increasing angles of incidence and the distributions, are concomitantly broader. Such a result is certainly reasonable where inelastic interactions dominate the scattering because the more grazing angles of incidence result in larger residence times in the neighborhood of the surface and smaller components of energy perpendicular to the surface.

In figure 8 the scattering of neon for the four fcc (111) surfaces is compared. Platinum scattering is more like that from silver than nickel, so that the correlation with the mean square displacement of the solid atoms is no longer followed. Instead the data seems to correlate better with the Debye temperature. Any inelastic dynamical interaction should be dominated by the ratio of the effective collision time to the natural frequency of the solid so that a simple correlation with the Debye temperature is expected. This correlation for the fcc (111) surfaces is also shown for argon scattering (figure 9).

The neon scattering from tungsten (110) is also shown in figure 8. It is much closer to the nickel data than the data from platinum, but the Debye temperature of tungsten ($\theta_D = 315°K$) is between that of platinum ($240°K$) and nickel ($440°K$). The tungsten data, however, were taken at an incidence angle of $45°$ while they are plotted as though they were taken at $50°$ for comparison with the fcc (111) data. Therefore, it is expected that actual data taken at $50°$ incidence angle would be slightly broader than the points plotted in figure 8. Also the temperature of the nickel surface ($700°K$) is greater than the temperature of the tungsten surface ($575°K$). Thus the scattering from nickel, at a temperature comparable to that of the tungsten surface, should be somewhat narrower than that shown in figure 8. It would appear then that the scattering from tungsten (110), at completely comparable conditions, would at least be consistent with the expected correlation with Debye temperature which is characteristic of this inelastic scattering regime.

The scattering of a room temperature argon beam from platinum (111) has been shown to exhibit only very minor thermal attenuation of the peak maximum (Stoll et al., 1971). Scattering of a high temperature ($700°C$) argon beam, however, exhibits thermal attenuation not unlike that shown for neon in figure 6. The scattering of krypton, however, shows a trend reversal in the intensity of the peak maximum with surface temperature. The peak maximum increases with increasing surface temperature while the shifts away from the specular angle are qualitatively the same (i.e. the scattering becomes more subspecular with increasing surface temperature). In the case of the high temperature ($700°C$) krypton beam normal thermal attenuation is observed. Xenon scattering, however, exhibits the trend reversal with surface temperature at both beam temperatures. It has been argued (Stoll et al., 1971) that when the surface temperature is raised the sum of two effects is observed. The first, thermal attenuation, tends to decrease the peak maximum. The second effect is a decrease in the initial trapping probability. Thus at higher surface temperatures atoms which would otherwise be trapped and emitted completely diffusely become part of the lobular scattering and tend to increase the intensity of the peak maximum. In room temperature argon scattering from platinum (111) these effects nearly compensate and there is little change in the intensity of the peak maximum. For the high temperature argon scattering the trapping

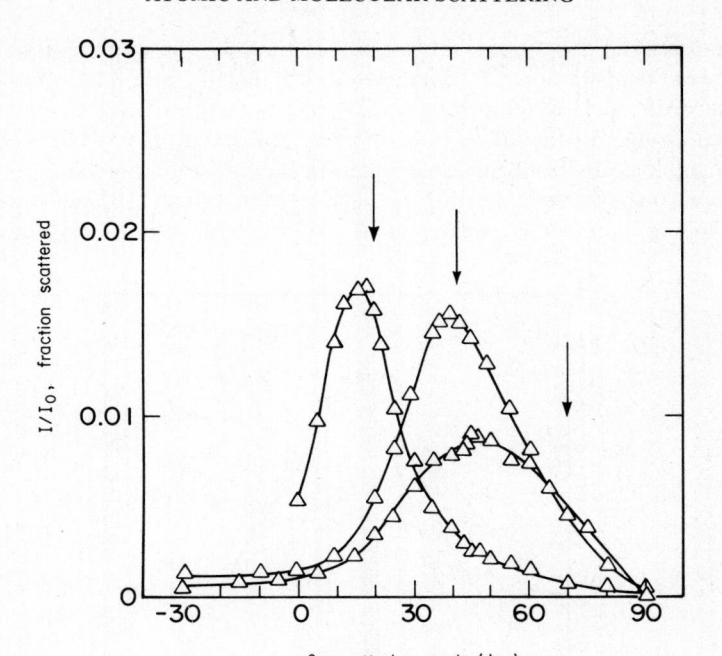

Fig. 7. Effect of incident angle on Ar scattering from Pt(111) 23°C beam.

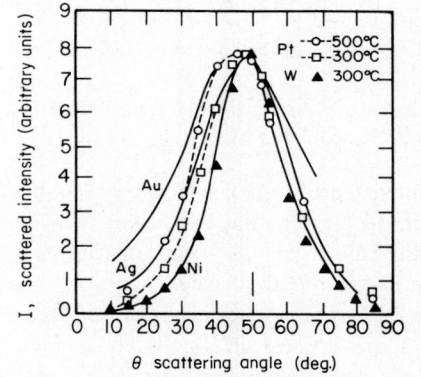

Fig. 8. Room temperature Ne scattering from W(110) at 575°K, Au(111) at 575°K, Ag(111) at 570°K, Ni(111) at 700°K and Pt(111) at 575°K and 775°K.

probability is reduced enough for all surface temperatures that changes in it cannot compensate for thermal attenuation. For low temperature krypton scattering the trapping probability is large enough to dominate, but again the higher beam temperature results in much lower trapping fractions for all surface temperatures and the thermal attenuation dominates. The trapping is so high at both beam temperatures for xenon scattering that its peak intensities always increase with surface temperature. Trapping so large that it dominates the observed changes in the intensity of the peak maximum, then defines the third phenomenologically distinguishable regime in scattering from metals.

Figure 10 shows the scattering of argon from tungsten (110) and is clearly in the trapping dominated regime, as is krypton and xenon scattering from

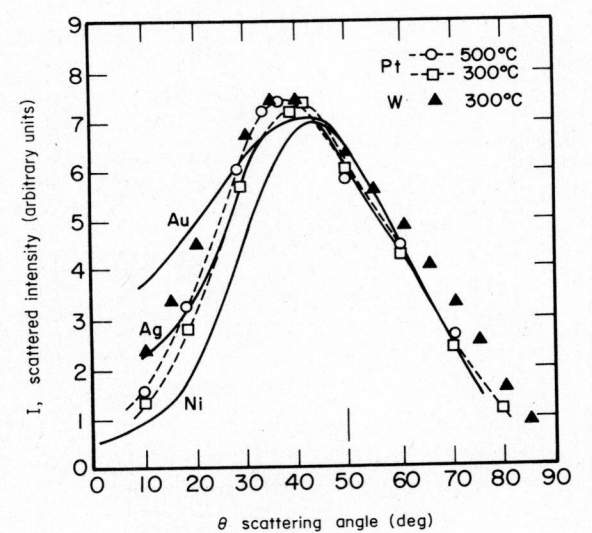

Fig. 9. Room temperature Ar scattering from W(110) at 575°K, Au(111) at 575°K, Ag(111) at 570°K, Ni(111) at 700°K and Pt(111) at 575°K and 775°K.

tungsten (110) (Weinberg and Merrill, 1971b). It has been shown empirically (Weinberg and Merrill, 1971) that the transitions from one regime to the next occur at characteristic values of the ratio of the neon gas energy to the adsorption interaction energy of the surface.

Points representing the argon scattering from tungsten (110) are also plotted in figure 9, which compares argon scattering from the fcc (111) surfaces. Clearly the tungsten data does not follow the correlation with Debye temperature exhibited by the scattering from fcc (111) surfaces and small corrections due to slightly different angles of incidence and surface temperature could not possibly be responsible for the lack of correlation. Rather it is believed that the width of the scattered beam is so much dominated by the large amount of trapping that the dynamical broadening of the untrapped portion of the scattering is obscured.

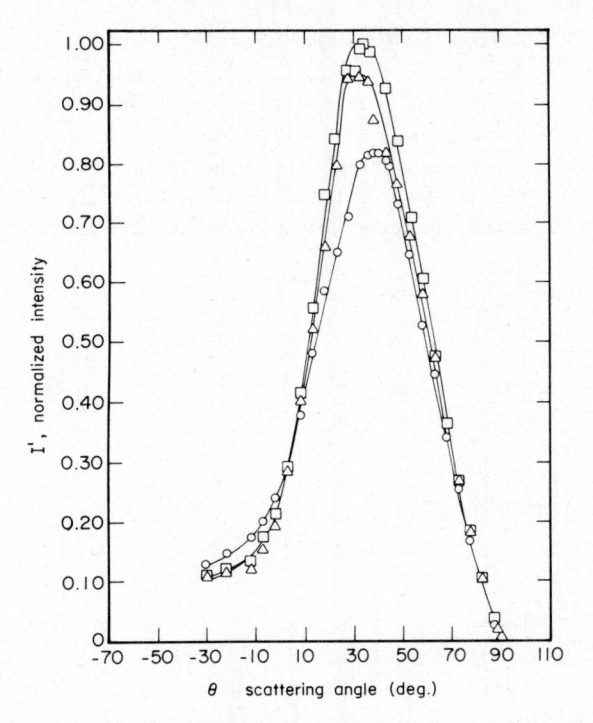

Fig. 10. Argon scattering from W(110) surface at $T_g = 295°K$ and $\theta_i = 45°$. $\bigcirc$ $T_S = 375°K$, $\triangle$ $T_S = 575°K$, $\square$ $T_S = 775°K$. Direct beam intensity is 80 000 pa, and the maximum reflected intensity at $T_S = 375°K$ is 840 pa.

III. DIFFRACTION

While diffraction was observed many years ago from lithium fluoride surfaces (Estermann and Stern, 1930) attempts to observe diffractive scattering from metal surfaces (Palmer *et al.,* 1970; Stoll *et al.,* 1971) has largely been unsuccessful. While preparing the tungsten (110) surface for the scattering work mentioned above (Weinberg and Merrill, 1972a) and discussed in this paper, a complex (3 x 5) LEED pattern (Weinberg and Merrill, 1972b) was observed. It was concluded, in agreement with the work of others (Stern, 1964; Chen and Papageorgopoulos, 1970), that the surface was due to carbon. In figure 11, helium scattering is shown from this (3 x 5) carbon structure which is oriented such that diffraction from a grating along the five spacing of the (3 x 5) unit cell could be observed within the principal scattering plane. The first and second order diffraction peaks are clearly evident and are present at very high intensities

(40% of the incident beam for the (00) beam) compared to the scattering usually observed from clean metals (15% to 25%). For comparison the scattering distribution from a clean tungsten surface is shown. It is substantially broader with a peak maximum of only 16% and exhibits no recognizable diffraction. Figure 12 shows diffraction features for angles of incidence of 20° and 70°. Deuterium diffraction is shown in figure 13 at incidence angles of 20° and 45°. As would be expected from the scattering of deuterium from clean metals

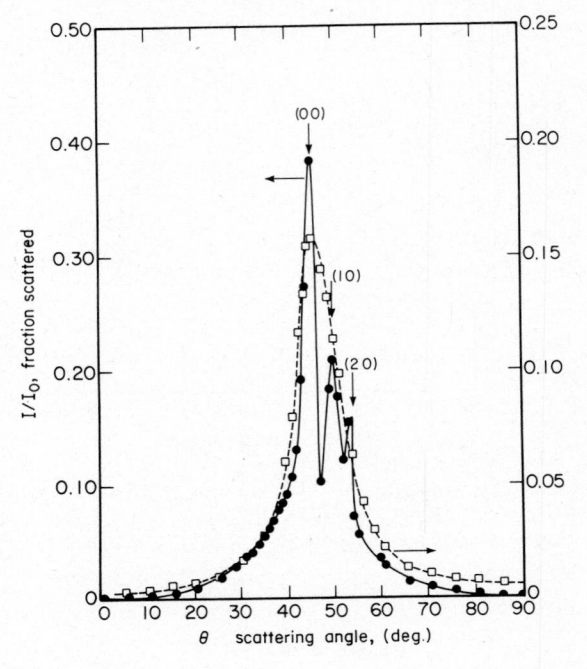

Fig. 11. Helium scattering from a W(110) surface. ● He diffraction pattern from R(3 x 5) carbide surface; □ He scattering from clean surface. $T_S = 375°$K, $T_g = 295°$K, $\theta_i = 45°$.

(Smith and Merrill, 1970; Weinberg and Merrill, 1972b) the intensities are lower (only 2% to 5% even for the (00) beams) and the peaks are considerably broader. This occurs because of the presence of rotational coupling and the higher interaction energy for deuterium compared to helium. Because of this broadness, however, it is possible to tilt the crystal in such a way that nearly all of the first and second order features are observed simultaneously. In figure 14 the neon scattering from the (3 x 5) carbon structure is shown. The scattering appears to be in the inelastic regime with no evidence of diffraction. Argon and xenon (Weinberg and Merrill, 1972b) are both in the trapping dominated regime.

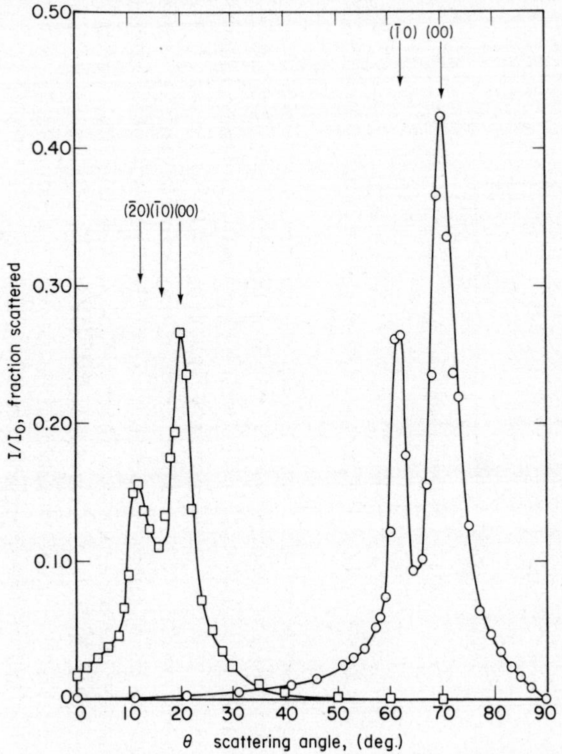

Fig. 12. Helium diffraction from a W(110) R(3 × 5) carbide surface. $\bigcirc \theta_i = 70°$, $\square \theta_i = 20°$, $T_S = 295°K$, $T_g = 375°K$.

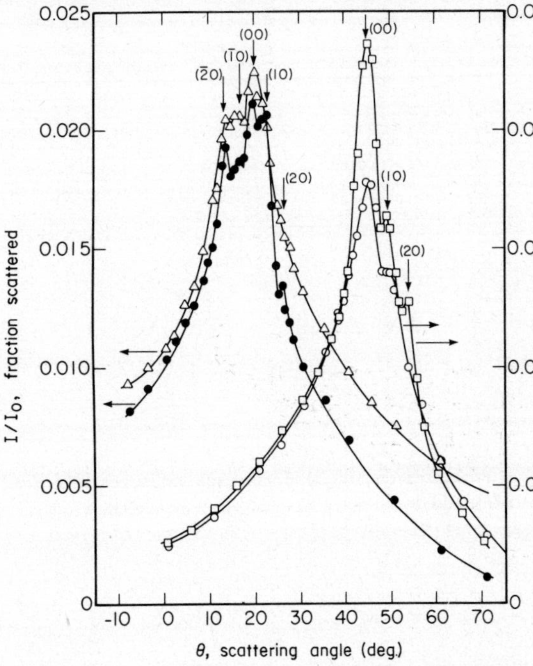

Fig. 13. Deuterium diffraction from a W(110) R(3 × 5) carbide surface. $\bullet$ $\theta_i = 20°$, $T_S = 625°K$, $T_g = 295°K$. $\triangle \theta_i = 20°$, $T_S = 775°K$, $T_g = 295°K$. $\square \theta_i = 45°$, $T_S = 625°K$, $T_g = 295°K$. $\bigcirc \theta_i = 45°$, $T_S = 775°K$, $T_g = 295°K$.

ATOMIC AND MOLECULAR SCATTERING

163

By prolonged heating above 2000°K the surface can be converted to a (1 x 1) LEED pattern but the helium scattering from this surface is remarkably different than helium scattering from the clean tungstem (110) surface. Though it is definitely in the quasi-elastic regime, the peak maximum is 55% compared to only 16% for the clean surface. This level of scattering in the specular direction is comparable to the diffraction maxima. Furthermore the clean surface can be

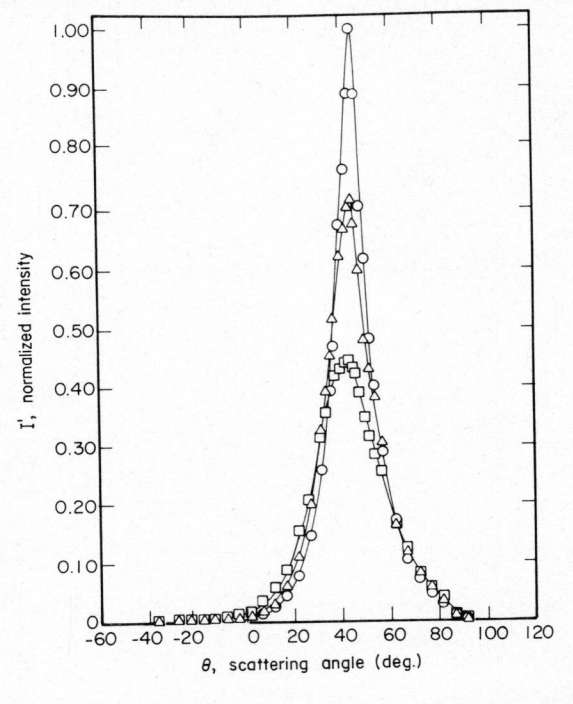

Fig. 14. Neon scattering from W(110) R(3 x 5) carbide surface $\theta_i = 45°$. $\circ$ T_S = 375°K, T_g = 295°K, $\triangle$ T_S = 775°K, T_g = 295°K. $\square$ T_S = 1300°K, T_g = 295°K.

produced by a high temperature oxygen treatment similar to that which converts the (3 x 5) carbide to clean tungsten (110). It is therefore concluded that the surface is a fully consolidated carbide with a (1 x 1) structure (i.e. in registry with and having the same lattice constant as the tungsten (110) surface). It may also be inferred that the (3 x 5) carbon structure is a nearly consolidated carbide rather than the dilute carbide suggested by Stern (1964) because the scattering is more directed than scattering from the metal the mean square displacement of the surface atoms must be less for the carbide (i.e. it is a stiffer lattice). From this it is possible to estimate crudely that the Debye temperature of the carbide is greater than 1200°K (Weinberg and Merrill, 1972b).

IV. STRUCTURAL INFORMATION
FROM NON-DIFFRACTIVE SCATTERING

When the scattering is in the quasi-elastic regime it is possible to derive some limited structural information from the scattering patterns. For instance, it has been possible to show that the adsorption of ethylene on clean platinum (111) is dissociative (Smith and Merrill, 1969), some evidence of roughness induced

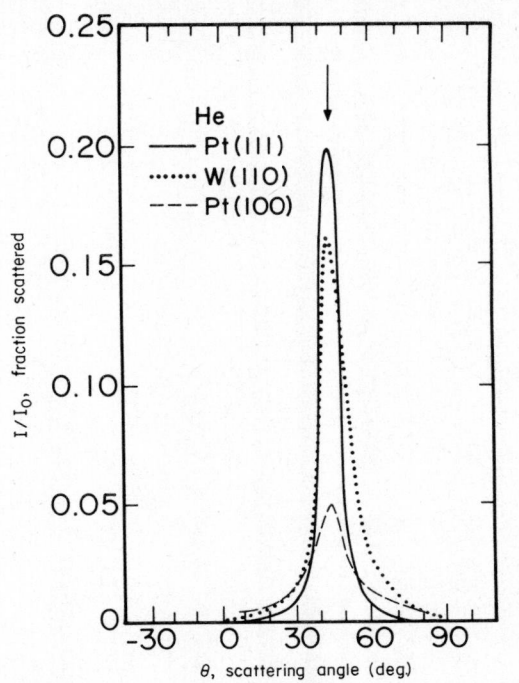

Fig. 15. Helium scattering from selected low index faces of Pt and W. $\theta_i =$ 45°, $T_S = 375°K$, $T_g = 295°K$. Pt(100) taken from West and Somorjai (1971).

rotational excitation has been found (Smith and Merrill, 1970) and marked differences between the helium scattering from tungsten (110) described here and that found by Yamamoto and Stickney (1970) have been ascribed to surface roughness (Weinberg and Merrill, 1972a). A more detailed assessment of the sensitivity of microscopic surface roughness can be made by comparing the scattering from the three surfaces depicted in figure 1. Geometrically these surfaces are very similar, as pointed out earlier. There is a regular progression to greater and greater roughness on a scale comparable to the diameter of a helium atom. In figure 15 the data from platinum (111), tungsten (110), and platinum (100) (West and Somorjai, 1971) are compared. As anticipated there is a marked effect on the peak maximum and the width of the distribution, with the platinum (111) the most specular, the platinum (100) the least specular and the tungsten (110) intermediate between the other two.

In figure 16 the arrangement of atoms on a diamond (111) surface is shown. In this surface the six "representative" atoms are not all in the same plane, as was the case for the metal surfaces shown in figure 1. Also, the atom corresponding to the central atom in the fcc (111) hexagonal array is missing. As a result, even though the carbon crystal radius is comparable to the helium radius, the surface is much rougher than the metal surfaces of figure 1. The

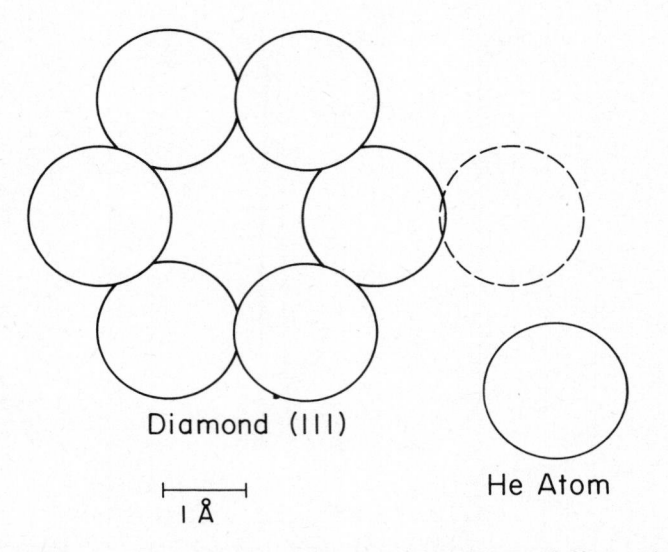

Diamond (III)

⊢——⊣
1 Å

He Atom

Fig. 16. Arrangement of atoms on the (111) face of diamond.

diamond (100) surface may be envisioned as a saw-tooth array of (111) surfaces and is rougher still. Figure 17 shows a helium scattering pattern taken from a (100) diamond surface. It is very diffuse, almost cosine. Such a scattering pattern could result from a number of situations. It could be that the surface was not properly cleaned, that helium is trapped with a near unity trapping probability, or that momentum exchange between carbon and helium is highly inelastic. The latter two seem unlikely because the heat of physical adsorption is not substantially different on carbon than it is on metals and scattering from the basal plane of graphite is far from diffuse (Merrill and Smith, 1970). The possibility of contamination cannot be ruled out but, even so, the increased surface roughness of the diamond surface seems adequate to explain the broad scattering shown in figure 17. This is particularly true in the light of the large changes shown in figure 15 for much smaller changes in the degree of surface roughness. If this explanation is correct it would be expected that neon and argon scattering from diamond might be more directed than helium, whereas they are always more diffuse on metals.

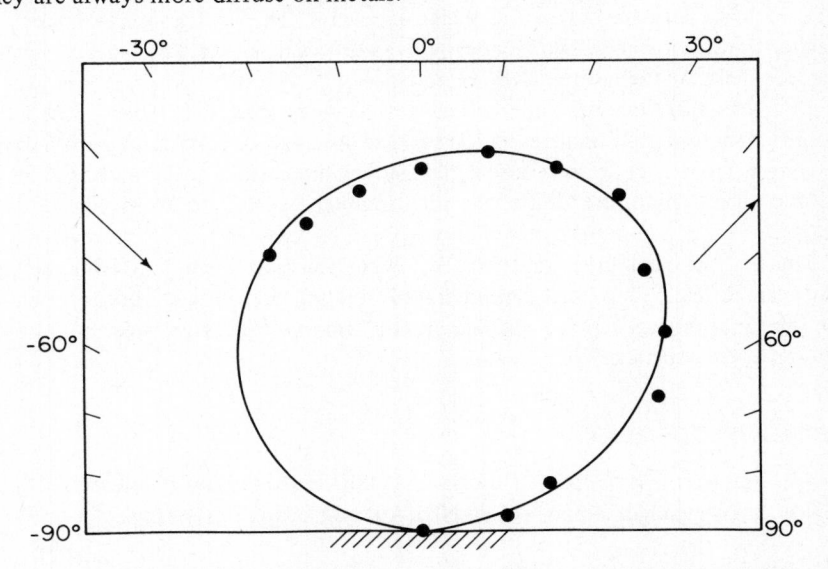

Fig. 17. Helium scattering from diamond (100) surface. $T_S = 295°K$, $T_g = 295°K$, $\theta_i = 45°$. Maximum reflected intensity is 0.0024 of incident beam. Calculated maximum intensity for diffuse reflection 0.0023 of incident beam.

V. CONCLUSIONS

The scattering of inert gases from three crystalographically similar surfaces has been discussed using data taken from well defined surfaces. Three regimes of scattering can be identified: quasi-elastic, inelastic, and trapping dominated.

The quasi-elastic regime is characterized by sharp distributions always found centered at the specular angle. They attenuate and broaden as the surface temperature is increased and become sharper and more directed at the more glancing angles of incidence. Comparison of the platinum (111) data with that from other fcc surfaces suggest roughening of the surface, due to increasing mean square displacements, is responsible for the thermal attenuation in this regime.

The inelastic regime is characterized by much broader distributions whose peak maxima can be either closer to the normal (subspecular) or further away from the normal (supraspecular) than the specular angle. Also the broadening is correlated with the natural lattice frequency (Debye temperature) rather than the mean square displacements of solid atoms. In this regime a thermal attenuation is also observed.

When the trapping fraction is great enough the peak maxima increase with increasing surface temperature, but the shifts away from the specular angle are similar to those observed in the inelastic regime. The widths of the scattering distributions, however, no longer correlate with either the mean square displacements or the natural lattice frequency.

Diffraction of helium and deuterium has been described from a (3 x 5) carbide structure on a tungsten (110) surface. Analysis of this, along with helium scattering from a (1 x 1) carbide, has shown the surface to be a consolidated carbide rather than the dilute carbide originally suggested from LEED data alone.

Finally, the structure sensitivity of even non-diffractive scattering in the quasi-elastic regime has been demonstrated. Helium scattering has been shown to become progressively more diffuse as the microscopic roughness of a well defined crystal surface is increased.

ACKNOWLEDGEMENT

This research was supported by the AFOSR. USAF, Grant Number 68-1409. W.H.W. acknowledges support of an NDEA Pre-Doctoral Fellowship.

REFERENCES

Chen, J. and Papageorgopoulos, C. A (1970). *Surface Sci.* **20**, 195.
Estermann, I. and Stern, O. (1930). *Z. Phys.* **61**, 95.
Logan, R. M. and Stickney, R. E. (1966). *J. Chem. Phys.* **44**, 195.
Logan, R. M. and Keck, J. C. (1968). *J. Chem. Phys.* **49**, 860.
Merrill, R. P. and Smith, D. L. (1970). *Surface Sci.* **21**, 203.
O'Keefe, D. R. and French, J. B. (1969). *Advan. Appl. Mech., Suppl.* 5, Vol. 2, 1279.
Palmer, R. L., Saltsburg, H. and Smith, J. N., Jr. (1969). *J. Chem. Phys.* **50**, 4661.
Palmer, R. L., O'Keefe, D. R., Saltsburg, H. and Smith, J. N., Jr. (1970). *J. Vac. Sci. Techol.* 7, 91.
Smith, D. L. and Merrill, R. P. (1969). *J. Chem. Phys.* **52**, 5861.
Smith, D. L. and Merrill, R. P. (1970). *J. Chem. Phys.* **53**, 3588.
Smith, J. N., Jr., Saltsburg, H. and Palmer, R. L. (1969). *Advan. Appl. Mech., Suppl.* 5, Vol. 2, 1141.
Stern, R. M. (1964). *Appl. Phys. Lett.* **5**, 218.
Stoll, A. G., Smith, D. L. and Merrill, R. P. (1971). *J. Chem. Phys.* **54**, 163.
Weinberg, W. H. and Merrill, R. P. (1970). *Phys. Rev. Lett.* **25**, 1198.
Weinberg, W. H. and Merrill, R. P. (1971). *J. Vac. Sci. Technol.* (In press).
Weinberg, W. H. and Merrill, R. P. (1972a). *J. Chem. Phys.* (In press).
Weinberg, W. H. and Merrill, R. P. (1972b). *J. Chem. Phys.* (In press).
West, L. A. and Somorjai, G. A. (1971). *J. Chem. Phys.* **54**, 2864.
Yamamoto, S. and Stickney, R. E. (1970). *J. Chem. Phys.* **53**, 1594.

SURFACE LIFETIMES OF THE ALKALIS AND HALOGENS ON MOLYBDENUM

M. D. SCHEER, R. KLEIN and J. D. McKINLEY

National Bureau of Standards,
Washington, USA

I. INTRODUCTION

The desorption of alkali and halogen atoms from a metal surface takes place as two parallel processes. In one channel a positive alkali or negative halogen ion is emitted, while in the other, a neutral atom desorbs into the gas phase. At a given surface temperature the ratio of rates of these two processes is determined by the metal affinity (ϕ) for the valence electron and the ionization potentials of the alkali atom (I) and negative halogen ion (A). A quantitative formulation of the gaseous ion-neutral ratios in a thermally equilibrated system consisting of a solid metal substrate and a metal vapor (Saha, 1920; Langmuir and Kingdon, 1925; Scheer, 1970) (either the same or different than the substrate) is given by the well-established Saha-Langmuir equation

$$\alpha_\pm = \frac{g_\pm}{g_0} \exp(\beta H) \tag{1}$$

H is $(\bar{\phi}_+ - I)$ for positive ions and $(A - \bar{\phi}_-)$ for negative ions, $\alpha_\pm$ is the degree of ionization, $(g_\pm/g_0)$ is the ratio of electronic partition functions for ion and neutral and β is $(kT)^{-1}$, k being the Boltzmann constant. The quantity $\alpha_\pm$ is the ratio of ions to neutrals in the gas phase, and may be expressed either in terms of number densities (N) or particle fluxes (i) to and from the substrate surface

$$\alpha_\pm = \frac{N_\pm}{N_0} = \frac{i_\pm}{i_0} \quad \text{since} \quad i_x = \frac{N_x}{(2\pi m \beta)^{1/2}} \tag{2}$$

The interaction of alkali metal atoms with incandescent refractory metal surfaces has been studied for over four decades. Much of the early work was concerned with the development of the Saha-Langmuir equation by determining ionization coefficients, and their temperature dependence, for such alkalis as potassium and cesium on tungsten. Desorption kinetics were first measured

169

directly by Moon and Oliphant (1932). Their technique was to impinge a flux of alkali ions with kilovolt energies on a tungsten target, with the surrounding collector electrically biased to suppress ion desorption. On sudden reversal of the electric field, the previously trapped alkalis desorb, and the temporal decay of the positive ion current was recorded on an oscillograph camera. Surface concentrations were determined separately from the current pulse obtained on flashing the tungsten to a high temperature. These experiments were repeated by Evans (1933), using thermal alkali atoms in the gas, and were extended to include potassium, rubidium and cesium lifetimes on tungsten. The first-order behavior of alkali desorption at low coverages was demonstrated, surface lifetimes were determined and, from their temperature dependence, ion desorption energies of 2.4, 2.1 and 1.8 eV were obtained for K, Rb, and Cs on tungsten. The modulated electric field method contains the essential feature of a mechanically chopped atomic-beam technique for surface lifetimes; the only significant difference is the means of producing the initial adsorbate concentration. Knauer (1948) and Starodubtsev (1949) reported the first successful applications of mechanically modulated atomic beams to surface lifetime measurements. A rotating shutter or vibrating reed was used to periodically interrupt a thermal beam of alkali atoms prior to desorption from the experimental surface. The period between pulses was longer than the characteristic lifetime of the adsorbate on the surface. The pulses of desorbed ions were detected and recorded synchronously. The time dependence of the pulses contained the kinetic information. With sensitive mass spectrometric detection, desorption data for specific ions and neutrals can be obtained over a large range of lifetimes. The early modulated beam data show generally poor agreement, probably due to the difficulty of maintaining clean surfaces in the absence of modern high-vacuum techniques. The effect on experimental desorption energies of surface contamination by impurity gases was illustrated in Scheer and Fine's (1962) data for cesium on tungsten ($\bar{l}$ = 2.0 eV) and on carbon monoxide covered tungsten ($\bar{l}$ = 1.5 eV).

The measurement of halogen surface lifetimes on metal surfaces has not been reported previously. In the present experiment a modulated beam of cesium halide was used to produce a dilute adlayer on a molybdenum surface. At temperatures above $1350°K$ such a dilute adlayer is almost completely dissociated. Under these conditions, halogen surface lifetimes are in the millisecond range while the cesium adatom lifetime is in the range of microseconds. Consequently within a millisecond after the beam shutter closes the surface layer consists almost entirely of the halogen which desorbs as neutrals and negative ions in a ratio given by the Saha-Langmuir equation.

II. EXPERIMENTAL DETERMINATION OF SURFACE LIFETIMES

Some of the features of the apparatus used in the present determinations of surface lifetimes has been described previously (Scheer and Fine, 1969). In outline, it consists of a chopped molecular beam source of either alkali atoms or cesium halide molecules, a metal ribbon target, and an ion collector consisting of either a Faraday cage or a mass spectrometer. In the present work, a mass spectrometer was used. A pulsed, low-duty cycle (10%) beam was directed to a polycrystalline molybdenum surface. Intensities were about 10^{12} cm^{-2} s^{-1} and surface temperatures were adjusted so that the mean surface lifetimes fell within the range 0.1 to 100 ms. Maximum coverage never exceeded 10^{-4} of a monolayer or approximately 10^{11} adsorbed particles per square centimeter.

The desorption flux of ions from the incandescent molybdenum surface was determined with a mass spectrometer tuned to the most abundant of the naturally occurring alkali or halogen isotopes. The ions were accelerated to 4 kV in the mass analyzer and detected with a 16-stage Cu-Be electron multiplier. Each ion impinging upon the first dynode produced a pulse at the sixteenth dynode 0.1 mV in height and 20 ns wide. These pulses were amplified and processed to yield uniform, triangular-shaped pulses 1 volt high 50 ns wide at half-maximum. With the beam shutter closed, the ion desorption flux produced an exponentially decreasing output pulse rate which was signal averaged using the multiscale mode of 1024 channel analyzer. In order to obtain ion desorption decay data with an acceptable signal to noise ratio, the outputs of several hundred pulsed beam cycles were stored and summed. A real-time triggering signal to the multiscaler was provided by the output of a photocell irradiated by a light beam passing through the molecular beam chopper. The data were retrieved from the analyzer memory on an X-Y plotter. The time constant for the ion flux decay was determined from a least-squares linear representation. Estimated uncertainties in the lifetimes were about 10%.

Vacuum conditions surrounding the molybdenum surface were maintained at about 2×10^{-8} Torr. At surface temperatures in excess of 900°K, no detectable accumulation of adsorbed residual gases were observed during the time required for a typical experiment. Prior to each lifetime determination, the molybdenum surface was flash heated to 2300°K for several seconds to establish a clean surface. Cooling of the surface to the experimental temperature and collecting the data required about 50 s. Reflashing the molybdenum surface to 2300°K at this time produced no observable pressure transient in a nearby vacuum gauge. It can be concluded that the adatom-molybdenum interaction under these conditions is free of perturbations due to residual background gas adsorption.

A polycrystalline molybdenum ribbon was mounted in an electric and magnetic field free region. A surface normal formed an angle of 45° with both

Table 1. *Alkali surface lifetimes on polycrystalline molybdenum*

Li		Na		K		Rb		Cs	
$T\,(K)$	$\bar{\tau}\,(ms)$	$T\,(K)$	$\bar{\tau}\,(ms)$	$T\,(K)$	$\bar{\tau}\,(ms)$	$T\,(K)$	$\bar{\tau}\,(ms)$	$T\,(K)$	$\bar{\tau}\,(ms)$
1150	14.8	986	49.0	994	114.4	968	119.3	928	80.6
1162	11.2	994	45.1	1007	71.8	980	79.4	942	51.8
1187	6.36	1002	36.8	1020	50.8	1007	42.0	954	52.1
1198	7.60	1020	18.0	1033	38.0	1020	24.6	968	27.2
1210	3.93	1020	17.9	1033	42.5	1033	19.9	980	19.3
1224	2.80	1033	11.1	1047	17.8	1047	16.7	994	12.3
1235	2.70	1047	7.80	1060	12.4	1060	10.2	1007	9.32
1249	1.14	1054	7.30	1072	13.0	1068	10.3	1020	6.90
1272	0.95	1060	5.02	1072	8.81	1072	7.32	1033	6.97
1296	0.48	1065	4.28	1084	8.91	1084	6.41	1047	5.03
1320	0.26	1072	3.50	1096	5.25	1096	3.91	1060	6.97
		1084	2.87	1110	3.90	1110	3.69	1072	2.75
		1096	2.43	1124	4.49	1136	2.98	1084	1.71
		1110	1.64	1136	3.03	1150	2.25	1096	1.74
		1136	0.95	1162	1.31	1162	1.38	1110	0.92
		1150	0.72	1175	1.27	1174	0.87	1136	0.61
								1162	0.49

the alkali beam axis and the entrance slit of an electrode assembly which accelerated and focused the positive ions into a 60°, 15 cm radius, magnetic sector. The entrance and exit slits of the direction focusing magnetic sector were 10 mm long and 0.25 and 0.60 mm wide respectively. The molybdenum ribbon target measured 25 x 2 x 0.05 mm. Only the central 10 mm was exposed to the impinging molecular beam. An optical pyrometer scan of the central 10 mm of the ribbon, heated to about 1150°K, showed no significant temperature gradients or temporal variation. During operation, the temperature was determined by measuring the electrical resistance of the ribbon using the four terminal method. The resistance-temperature scale was determined with a calibrated optical pyrometer focused on the central portion of the ribbon. Temperature differences at 1150°K were known to within several tenths of a degree, while the absolute accuracy of the resistance temperature scale was about ±3°K in the 950 to 2000°K range.

III. ALKALI AND HALOGEN SURFACE LIFETIMES ON POLYCRYSTALLINE MOLYBDENUM

Table 1 gives the surface lifetimes for Li, Na, K, Rb, and Cs on molybdenum in the temperature range 930 to 1330°K. Table 2 gives the lifetimes obtained for F, Cl, Br, and I on the same molybdenum surface in the 1350 to 1830°K range. The Arrhenius desorption parameters were obtained from a linear least-squares

Table 2. *Halogen surface lifetimes on polycrystalline molybdenum*

F		Cl		Br		I	
$T(K)$	$\bar{\tau}\,(ms)$	$T(K)$	$\bar{\tau}\,(ms)$	$T(K)$	$\bar{\tau}\,(ms)$	$T(K)$	$\bar{\tau}\,(ms)$
1700	21.3	1469	36.4	1431	24.8	1350	8.06
1715	15.9	1488	23.7	1440	20.0	1369	4.93
1731	11.8	1506	15.9	1469	12.0	1400	2.95
1748	8.95	1517	12.8	1489	9.05	1421	2.04
1765	6.96	1525	12.5	1499	6.71	1431	1.72
1775	5.75	1561	5.50	1516	4.64	1460	1.04
1796	4.20	1570	4.32	1516	4.60	1460	0.96
1800	3.79	1570	4.19	1535	3.53	1480	0.72
1834	1.96	1579	4.18	1543	2.90	1500	0.51
		1596	3.08	1561	2.17		
		1613	2.18	1561	2.10		
		1613	2.11	1579	1.55		
		1615	1.99	1587	1.24		
		1630	1.47	1623	0.73		
		1647	0.94				

fit to the function $\ln \bar{\tau} = \ln \bar{\tau}^\circ + A/T$ where the slope A is $(\bar{l}/k)$, $\bar{l}$ being the average desorption energy and $\bar{\tau}^\circ$ the average temperature insensitive pre-exponential. Table 3 gives the results of these calculations with the listed uncertainties equal to the standard error in a least-squares average. The $\bar{\tau}^\circ$'s increase from 3×10^{-16} to 3×10^{-13}s for Li through Cs. The mean binding energies, on the other hand, decrease from 3.14 to 2.10 eV for the same alkali sequence. The binding energies of the halogens on the other hand, decrease from 4.63 to 3.15 eV for F through I while the $\bar{\tau}^\circ$'s increase from 3×10^{-16} to 1×10^{-14} s.

Table 3. *Experimental desorption parameters for the alkalis and halogens on molybdenum*

	$\bar{\tau}$ (s)	$\bar{l}$ (eV)
Li	$(3 \times 10^{-16}) \exp([36,400 \pm 1900]/T)$*	3.14 ± 0.17
Na	$(3 \times 10^{-15}) \exp([30,200 \pm 700]/T)$	2.60 ± 0.06
K	$(2 \times 10^{-14}) \exp([29,400 \pm 1100]/T)$	2.53 ± 0.10
Rb	$(1 \times 10^{-13}) \exp([26,900 \pm 500]/T)$	2.31 ± 0.05
Cs	$(3 \times 10^{-13}) \exp([24,500 \pm 600]/T)$	2.10 ± 0.06
F	$(4 \times 10^{-16}) \exp([53,800 \pm 1200]/T)$	4.63 ± 0.10
Cl	$(3 \times 10^{-16}) \exp([47,800 \pm 900]/T)$	4.12 ± 0.08
Br	$(2 \times 10^{-15}) \exp([43,000 \pm 600]/T)$	3.70 ± 0.05
I	$(1 \times 10^{-14}) \exp([36,500 \pm 500]/T)$	3.15 ± 0.04

* The least-squares standard error in all of the pre-exponentials is a factor of 2-3.

THE WORK FUNCTION OF MOLYBDENUM

The value of $\bar{\phi}_\pm$ for the experimental molybdenum surface was obtained from a determination of the temperature dependence of the steady-state surface ionization of lithium and iodine. From a consideration of equations (1) and (2), and noting that $I(\mathrm{Li}) > \phi(\mathrm{Mo}) > A(\mathrm{I})$, it can be concluded that the desorption of neutral lithium and iodine atoms are the overwhelmingly predominant processes. With a constant unmodulated beam, the steady-state neutral desorption flux (i_0) is essentially equal to the impinging beam intensity (R) so that the Saha-Langmuir equation may be written

$$i_\pm = R\frac{g_\pm}{g_0} \exp(\beta H) \quad \text{where} \quad H = \begin{cases} \bar{\phi}_+ - I & \text{for } PSI \\ A - \bar{\phi}_- & \text{for } NSI \end{cases} \tag{3}$$

A determination of the $^7\mathrm{Li}^+$ and $^{127}\mathrm{I}^-$ currents as a function of β ($= 1/kT$) are shown in figures 1 and 2. The least squares values for the slopes are $(\bar{\phi}_+ - I) = -1.11 \pm 0.05$ eV and $(A - \bar{\phi}_-) = -1.23 \pm 0.02$ eV respectively. Since $I(\mathrm{Li}) = 5.39$ eV; $\bar{\phi}_+(\mathrm{Mo}) = 4.28 \pm 0.05$ eV in good agreement with thermionic values

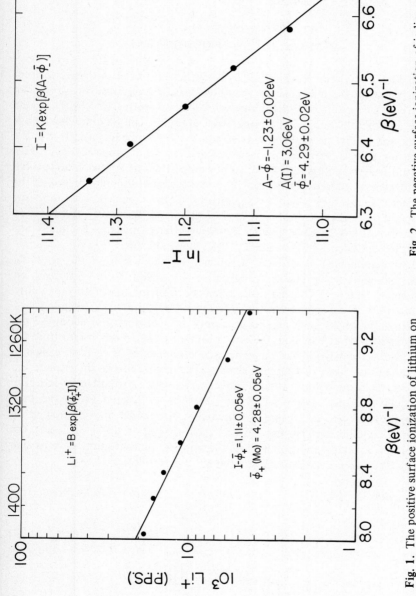

Fig. 2. The negative surface ionization of iodine on molybdenum.

$$I^- = K \exp[\beta(A - \bar{\phi}_-)]$$

$A - \bar{\phi} = -1.23 \pm 0.02 \, \text{eV}$
$A(I) = 3.06 \, \text{eV}$
$\bar{\phi}_- = 4.29 \pm 0.02 \, \text{eV}$

Fig. 1. The positive surface ionization of lithium on molybdenum.

$$Li^+ = B \exp[\beta(\bar{\phi}_+ - I)]$$

$I - \bar{\phi}_+ = 1.11 \pm 0.05 \, \text{eV}$
$\bar{\phi}_+ (Mo) = 4.28 \pm 0.05 \, \text{eV}$

given in the literature (Fomenko, 1966). Similarly since $A(I) = 3.06$ eV, $\bar{\phi}_- = 4.29 \pm 0.02$ eV. This excellent agreement between $\bar{\phi}_+$ and $\bar{\phi}_-$ indicates that the experimental molybdenum surface was isotropic insofar as positive and negative surface ionization are concerned.

V. THE KINETICS OF DESORPTION

The adsorption and desorption of an electropositive or electronegative atom from a metal surface may be described in two ways. Either a one- or two-species model can be postulated. In the one-species model only $(A_s{}^*)$ exists on the surface and desorbs as either a neutral atom or an ion. On the other hand, both adatoms and adions are assumed to be in equilibrium on the metal surface in the two-species model. Some consideration of the nature of a surface layer on a metal substrate is required. In 1935 Gurney noted that the approach of an alkali atom to a metal surface would result in a broadening of the valence level of the atom and that this level would be depleted by electron tunnelling to an energy corresponding to that of the Fermi level of the metal. A number of theoretical papers quantifying and expanding these ideas have appeared (Schmidt and Gomer, 1966; Gadzuk, 1967; Gadzuk, 1969; Remy, 1970).

Electron tunnelling is strongly dependent on the width of the potential barrier and should be especially favorable for alkali atoms adsorbed on molybdenum. Remy (1970) has made calculations of the transition probability per unit time of the bound electron for the system lithium on rhenium, and indicated it to be 10^{15} s^{-1} up to an atom-metal separation of about 0.3 nm decreasing to 10^{10} s^{-1} at about 0.9 nm. A lithium atom 0.9 nm distant from the surface would be beyond the range of surface interaction if electron exchange were absent. Because of the rapid electron exchange between the surface and adsorbed alkali, the concept of an adsorbed atom with a partial charge is useful. A partially charged adatom $(A_s{}^*)$ may then be considered to be the only adsorbed species on the surface. Alternatively, the model may be that of an ion-neutral equilibrium on the surface such that $K = \sigma_\pm/\sigma_0$, K being the equilibrium constant and $\sigma_\pm$ and σ_0 the surface concentration of ions and neutrals. The two mechanisms may be symbolized as follows:

One-Species Model (I)   *Two-Species Model* (II)

$$A_g \underset{-1}{\overset{1}{\rightleftarrows}} A_s^* \pm e_s^*$$

$$e_s^* \pm A_s^* \overset{2}{\longrightarrow} A_g^\pm \pm e_s$$

$$A_g \underset{-1}{\overset{1}{\rightleftarrows}} A_s$$

$$A_s \rightleftarrows A_s^\pm \pm e_s$$

$$A_s^\pm \overset{2}{\longrightarrow} A_g^\pm$$

where $e_s{}^*$ represents the fractional charge gained or lost by the adatom to the metal and e_s is an electron in the Fermi sea of the metal. The parallel desorption processes are labelled -1 and 2 in both mechanisms, and 1 represents the adsorption of an atom from the impinging atom beam. Assuming a unit sticking coefficient, the rate of 1 is given by the alkali beam intensity (R). The rates of -1 and 2 are given by the neutral and ionic desorption fluxes i_0 and $i_\pm$ respectively. The surface concentrations of $A_s{}^*$, A_s and $A_s{}^\pm$ are σ^*, σ_0 and $\sigma_\pm$ respectively. Their rates of change are:

$$\frac{d\sigma^*}{dt} = R - i_\pm - i_0 \tag{4}$$

for mechanism I, and

$$\frac{d(\sigma_0 + \sigma_\pm)}{dt} = R - i_\pm - i_0 \tag{5}$$

for mechanism II. The desorption fluxes $i_\pm$ and i_0 may be written in terms of the σ's and lifetimes $\tau_\pm$ and τ_0 for processes 2 and -1. In the case of mechanism I

$$i_\pm = \frac{\sigma^*}{\tau_\pm} \quad \text{and} \quad i_0 = \frac{\sigma^*}{\tau_0} \tag{6}$$

and for mechanism II

$$i_\pm = \frac{\sigma_\pm}{\tau_\pm} \quad \text{and} \quad i_0 = \frac{\sigma_0}{\tau_0} \tag{7}$$

The degree of ionization $\alpha_\pm$ is

$$\alpha_\pm = \frac{i_\pm}{i_0} = \frac{\tau_0}{\tau_\pm} \quad \text{and} \quad \alpha_\pm = \frac{\sigma_\pm}{\sigma_0}\frac{\tau_0}{\tau_\pm} = K\frac{\tau_0}{\tau_\pm}\,. \tag{8}$$

for mechanisms I and II respectively. In terms of the equilibrium constant K the fraction of the adlayer ionized is $\sigma_\pm/(\sigma_\pm + \sigma_0) = K/(K+1)$. This quantity is equivalent to the fractional charge Z associated with an adsorbed atom.

With the beam shutter closed $(R = 0)$, one finds on substituting (6) and (7) into (4) and (5) and then taking logarithmic derivatives that

$$-\frac{d\ln i_\pm}{dt} = -\frac{d\ln i_0}{dt} = \frac{1}{\tau_0} + \frac{1}{\tau_\pm} = \frac{1}{\bar{\tau}} \tag{9}$$

and

$$-\frac{d\ln i_\pm}{dt} = -\frac{d\ln i_0}{dt} = \frac{K}{K+1}\left(\frac{1}{\tau_\pm} + \frac{1}{K\tau_0}\right) = \frac{z}{\tau_\pm} + \frac{1-z}{\tau_0} = \frac{1}{\bar{\tau}} \tag{10}$$

for mechanisms I and II, where $K = Z/(1 - Z)$. For mechanism I τ_0 and $\tau_\pm$ can be expressed in terms of the experimental $\bar{\tau}$ and the degree of ionization $\alpha_\pm$ as follows

$$\tau_0 = \bar{\tau}(1 + \alpha_\pm) \quad \text{and} \quad \tau_\pm = \bar{\tau}(1 + 1/\alpha_\pm) \tag{11}$$

while for the two-species model, the result is

$$\tau_0 = (1 - z)\bar{\tau}(1 + \alpha_\pm) \quad \text{and} \quad \tau_\pm = z\bar{\tau}(1 + 1/\alpha_\pm) \tag{12}$$

The quantity $\alpha_\pm$ can be obtained from a knowledge of ϕ, I and A and the Saha-Langmuir equation. The partial charge Z on the adatom was not determined in the present study so that the individual lifetimes τ_0 and $\tau_\pm$ can be obtained only by applying mechanism I and equation (11).

VI. THE CALCULATION OF $\bar{\tau}_+{}^\circ, \bar{\tau}_-{}^\circ, \bar{\tau}_0{}^\circ, l_+, l_-,$ AND l_0 WITH MECHANISM I

A formulation for the desorption process, by Dobretsov (1952) assumes a critical distance from the surface so that at greater distances the partial charge characteristic of the adatom is lost and separation into ions and neutrals occurs. Such a model corresponds to mechanism I. Using this model, equation (11), and the Saha-Langmuir equation (1)

$$\tau_0 = \bar{\tau}\left\{1 + \frac{g_\pm}{g_0} \exp(\beta H)\right\}$$

and

$$\tau_\pm = \bar{\tau}\left\{1 + \frac{g_0}{g_\pm} \exp(\beta H)\right\} \tag{13}$$

where $g_\pm = 1$, g_0 (alkalis) $= 2$, g_0 (halogens) $= 4 + 2 \exp(-\beta\epsilon)$; ϵ being the energy difference between the $^2P_{3/2}$ and $^2P_{1/2}$ states of the halogen atom. These quantities have an Arrhenius temperature dependence over the experimental temperature range

$$\tau_0 = \tau_0{}^\circ \exp(\beta l_0) \quad \text{and} \quad \tau_\pm = \tau_\pm{}^\circ \exp(\beta l_+) \tag{14}$$

These are consistent with a ratio of $g_\pm/g_0$ for the pre-exponential and the well-known Schottky (1920) relations

$$\bar{\phi}_+ - I = l_0 - l_+ \quad \text{and} \quad A - \bar{\phi}_- = l_0 - l_- \tag{15}$$

The results of these calculations are given in table 4.

Table 4. *Desorption parameters for the alkalis and halogens on polycrystalline molybdenum, based on the single specie system*

From steady-state surface ionization of Li and I: $\bar{\phi}_{\pm}$ = 4.28 eV
I (Li, Na, K, Rb, Cs) = 5.39, 5.14, 4.34, 4.18, 3.89 eV
A (F, Cl, Br, I) = 3.45, 3.61, 3.36, 3.06 eV

		Experimental			Calculated		
	T range (K)	l (eV)	$\bar{\tau}^{\,\circ}$ (s)	l_0 (eV)	$\tau_0^{\,\circ}$ (s)	$l_{\pm}$ (eV)	$\tau_{\pm}^{\,\circ}$ (s)
Li	1150–1320	3.14	3×10^{-16}	3.14	3×10^{-16}	4.25	6×10^{-16}
Na	985–1150	2.60	3×10^{-15}	2.59	3×10^{-15}	3.45	6×10^{-15}
K	995–1175	2.53	2×10^{-14}	2.49	2.5×10^{-14}	2.55	5×10^{-14}
Rb	970–1175	2.31	1×10^{-13}	2.39	1.5×10^{-13}	2.29	3×10^{-13}
Cs	930–1160	2.10	3×10^{-13}	2.48	2×10^{-13}	2.09	4×10^{-13}
F	1700–1835	4.63	4×10^{-16}	4.63	4×10^{-16}	5.46	2×10^{-15}
Cl	1470–1650	4.12	3×10^{-16}	4.12	3×10^{-16}	4.78	2×10^{-15}
Br	1430–1625	3.70	2×10^{-15}	3.70	2×10^{-15}	4.63	9×10^{-15}
I	1350–1500	3.15	1×10^{-14}	3.15	1×10^{-14}	4.38	5×10^{-14}

Similar results would have been obtained if mechanism II and equation (12) were used provided $dZ/d\beta$ = 0. It will be shown that this requirement is met for the alkalis on molybdenum so that the only difference arising from the application of mechanism II is that one would find the ratio of pre-exponentials ($\tau^{\circ}/\tau_{\pm}^{\circ}$) to be $(g_{\pm}/g_0)(1 - Z)/Z$ instead of $g_{\pm}/g_0$.

VII. ADLAYER MOBILITY AND THE ARRHENIUS PRE-EXPONENTIAL FACTOR

By assuming a dynamic equilibrium between gaseous particles and their adsorbed counterparts, an expression for the mean surface lifetime can be derived. Such an equilibrium assumption requires that the gas and surface temperatures be equal and that the number of atoms adsorbed per unit time be equal to the rate of desorption. To identify experimental surface lifetimes with equilibrium adsorption lifetimes the observed rate of desorption is equated to the desorption rate in a system with gaseous and surface species in thermal equilibrium. This requirement is met when the surface residence times are long enough for the adatoms to equilibrate with the surface. The experimental surface temperatures in this study were adjusted so that these times were in the range of milliseconds. As a result, there was sufficient time for the adatoms to attain an energy distribution characteristic of the surface temperature.

At thermal equilibrium, the absolute activity of the gas (λ_g) is equal to that of the adlayer (λ_s). These activities are defined by (Fowler and Guggenheim, 1956)

$$\lambda_g = \exp(\beta\mu_g) \quad \text{and} \quad \lambda_s = \exp(\beta\mu_s) \tag{16}$$

where μ_g and μ_s are the gaseous and adlayer chemical potentials. The gaseous and adlayer Helmholtz free energies $(F_g$ and $F_s)$ are written in terms of their respective partition functions (Q) for n-particle assemblies as follows

$$F_g = \frac{1}{\beta} \ln \frac{1}{Q(n_g)} \quad \text{and} \quad F_s = \frac{1}{\beta} \ln \frac{1}{Q(n_s)} \tag{17}$$

The chemical potentials in equation (16) are defined in terms of these free energies by

$$\mu_g = \left(\frac{\partial F_g}{\partial n_g}\right)_{\beta,V} \quad \text{and} \quad \mu_s = \left(\frac{\partial F_s}{\partial n_g}\right)_{\beta,A} \tag{18}$$

where V and A are the gaseous volume and surface area respectively. For a monatomic gas consisting of n_g atoms; each of mass m with three translational degrees of freedom, the partition function is

$$Q(n_g) = \frac{1}{n_g!} \left(\frac{2\pi m}{\beta h^2}\right)^{3n_g/2} V^{n_g} \tag{19}$$

Using the ideal gas law $(n/V = \beta p)$, the approximation $\ln n! = n \ln n - n$, and equations (19), (18), (17) and (16), the gaseous absolute activity for atoms of mass m is found to be

$$\lambda_g = \beta_p \left(\frac{\beta h^2}{2\pi m}\right)^{3/2} \tag{20}$$

Consider an assembly of n_s adsorbed atoms. A fraction f may be assumed to be localized in a potential well of mean depth $\bar{l}$. The remaining adatoms $(1 - f)n_s$ are thermally populated into a potential well whose depth is $(\bar{l} - \bar{L})$ where $\bar{L}$ is the energy barrier between localized adsorption and a two-dimensional gas on the surface. The fraction $(1 - f)$ of the latter is given by

$$\frac{1-f}{f} = \exp(-\beta\bar{L}) \tag{21}$$

assuming that the two state densities are equal. The mobile fraction of adatoms is characterized by two degrees of translational freedom and a vibration normal to the surface. A partition function (Q_{sm}) for such an assembly of $(1 - f)n_s$ mobile adatoms is

$$Q_{sm} = \frac{1}{[(1-f)n_s]!} \left(\frac{2\pi m}{\beta h^2} A Q_v \right)^{(1-f)n_s} \exp\left[(1 - f)n_s \beta(\bar{l} - \bar{L}) \right] \tag{22}$$

where Q_v is the partition function for the vibration normal to the surface and A is the surface area. The localized part of the adlayer consists of fn_s adatoms adsorbed on X_s distinguishable sites. All of the translational degrees of freedom are lost in the process of adsorption and converted into three perpendicular vibrational modes. The vibrational quanta associated with these degrees of freedom will be assumed to be $>kT$ so that thermal excitation to vibrational levels above the ground state may be neglected and the partition functions for these vibrational modes set equal to unity. Since the number of ways of distributing the fn_s particles on X_s sites is $X_s!/(fn_s)!(X_s - fn_s)!$, the partition function (Q_{sim}) for the immobile fraction of the adlayer is given by

$$Q_{sim} = \frac{X_s!}{(fn_s)!(X_s - fn_s)!} \exp(fn_s \beta \bar{l}) \tag{23}$$

The total partition function for the assembly of n_s adatoms, $Q(n_s)$ is the product of equations (22) and (23)

$$Q(n_s) = \frac{X_s! \left[\frac{2\pi m}{\beta h^2} A Q_v \right]^{(1-f)n_s}}{(fn_s)!(X_s - fn_s)![(1-f)n_s]!} \exp\{h_s \beta[\bar{l} - (1-f)\bar{L}]\} \tag{24}$$

Using equations (16), (17), (18), (21) and (24) the following expression is obtained for the absolute activity (λ_s) of the adlayer

$$\lambda_s = \left[\frac{fn_s}{X_s - fn_s} \right]^f \left[\frac{n_s}{A} \left(\frac{\beta h^2}{2\pi m} \right) \frac{f}{Q_v} \right]^{1-f} \exp(-\beta \bar{l}) \tag{25}$$

Equilibrium between the gaseous and adsorbed particles is expressed by equating their absolute activities, equations (20) and (25) respectively

$$\beta p \left(\frac{\beta h^2}{2\pi m} \right) = \left[\frac{fn_s}{X_s - fn_s} \right]^f \left[\frac{n_s}{A} \left(\frac{\beta h^2}{2\pi m} \right) \frac{f}{Q_v} \right]^{1-f} \exp(-\beta \bar{l}) \tag{26}$$

The flux of atoms to the metal surface is $p(\beta/2\pi m)^{1/2}$ and the rate of desorption is $n_s/A\bar{\tau}$, where $\bar{\tau}$ is the mean surface lifetime of the adatom. At equilibrium these two quantities are equal so that the left-hand side of (26) becomes $(\beta h n_s / A \bar{\tau})(\beta h^2 / 2\pi m)$. The resulting expression for $\bar{\tau}$ is

$$\bar{\tau} = \frac{\beta h}{f} Q_\nu^{1-f} \left[\frac{X_s - f n_s}{A} \left(\frac{\beta h^2}{2\pi m} \right) \right]^f \exp(\beta \bar{l}) \tag{27}$$

An expression similar to this for an immobile adlayer ($f = 1$) was first given by DeBoer (1952). The limit of an infinitely dilute adlayer is pertinent to the present experimental results. The equation for the mean surface lifetime of an adatom in such a dilute adlayer is [letting $n_s \to 0$ in (27)]

$$\bar{\tau} = \frac{\beta h}{f} Q_\nu^{1-f} \left[\frac{X_s}{A} \left(\frac{\beta h^2}{2\pi m} \right) \right]^f \exp(\beta \bar{l}) \tag{28}$$

The mean surface lifetime of a *completely localized* adatom ($f = 1$) at infinite dilution is given by

$$\bar{\tau} = \beta h \left[\frac{X_s}{A} \left(\frac{\beta h^2}{2\pi m} \right) \right] \exp(\beta \bar{l}) \tag{29}$$

Since βh is of the order of 10^{-14} s at $1150°K$ and the number of adsorption sites per unit area (X_s / A) is about 10^{15} cm^{-2}, the pre-exponential in (29) is approximately 10^{-17} s. Lithium adsorbed on molybdenum (table 3) approximates this limit. The cesium adlayer on the other hand, approaches the other extreme of a two dimensional gas. Rewriting (28) with $f = 1/2$ (when $\beta \bar{L} = 0$) the following expression for the mean surface lifetime of a particle in an adlayer consisting of equal numbers of adatoms in the two surface states is obtained.

$$\bar{\tau} = 2\beta h \left[Q_\nu \frac{X_s}{A} \left(\frac{\beta h^2}{2\pi m} \right) \right]^{1/2} \exp(\beta \bar{l}) \tag{30}$$

If the mobile vibrational quanta are assumed to be small ($h\nu \ll kT$) so that $Q_\nu = [1 - \exp(-\beta h\nu)]^{-1} \sim (\beta h\nu)^{-1}$. The pre-exponential in equation (28) may therefore be written as

$$\bar{\tau}^0 = \frac{1}{f\nu} \left[\beta h\nu \frac{X_s}{A} \left(\frac{\beta h^2}{2\pi m} \right) \right]^f \tag{31}$$

It is interesting to note that in the case of a 'supermobile' adlayer (Kemball, 1950) consisting only of two-dimensional gas state, one would let $f = \bar{L} = 0$ in the partition function given in equation (22) and

$$\bar{\tau}^0 = \beta h Q_\nu \xrightarrow{\beta h\nu < 1} \frac{1}{\nu} \tag{32}$$

which is a result first obtained by Frenkel. For Cs on Mo $\beta\bar{L}$ may be assumed to be small, ~ 0.1 (i.e. $\bar{L} \sim 0.1$ eV at $1150°$K), with f equal to 0.52 from equation (21). Using this f value and 3×10^{-13} for $\bar{\tau}°$ in (31), $\nu \sim 1.2 \times 10^8$ s^{-1}. Since ν is proportional to $(m)^{1/2}$, the corresponding vibrational frequencies for the other alkali atoms may be readily obtained. These values for ν are then used with the corresponding $\bar{\tau}°$'s for the other alkalis to obtain a set of f values which with equation (21) yields the surface state energy difference $(\bar{L})$, at infinite dilution. The results are shown in table 5 for Rb, K, Na, and Li. Note that even in the case of Li where only 3% of the adlayer is mobile, the barrier $\bar{L}$ is 0.35 eV or only about 1/10 of the desorption energy $\bar{l}$.

Table 5. *The fractional mobilities* $(1 - f)$ *and barriers between localized and mobile surface states* $(\bar{L})$ *for infinitely dilute alkali adlayers on molybdenum at* $1150°$K $(\beta h = 4 \times 10^{-14}$ s *and* $kT = 0.1$ eV)

$\bar{\tau}°$ (s)	$X_s \beta h^2 / 2\pi A m$	$1 - f$	$\bar{L}$ (eV)
Cs 3×10^{-13}	2.1×10^{-4}	0.48	0.01
Rb 1×10^{-13}	3.3×10^{-4}	0.42	0.03
K 2×10^{-14}	7.1×10^{-4}	0.32	0.08
Na 3×10^{-15}	1.2×10^{-3}	0.21	0.13
Li 3×10^{-16}	4.6×10^{-3}	0.03	0.35

A further slight modification of the Arrhenius pre-exponential may be necessary due to the temperature dependence of $\bar{l}$. This effect is undoubtedly small and can be assumed to be linear over the experimental temperature range. It may be described by $\bar{l} = \bar{l}° + aT$. The pre-exponential in the surface lifetime equation will include a factor $e^{a/k}$. This behavior of $\bar{l}$ may be directly related to its dependence on the metal work function (ϕ) since $\bar{l} = l_\pm + [1/(1 + \alpha)](H) = l_0 + [\alpha/(1 + \alpha)](H)$ (where $H = (\bar{\phi}_+ - I)$ and $(A - \bar{\phi}_-)$ for positive and negative surface ionization respectively). The temperature coefficients for metal work functions are usually found to be between k and $2k$. The range of values for the $\bar{\tau}°$'s in table 3 is seen to be about 10^3 or e^7. The factor $e^{a/k}$ cannot account for more than a small fraction of the observed pre-exponential variation but should be included for completeness. It was not done here because a reliable value for a/k is not available for molybdenum.

VIII. THE SURFACE IONIZATION POTENTIAL

The calculation of an effective adatom charge Z has been treated by Gadzuk (1969). It is given by

$$Z = 1 - \int_{-\infty}^{\infty} \rho(E) F(E) \, dE \tag{33}$$

where $\rho(E)$ is the charge density per unit energy range of the perturbed valence level of an alkali adatom and $F(E)$ is the Fermi function for the metal electrons. The symbol E represents the difference between the electron energy and the Fermi energy. Gadzuk gave $\rho(E)$ as a Lorentzian of the form $\Gamma/\{(E + A)^2 + \Gamma^2\}$ and calculated Z in the zero temperature approximation. To determine Z as a function of the temperature, the Gaussian form

$$\rho(E) = \frac{1}{\pi\Gamma} \exp\left\{-\frac{\ln 2}{\Gamma^2} (E + A)^2\right\} \tag{34}$$

is more convenient. The correspondence between the two charge density forms with regard to the maximum, position of the maximum, and width at half-maximum are maintained. The quantity A is $(I - \bar{\phi}_+ - \Delta E)$ where ΔE is the shift in the alkali valence energy level from the gas phase to the metal surface. This value is given as 0.7 eV for all of the alkalis on transition metal surfaces (Gadzuk, 1967). The quantity Γ is the bandwidth of the alkali valence energy level. This width is extremely narrow in the gas and increases as electron tunnelling becomes more intense near the surface. Γ was found to vary about 0.75 eV depending on the alkali adatom.

It has been suggested (Becker, 1954; Hughes, 1959) that the Schottky cycle ($\phi - I = l_0 - l_+$) does not close for alkali desorption from metal surfaces. A quantity called the surface ionization potential, I_s, was postulated as an additional term in this equation. This energy is given by $(-d \ln K/d\beta)$ where $K = \sigma_+/\sigma_0 = Z/(1 - Z)$; therefore

$$-I_s = \frac{1}{Z(1 - Z)} \frac{dZ}{d\beta} \tag{35}$$

The calculation of $dZ/d\beta$ for the determination of I_s can be made by using (34) and $[1 + \exp(\beta E)]^{-1}$ for $F(E)$ in (33).

$$\frac{dZ}{d\beta} = -\frac{1}{\pi\Gamma} \int_{-\infty}^{\infty} \frac{\exp\left\{-\dfrac{\ln 2}{\Gamma^2} (E + A)^2\right\} E \exp(\beta E)\, dE}{[1 + \exp(\beta E)]^2} \tag{36}$$

The factor $(\ln 2)/\Gamma^2$ can be assumed to be equal to 1, so that

$$\frac{dZ}{d\beta} = \frac{1}{\pi\Gamma} \int_{0}^{\infty} \frac{\{\exp[-(A - E)^2] - \exp[-(A + E)^2]\} E \exp(\beta E)\, dE}{[1 + \exp(\beta E)]^2} \tag{37}$$

This integral was evaluated numerically and found to be $\sim 5 \times 10^{-4}$ eV. It is insensitive to the choice of A or β within the limits of the surface ionization studies reported here. The value of $1/Z(1 - Z)$ is largest for lithium and cesium, but does not exceed a value of 6. Therefore I_s is, to within less than 3×10^{-3} eV, approximately equal to zero for the alkali atoms adsorbed on a metal surface.

A reasonable potential representation for the desorption of alkali and halogen atoms may be obtained from a combination of the potential curves for neutral and ionic species. Figures 3, 4, and 5 show potential energy diagrams for cesium, lithium, and iodine on molybdenum where r is the distance to the surface. Morse (1929) and Born and Mayer (1932) potential functions were used to calculate the adatom and adion curves. The long range Born-Mayer ionic potential was adjusted so that it approached an attractive image potential $(e^2/4r_i)$ at large distances and a Lennard-Jones (r^{-12}) repulsive potential at short distances. It has been shown (Remy, 1970) that the characteristic time for electron exchange between an adsorbed alkali and a metal surface is of the order of 10^{-15} s. Figures 3, 4, and 5 show that in order to go from a full adion to an adatom the alkali nucleus must move through distances of the order 10^{-8} cm so that a velocity of 10^7 cm/s is required. This is orders of magnitude greater than a mean thermal velocity. Electron exchange must occur with the adatom nucleus at

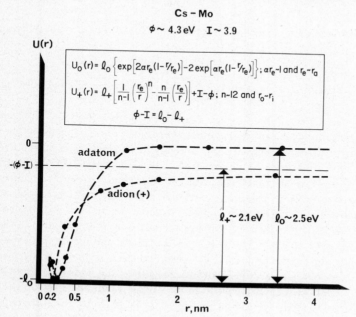

Fig. 3. Schematic potential energy diagram for cesium on molybdenum.

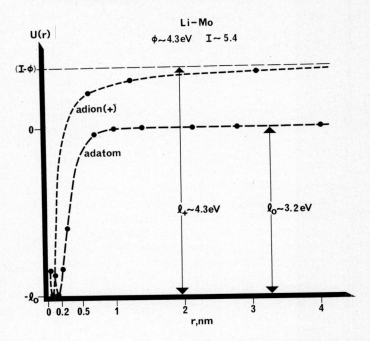

Fig. 4. Schematic potential energy diagram for lithium on molybdenum.

some fixed distance from the surface, so that the concept of a partially charged adatom is appropriate.

The surface charge of a cesium adatom (Gadzuk, 1967) on tungsten was found to be about $0.8e$ so that its potential energy approximates the adion more closely than it does the adatom curve in figure 3. The minima in both curves are shown at the same level $(-l_0)$ consistent with a negligible surface ionization energy. Figure 4 is an analogous diagram for lithium on molybdenum. The adatom charge in this case is about $0.2e$ and hence its potential energy approximates the adatom curve most closely. Iodine ($A = 3.06$ eV) on molybdenum undoubtedly approximates the adatom potential curve in figure 5 at distances close to the surface. For all the alkalis and halogens, however, at distances greater than about $10r_i$, the desorbed particle is either a neutral or fully charged ion and would conform to either the Morse or Born-Mayer potential curves. At such distances from the surface where the electron exchange processes have ceased, the ion to neutral ratio is given by the Saha-Langmuir equation.

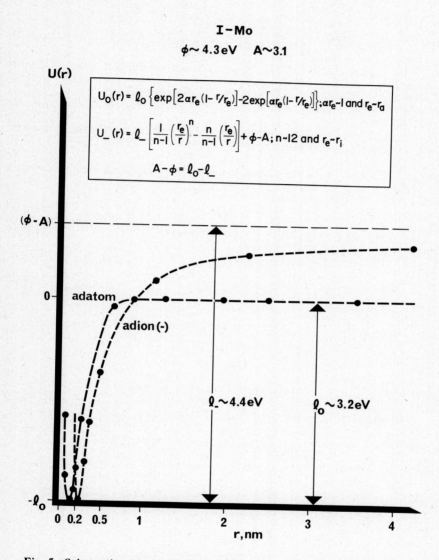

Fig. 5. Schematic potential energy diagram for iodine on molybdenum.

REFERENCES

Becker, J. A. (1954). *Ann. N.Y. Acad. Sci.* **58**, 723.

Born, M. and Mayer, J. E. (1932). *Z. Phys.* **75**, 1.

DeBoer, J. H. (1952). "The Dynamical Character of Adsorption", Chapter III, The Clarendon Press, Oxford.

Dobretsov, L. N. (1952) "Elektronnoya i Ionnaya Emissiya" State Publishing House for Technical and Theoretical Literature, Moscow-Leningrad.

Evans, R. C. (1933). *Proc. Roy. Soc., London Ser. A*, **139**, 604.

Fomenko, V. S. (1966). "Handbook of Thermionic Properties" Plenum Press, New York.

Fowler, R. H. and Guggenheim, E. A. (1956). "Statistical Thermodynamics", Cambridge University Press.

Gadzuk, J. W. (1967). *Surface. Sci.* **6**, 113.

Gadzuk, J. W. (1969). *In* "Proceedings of the 4th International Materials Symposium", John Wiley, New York.

Gurney, R. W. (1935). *Phys. Rev.* **47**, 429.

Hughes, F. L. (1959). *Phys. Rev.* **113**, 1036.

Kemball, C. (1950). *In* "Advances in Catalysis" Vol. 2, 233, Academic Press, New York and London.

Knauer, F. (1948). *Z. Phys.* **125**, 278.

Langmuir, I. and Kingdon, K. H. (1925). *Proc. Roy. Soc. London Ser. A*, **107**, 61.

Moon, P. B. and Oliphant, M. L. E. (1932). *Proc. Roy. Soc. London Ser. A*, **137**, 463.

Morse, P. M. (1929). *Phys. Rev.* **34**, 57.

Pauling, L. (1960). "The Nature of the Chemical Bond", Cornell University Press, New York.

Remy, M. (1970). *J. Chem. Phys.* **53**, 2487.

Saha, M. N. (1920). *Phil. Mag.* **40**, 472.

Scheer, M. D. (1970). *J. Res. Nat. Bur. Stand.* **74A**, 37.

Scheer, M. D. and Fine, J. (1962). *J. Chem. Phys.* **37**, 107.

Scheer, M. D. and Fine, J. (1969). "Proceedings of the 6th International Symposium on Rarefied Gas Dynamics", Vol. 2, p. 1469. Academic Press, New York and London.

Schmidt, L. J. and Gomer, R. (1966). *J. Chem. Phys.* **45**, 1605.

Schottky, W. (1920). *Ann. Phys.* **62**, 113.

Starodubtsev, C. B. (1949). *Z. Eksp. Teor. Fiz.* **19**, 215.

THE INTERACTION OF HYDROGEN AND DEUTERIUM BEAMS WITH METAL SURFACES

R. CHAPPELL and D. O. HAYWARD

Department of Chemistry, Imperial College London, England

I. INTRODUCTION

The scattering of light atoms and molecules from metal surfaces has excited considerable interest in recent years. Beams of helium, hydrogen and deuterium have been scattered from polycrystalline nickel foils (Smith and Fite, 1963; Smith, 1964), polycrystalline and single-crystal platinum (Datz *et al.,* 1963; Moore *et al.,* 1966; Hinchen and Foley, 1966; Hinchen and Shepherd, 1967; Smith and Merrill, 1970) and epitaxially grown films of gold and silver (Smith and Saltsburg, 1964, 1966; Saltsburg and Smith, 1966; Saltsburg, *et al.* 1967). Much of this work has been performed under poor vacuum conditions with inadequate facilities for outgassing of the metal, and there seems little doubt that many of the surfaces used in these experiments were contaminated. Only in the work of Saltsburg and Smith, who investigated the scattering of hydrogen, deuterium and helium from gold and silver films, and in the work of Smith and Merrill, who studied the reflection of helium and deuterium from a platinum (111) surface, can one be reasonably certain that clean surfaces were being investigated. With the exception of deuterium on silver, all the scattering phenomena reported by these workers showed a very sharp specular peak superimposed on a broader lobe, and this appears to be typical of scattering from clean surfaces generally.

The work to be described in this paper was started with the object of examining this type of scattering in greater detail. The systems chosen for study were the reflection of hydrogen and deuterium beams at the surfaces of polycrystalline platinum, and polycrystalline and single-crystal molybdenum. All experiments have been performed in ultra-high vacuum so that contamination of the surface by adsorption from the background gases has not been a problem. A novel feature of the work has been the observation of changes, in the spacial distributions of scattered gas, as molecules from the beam adsorb on the surface. Previous work has always been done under steady-state conditions where the surface is saturated by adsorption before measurements are taken.

189

II. APPARATUS

The molecular beam apparatus was constructed out of Pyrex glass and is capable of attaining vacua of the order of 10^{-10} Torr. The essential features of the beaming section are shown in figure 1. Hydrogen (or deuterium) is purified by diffusion through a heated palladium thimble and then passes into a storage vessel where the pressure is maintained constant at about 10^{-2} Torr. The beam is formed by four accurately aligned capillary tubes of internal diameter 2 mm. The first two beam chambers are pumped by mercury diffusion pumps and the last one by a titanium film. The beam can be closed off by a magnetically operated shutter attached to the second chamber. This system provides a fairly intense beam with an angular divergence of less than two degrees. The diffuse

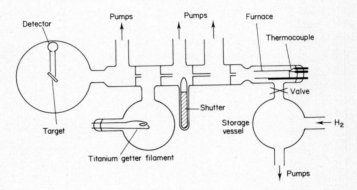

Fig. 1. Schematic diagram of beaming section of apparatus.

component is between 1 and 10% of the total beam flux. The temperature of the beam can be varied by using the furnace situated just in front of the first capillary. This is constructed from molybdenum sheet and can be heated to a temperature of about 1300°K by a radio-frequency coil which is placed around the glass envelope.

The main cell consists of a 1 litre bulb with various side-arms attached and is illustrated in figure 2. The metal target is supported through a ground-glass sleeve and can be rotated with the aid of a magnet about an axis at right angles to the beam and the surface normal. It can also be moved laterally about its axis of rotation. The target is heated by electron bombardment from a hot tungsten filament (not shown in figure 2) situated just behind its rear surface and out of sight of the beam. With a potential difference of 2000 volts applied between target and filament the temperature of the metal could easily be raised above 2000°K. The spacial distribution of gas molecules reflected from the metal surface is monitored by a rotable detector arm which is mounted on a ground-glass sleeve similar to the one used with the target assembly. The arm terminates in a small glass bulb which has a 2 mm diameter hole bored in it through which molecules reflected from the surface can enter the detection system. The pressure generated by these molecules can be monitored either by an ionization gauge or by a mass spectrometer, both of which contain rhenium cathodes coated with lanthanum hexaboride (Hayward and Taylor, 1966).

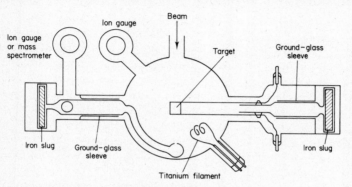

Fig. 2. Schematic diagram of main cell.

In order to keep the pressure in the cell as low as possible when the beam is impinging on the metal surface, a titanium film is deposited on to the walls of the cell from the filament shown in figure 2. The metal target is shielded during evaporation by a piece of molybdenum sheet, and deposition of film in the detector arm is prevented by moving a nickel cylinder across the detector orifice. The latter precaution is most important as any uptake of gas in this part of the system would lead to erroneous results.

III. TARGET PREPARATION

The polycrystalline platinum and molybdenum targets were cut from polished sheet in such a way that their long edges followed the direction of rolling of the original sheet. The platinum used was quoted as having less than 10 parts per million of metallic impurity; the figure for molybdenum was 150 parts per million. The targets were attached to their holders by small tantalum flags spot-welded to each corner.

Three single crystals of molybdenum were also used, cut respectively to expose the (100), (110) and (111) faces. They were polished mechanically using diamond paste until they were optically smooth and highly reflecting. After this they were given a brief (about 5 seconds) chemical etch in an aqueous solution of potassium ferricyanide and sodium hydroxide to remove the top surface layers, which were thought to be highly fragmented after the mechanical polishing and to contain embedded particles of the polishing material.

IV. EXPERIMENTAL PROCEDURE

When the pressure in the apparatus had fallen to between 2 and 3×10^{-10} Torr and all metal parts had been thoroughly outgassed, titanium film was deposited in the main cell and also in the getter bulb attached to the third stage of the beaming system. At this point hydrogen was introduced into the storage bulb at a pressure of about 10^{-2} Torr, this being sufficient to give a beam flux of around 10^{13} molecules per second. The detector orifice was then placed at a suitable angle to the surface and the target flashed to remove adsorbed molecules. When the surface had cooled to the requisite temperature the beam was introduced into the cell and the pressure generated in the detector system was recorded as a function of time. By working in this way with the detector orifice at different angles to the surface it was possible to obtain scattering distributions corresponding to different surface coverages of adsorbed hydrogen. It usually took between 30 and 100 seconds for the detector signal to build up to its steady value, a time which is long compared to the response time of the system.

When diffuse reflection was being investigated the maximum pressure rise in the detector was typically about 9×10^{-11} Torr whereas with specular reflection the rise was sometimes as high as 6×10^{-10} Torr. Such pressure rises could be measured accurately only when the residual positive current from the ion gauge was backed-off at the electrometer. This proved to be quite satisfactory provided the residual current remained constant, which it did in the majority of experiments.

Introduction of the beam into the cell caused the background pressure to rise by about 1×10^{-11} Torr and this value had to be subtracted from the total pressure rise measured by the detector in order to obtain the correct signal for gas scattered from the metal target.

V. SCATTERING FROM PLATINUM SURFACES

Three types of scattering patterns have been observed with polycrystalline platinum sheet. First, a broad reflected lobe was observed after the surface had been heated *in vacuo* for ten hours at a temperature of 1500°K. The maximum intensity of this lobe occurred at the specular angle when the surface and the gas were at the same temperature but moved towards the surface normal when the surface was at a higher temperature than the beam. This behaviour is probably caused by momentum accommodation normal to the surface but not parallel to it (Logan and Stickney, 1966; Logan et al., 1967). Similar observations have been made by other workers while investigating the reflection of light gas atoms and molecules from metal surfaces (Smith, 1964; Datz et al., 1963; Hinchen and Foley, 1966; Hinchen and Shepherd, 1967). It seems evident from recent work (Merrill and Smith, 1970) however, that surfaces producing this type of scattering are not clean, the most likely contaminant being carbon.

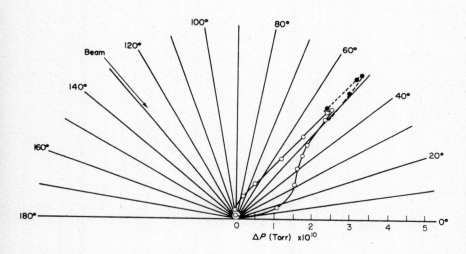

Fig. 3. Hydrogen refection from platinum with surface and beam at 300°K. Open circles represent final signal from hydrogen-covered surface; filled circles represent initial signal from clean surface.

In our experiments an attempt was made to remove residual impurities from the surface by heating the platinum to a temperature of 650°K for two hours in 1 Torr pressure of oxygen. This was followed by annealing at 1900°K for eight hours in vacuum. This treatment was subsequently found to have produced large crystallites about 1 mm across with the (111) face exposed in the surface. When hydrogen was scattered from this surface at 300°K a very sharp specular peak was observed superimposed on a broader lobe, as shown in figure 3. It was also noticed that when the detector orifice was at or close to the specular angle and the beam was admitted to the clean surface, the pressure rise within the detector was initially large but gradually decreased to a lower constant value. If the detector orifice was moved away from the specular angle, however, the pressure rise within the detector slowly increased after the beam was admitted and reached its maximum value only after about 30 seconds. This is illustrated in figure 4.

The observed behaviour, away from the specular lobe, is not unexpected since initially some hydrogen is chemisorbed on the platinum surface, and the total number of molecules scattered must therefore rise as the surface becomes saturated with hydrogen and chemisorption ceases. On the other hand, the behaviour at the specular angle is in the opposite direction and shows that less hydrogen is specularly reflected from a hydrogen-covered surface than from a

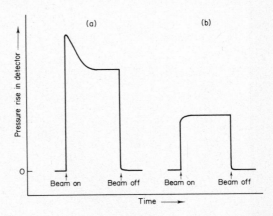

Fig. 4. Pressure rise in detector plotted as a function of time after beam is admitted to clean surface with detector (a) at specular angle, and (b) away from specular angle.

clean surface. This can be explained in terms of energy accommodation; when a light molecule such as hydrogen impinges on a clean surface consisting of heavy metal atoms, energy transfer between the hydrogen and the surface is difficult and a large proportion of the gas is elastically reflected, but when chemisorbed hydrogen is present energy transfer becomes much easier and the proportion of elastic collisions is reduced.

Beam reflections were also studied from heated surfaces. As the temperature increased the specular peak became less well defined and the broad lobe appeared to increase in intensity. At 550°K the specular peak was still visible but at 750°K it had almost completely merged into the broad lobe. The result is not easily explained. It may be connected with increasing thermal disorder as the temperature is raised.

The reflection of deuterium from this surface was also investigated and the results are shown in figure 5 for temperatures of 300, 550, and 700°K. The sharp specular peak is seen to be absent and increasing the temperature has little

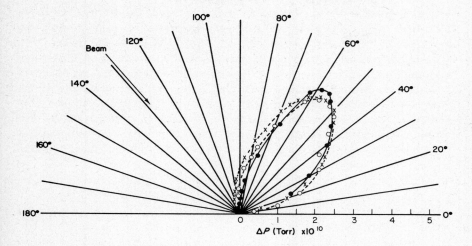

Fig. 5. Deuterium reflection from platinum with beam at $300°K$. Filled circles—surface at $300°K$; open circles—surface at $550°K$; crosses—surface at $700°K$.

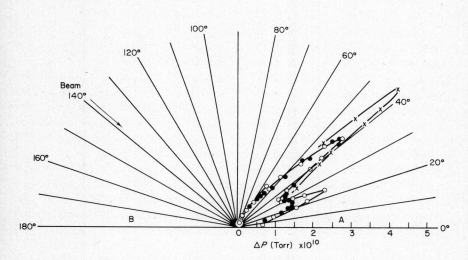

Fig. 6. Hydrogen reflection from platinum with surface and beam at $300°K$. Open circles—surface flashed once and detector arm taken from A to B. Closed circles—detector arm taken in reverse direction after flashing. Crosses—signal from clean surface at each angle.

effect on the shape of the broad lobe that is obtained. Saltsburg *et al.* (1967) have also observed differences between the scattering of hydrogen and deuterium in their experiments using the (111) face of silver. They believe that transfer of rotational energy occurs between deuterium and the metal lattice, resulting in inelastic collisions, whereas with hydrogen this process is thought to be prevented because of the greater size of the rotational quanta. However, Smith and Merrill (1970) have recently observed a sharp specular peak for deuterium scattered from the (111) face of a platinum single crystal so that elastic collisions can apparently occur with deuterium under appropriate conditions. The reason for the difference between our results and those of Smith and Merrill is not known.

At this stage it was decided to look for evidence of diffraction in the lobe structure since some of the earlier results had pointed to the existence of a diffraction effect. It was realised, however, that diffraction effects might be obscured by the presence of hydrogen on the surface and it was therefore necessary to investigate the lobe structure when the hydrogen coverage was as low as possible. This was done by repeatedly flashing the target at $1950°K$ while it was completely out of the path of the very small residual beam which still exists when the shutter is closed. The beam was then admitted to the clean surface for only 5 sec at each angle of the detector orifice so that the amount of hydrogen adsorbed would be small. This was less tedious than having to flash the surface before each measurement, although this was done at angles close to the specular angle. The results of these experiments are shown in figure 6.

A definite secondary peak is apparent at an angle of about $24°$ and this is most pronounced in the experiment where measurements were taken, starting with the detector in the position marked A and working round to the position B. In this process the secondary peak is reached while the surface is still relatively clean whereas in the opposite experiment, in which the detector is moved from B to A, the surface is nearing saturation by the time this point is reached. If the secondary peak is a diffraction effect one would expect to find a similar peak on the other side of the specular angle. Although this is not observed there is indeed a slight shoulder which occurs at approximately the correct angle.

In a further series of experiments a different polycrystalline platinum sheet was used. This was heated *in vacuo* to a temperature of $1950°K$ for eight hours to outgas it and to promote crystal growth. It was then heated to $1800°K$ for four hours in oxygen at a pressure of 10^{-6} Torr to oxidise residual carbon from the surface, and this was followed by a short period of heating in hydrogen. As with the previous sample, this treatment was found to produce large crystallites with the (111) face exposed in the surface. Finally the metal was flashed *in vacuo* and the beam admitted. Figure 7 shows the distribution of gas reflected from the clean surface. A subspecular peak is again observed at an angle of about $15°$ but there is also an equally prominent peak at about $61°$. It is interesting to note that all the structural features seen in figure 7 completely disappear when the surface temperature is raised above about 500°K.

If these peaks are really first order diffraction maxima the distance, d, between diffracting rows of atoms in the surface can be calculated. Using a wavelength for hydrogen molecules at room temperature of 0.80 Å, which is calculated for the most probable velocity, d is found to be 5.03 Å for the peak shown in figure 6 and 4.77, 2.54 Å respectively for the subspecular and superspecular peaks shown in figure 7. The distance between close-packed rows of atoms in the platinum (111) surface is 2.46 Å, which is in moderately close

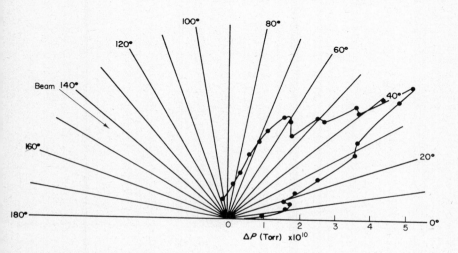

Fig. 7. Hydrogen reflection from clean platinum with surface and beam at 300°K. The surface was flashed at each angle of the detector arm and the initial pressure rise is plotted.

agreement with the value calculated for the superspecular peak. The corresponding peak on the other side of the specular angle would not be seen because calculation shows that it lies within the surface. The other peaks observed appear to correspond roughly to a surface structure having a unit mesh twice the size of that for the normal metal lattice.

An alternative explanation of these peaks is in terms of surface facetting. However, photographs taken with an electron scanning microscope just after this series of experiments were completed, showed a smooth, fairly flat surface. Also, if facetting is the explanation, it is difficult to account for the disappearance of the peaks on adsorbing hydrogen or on increasing the temperature of the surface.

VI. SCATTERING FROM MOLYBDENUM SURFACES

The initial experiments with molybdenum were performed on a poly-crystalline sheet which had been outgassed at a temperature of 1800°K for several hours. The spacial distributions of hydrogen scattered from both the clean and the hydrogen-covered surfaces were obtained by flashing the target briefly to 2000°K at different angles of the detector orifice and then following the pressure rise in the detector as a function of time after the beam was introduced. The results of this experiment are shown in figure 8. The

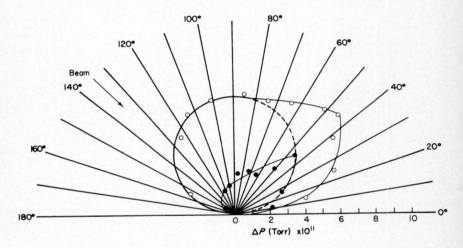

Fig. 8. Hydrogen reflection from polycrystalline molybdenum at 300°K. Filled circles—initial signal from clean surface. Open circles—final signal from hydrogen-saturated surface.

distribution of scattered gas can be resolved into a diffuse and a specular component. It is found that, as the surface becomes saturated with hydrogen, the amount of diffuse scattering increases but the specular component remains approximately unaltered. By integrating the detector signals over all angles it is possible to calculate the amount of hydrogen contributing to any particular component. For the clean molybdenum surface approximately 8% of the incident molecules are reflected specularly, 24% are scattered diffusely according to the cosine law, and the remaining 68% must therefore be chemisorbed. When the surface is saturated with hydrogen the specular component still remains at approximately 8% but the diffuse component has now risen to 92% of the total gas impinging on the surface.

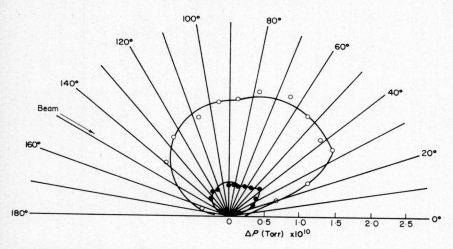

Fig. 9. Hydrogen reflection from the (100) face of molybdenum at 300°K. Filled circles—initial signal from clean surface. Open circles—final signal from hydrogen-saturated surface.

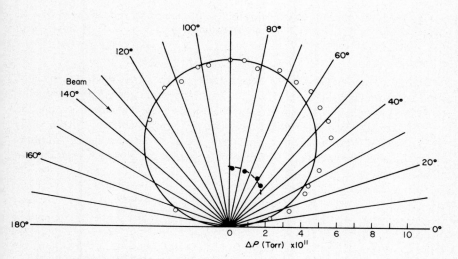

Fig. 10. Hydrogen reflection from the (111) face of molybdenum at 300°K. Filled circles—initial signal from clean surface; open circles—final signal from hydrogen-saturated surface.

To test whether the two types of scattering were occurring from different crystal faces of the polycrystalline sheet, molybdenum single crystals were used in subsequent work. The results of these experiments are shown in figures 9, 10, and 11, where it can be seen that the (100) and (111) faces give mainly diffuse scattering whereas the (110) face gives specular reflection. Clearly the orientation of the crystal is all important in determining the trapping efficiencies

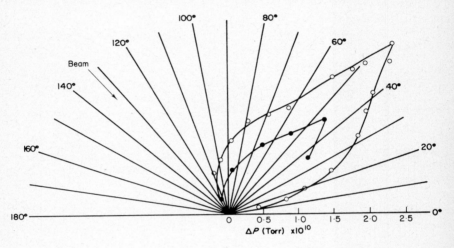

Fig. 11. Hydrogen reflection from the (110) face of molybdenum at 300°K Filled circles—initial signal from clean surface; open circles—final signal from hydrogen-saturated surface.

of the hydrogen molecules at the metal surface. Nearly all the molecules incident upon the clean (100) and (111) faces became trapped initially. Of these it can be calculated that about 70% are strongly chemisorbed and the remainder must be only weakly adsorbed since they finally escape from the surface. On the (110) face, however, weak adsorption does not appear to be very important since molecules are either strongly adsorbed or reflected straight back into the gas phase.

REFERENCES

Datz, S., Moore, G. E. and Taylor, E. H. (1963). *In* "Rarified Gas Dynamics" (J. A. Laurmann, ed.), Vol. 1, p. 347. Academic Press. New York and London.
Hayward, D. O. and Taylor, N. (1966). *J. Sci. Instrum.* **43**, 762.
Hinchen, J. J. and Foley, W. M. (1966) *In* "Rarified Gas Dynamics" (J. H. de Leeuw, ed.), Vol. 2, p. 505. Academic Press, New York and London.

Hinchen, J. J. and Shepherd, E. F. (1967). *In* "Rarified Gas Dynamics" (C. L. Brundin, ed.), Vol. 1, p. 239. Academic Press, New York and London.

Logan, R. M. and Stickney, R. E. (1966). *J. Chem. Phys.* **44**, 195.

Logan, R. M., Keck, J. C. and Stickney, R. E. (1967). *In* "Rarified Gas Dynamics" (C. L. Brundin, ed.), Vol. 1, p. 49. Academic Press, New York and London.

Merrill, R. P. and Smith, D. L. (1970). *Surface Sci.* **21**, 203.

Moore, G. E., Datz, S. and Taylor, E. H. (1966). *J. Cataly.* **5**, 218.

Saltsburg, H. and Smith, J. N. (1966). *J. Chem. Phys.* **45**, 2175.

Saltsburg, H., Smith, J. N. and Palmer, R. L. (1967). *In* "Rarified Gas Dynamics" (C. L. Brundin, ed.), Vol. 1, p. 223. Academic Press, New York and London.

Smith, J. N. (1964). *J. Chem. Phys.* **40**, 2530.

Smith, J. N. and Fite, W. L. (1963). *In* "Rarified Gas Dynamics" (J. A. Laurmann, ed.), Vol. 1, p. 430. Academic Press, New York and London.

Smith, D. L. and Merrill, R. P. (1970). *J. Chem. Phys.* **53**, 3588.

Smith, J. N. and Saltsburg, H. (1964). *J. Chem. Phys.* **40**, 3585.

Smith, J. N. and Saltsburg, H. (1966). *In* "Rarified Gas Dynamics" (J. H. de Leeuw, ed.), Vol. 2, p. 491. Academic Press, New York and London.

INTERACTION OF IONS AND ELECTRONS WITH ADSORBED GASES

R. CLAMPITT

Culham Laboratory, Abingdon, England

I. INTRODUCTION

Collisions between electrons or ions, with atoms or molecules weakly bound to a surface, are often similar to those occurring in the gas phase in isolation. For example, electrons may excite the adsorbed species to radiative, forbidden or ionising states, giving, respectively, radiation, excited particles or ions. Ions may suffer elastic or inelastic scattering; be neutralised by charge-exchange to a radiative or forbidden state; undergo an ion-molecule reaction; or may simply induce a dipole in an otherwise non-reactive adsorbed atom.

II. APPARATUS

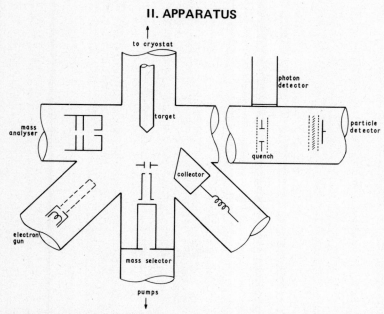

Fig. 1. Apparatus for studying ion/electron interactions with surfaces.

The apparatus used in all the studies reported here is shown in figure 1. The surface can be bombarded with beams of electrons or mass-selected ions. Ejected and reflected ions are analysed with a mass spectrometer 1 cm away. Excited neutral species pass along a flight tube of 50 cm and are detected by electron ejection at a particle multiplier. A quench field and U.V. photomultiplier are located along the flight tube. These are discussed below. The target can be cooled to 3°K or heated thermally.

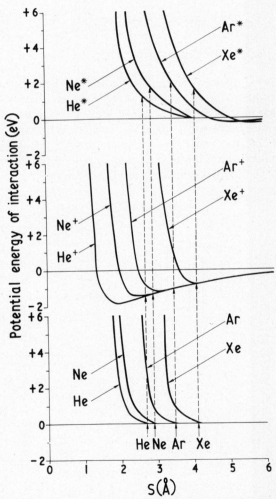

Fig. 2. Potential-energy curves of the interactions between a surface and rare-gas atoms, excited atoms and ions.

III. ELECTRON INTERACTIONS WITH ADSORBED ATOMS

Some potential-energy curves for rare-gas atoms, excited atoms and ions close to a surface are shown in figure 2 (Hagstrum, 1954). When an incoming electron promotes an atom to an ionic state, the ion so formed remains bound to the surface. Excitation to a non-ionising level is generally repulsive, so that, if the state is non-radiative an excited atom will be ejected from the surface, and with some translational energy. In addition, some emerging atoms may be resonance-ionised (Hagstrum, 1954) when the condition $E^+ - E^* \leq \phi$ is fulfilled, where E^+ is the ionisation potential of the atom, E^* is the excitation potential of the

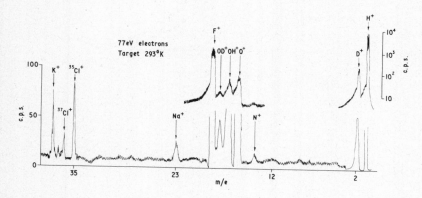

Fig. 3. Mass spectrum of ions ejected from unclean copper by electron impact.

newly-formed state, and ϕ is the work function of the surface. The ions also will possess kinetic energy.

The ejection of ions by electron impact of adsorbed gases is commonly known. A typical ion spectrum from unclean copper is shown in figure 3. Estimates of the kinetic energy of the ions can be made by retarding them electrically (figure 4). The rare-gas ions do indeed possess kinetic energy, as do protons formed from ionisation of chemisorbed H atoms. If we now physisorb molecular H_2 onto chemisorbed H atoms, another process, that of dissociative ionisation of H_2, contributes protons to the ejected proton current. Depending upon the incident electron energy these may have more, or less, kinetic energy than the protons of chemisorbed H origin. Looking for excited atoms from the surface is a more difficult matter. These must be time-resolved from photons also produced by electron impact. Figure 5 shows excited argon atoms resolved from photons during pulsed electron excitation of argon adsorbed on copper.

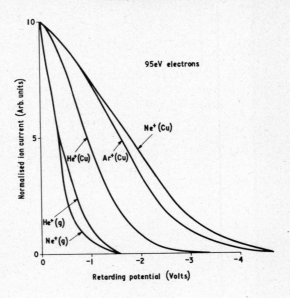

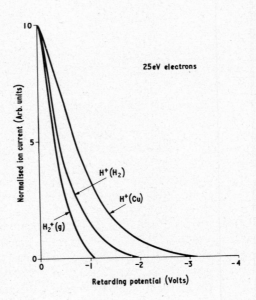

Fig. 4. Retarding-potential curves of ions formed by electron impact. The symbols in brackets refer to the origin or the ion: (Cu) is the surface; (g) is the gas phase; (H_2) is gas phase hydrogen.

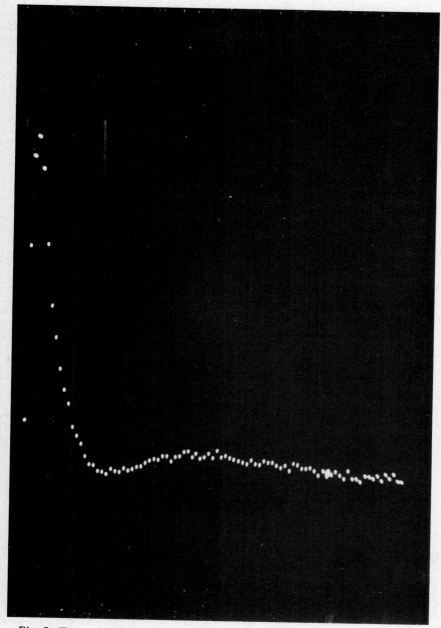

Fig. 5. Time-of-flight resolution of metastable Ar atoms ejected from copper at 3°K: 23 eV electrons; 20 μs full scale.

The flight time gives a mean kinetic energy of ~4 eV which compares reasonably with that of the ion (figure 4). The time-of-flight method is discussed later.

Electron excitation of a physisorbed monolayer, such as H_2, leads to the ejection of H_2^+, and H^+ by dissociative ionisation. Excitation of multilayers results initially in H_3^+ mainly from the reaction: $H^+ + 2H_2 \rightarrow H_3^+ + H_2$. As this ion emerges from the surface, neighbouring H_2 molecules are polarised and cluster around it. The ejected ionic species is of the type $H_3^+(H_2)_n$ with the ion $H_3^+(H_2)_6$ especially prominent (Clampitt and Gowland, 1969).These are a new type of species: ion/induced-dipole clusters.

IV. ION INTERACTIONS WITH SURFACES

Oliphant (1929) and Hagstrum (1961) reported the conversion of ions to excited atoms at metal surfaces. The atoms are usually detected by electron ejection at a metal surface. To distinguish unequivocally between ultra-violet radiation and metastable excited atoms we have used the time-of-flight technique (Clampitt, 1969). Ions or electrons are pulsed onto the target. Photons and excited atoms pass along a flight tube to be detected by a particle multiplier and processed by a time-to-height converter and pulse-height analyser. Charged particles are removed by a magnetic field.

Figure 6 shows the time-of-flight distribution of H_2^+ (narrow peak) and metastable H_2 $(c^3\pi)$ molecules (broad peak) reflected from molybdenum. The reflected H_2^+ distribution is shown to contrast its sharp energy distribution with the broad one of excited H_2 molecules. The laboratory angle of scattering is $90°$. Considerations of conservation of energy and momentum show that for H_2^+ scattering on Mo, the reflected energy E_1 lies in the range of $0.984 \geqslant E_1/E_0 > 0.96$, where E_0 is the incident energy. That the energy distributions differ suggests that scattering occurs after conversion of the ion to H_2^*. That is, ions are converted to excited molecules on the incoming trajectory.†

Above a certain velocity, unexcited atoms impinging on a surface can eject an electron by kinetic emission (Hagstrum, 1956). In our experiments with a Cu/Be detector this is ~100 eV for ground-state H atoms. Above this energy the method of detecting excited atoms by electron ejection from a surface is suspect. In one specific case, the conversion of protons to excited H atoms, Stark-effect quenching can be employed (Lamb and Retherford, 1950) whereby Lyman $-\alpha$ photons are liberated by the process $H(2S) \rightarrow H(1S) + h\nu$.

These are detected with a photomultiplier and the signal processed as before. In this way we have proved that protons are converted to H(2S) atoms at a metal surface. We could not estimate the efficiency of the process.

† Evidence for the $H_2^+ \rightarrow H_2^*$ conversion in a metal vapour has been reported recently by Morgan et al. (1971).

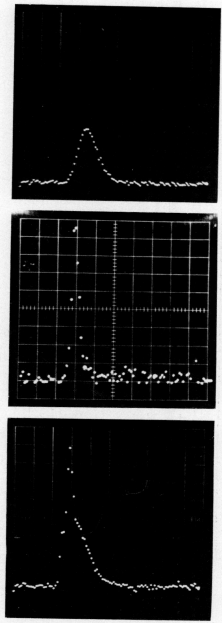

Fig. 6. Time-of-flight distribution of metastable $H_2(c^3\pi)$ molecules (top), H_2^+ (middle) and both species (bottom), reflected from molybdenum.

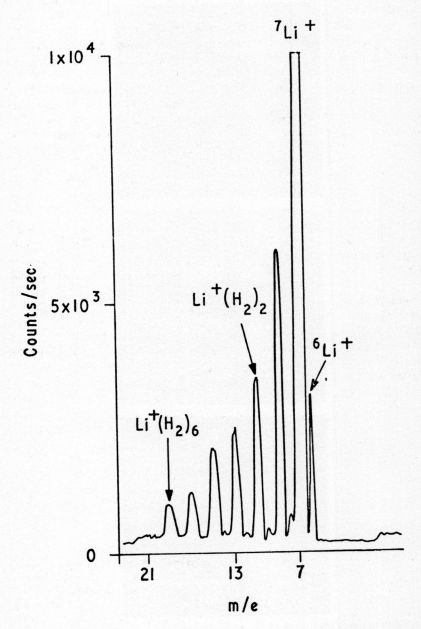

Fig. 7. Cluster ions produced by Li^+ impact on H_2^+ layers.

The efficiency of conversion of 200 eV H_2^+ ions to excited H_2 molecules was obtained as follows. Reflected H_2^+ and H_2^* were detected by potential emission from the earthed cathode of the particle multiplier. In this condition the secondary electron coefficients are essentially equal (Hagstrum, 1954). We find that the measured angular distributions of H_2^+ and H_2^* are identical; so the ion count rate and neutral count rate (obtained by deflecting the ions) give directly the ratio of H_2^+ to H_2^*. This value is $\leqslant 0.025$. Now the total fraction of H_2^+ ions reflected as H_2^+ from unclean Mo was found, using a Faraday cup, to be 0.015 at 200 eV. Hence $>60\%$ of all incident H_2^+ ions are reflected as H_2^* molecules. This process is by far the most dominant one in the interaction of low-energy ions with metal surfaces. Other processes such as electron ejection ($\sim 10^{-1}$/ion) (Hagstrum, 1954), ion reflection ($\sim 10^{-2}$/ion) (Hagstrum, 1961), photon production ($\sim 10^{-3}$/ion) (Clampitt and Jefferies, 1971) and ion dissociation ($< 10^{-4}$) (Clampitt and Jefferies, 1971) are negligible by comparison.†

When a gas is physisorbed on the surface, the interaction is now between it and the ion. The potential of interaction is proportional to $\alpha \,.\, r^{-4}$ where α is the average polarisability of the adsorbed molecule and r is the ion-molecule internuclear separation. A dipole moment is induced in the adsorbed molecule which nucleates or clusters around the ion. The reflected species are now ion/induced-dipole clusters of the type $A^+(B)_n$ (figure 7). We have reported these previously (Clampitt and Jefferies, 1970). The surface is, in this case, relatively inert: it simply provides for concentrating target molecules to a density not readily attainable in the gas phase. Thus atomic and molecular processes that are almost inaccessible to study by gas phase methods can be explored at surfaces.

---

† This result, together with those of the earlier workers, Oliphant (1929) and Hagstrum (1961), casts doubt on a reported measurement of photon emission by low-energy He^+ and Ne^+ impact on tungsten (Bohmer and Luscher, 1963). They detected what was thought to be ultra-violet radiation with a windowless particle multiplier. By time-resolving these species, we have shown that photon production is negligible compared to conversion to He^* and Ne^*.

REFERENCES

Böhmer, H. and Lüscher, E. (1963). *Phys. Lett.* **5**, 240
Clampitt, R. (1969). *Phys. Lett. A*, **28**, 581.
Clampitt, R. and Gowland, L. (1969). *Nature (London)* **223**, 815.
Clampitt, R. and Jefferies, D. K. (1970). *Nature (London).* **226**, 141.
Clampitt, R. and Jefferies, D. K. (1971). In press.
Hagstrum, H. D. (1954). *Phys. Rev.* **96**, 336.
Hagstrum, H. D. (1956). *Phys. Rev.* **104**, 1516.
Hagstrum, H. D. (1961). *Phys. Rev.* **123**, 758.
Lamb, W. E. and Retherford, R. C. (1950) *Phys. Rev.* **79**, 549.
Morgan, T. J., Berkner, K. H. and Pyle, R. V. (1971). *Phys. Rev. Lett.* **26**, 602.
Oliphant, M. L. E. (1929). *Proc. Royal Soc. Ser. A,* **124**, 228.

III. CHEMISORPTION OF GASES BY METALS

ANDERSON'S HAMILTONIAN AND THE THEORY OF THE CHEMISORPTION OF ATOMS BY METALS

T. B. GRIMLEY

The Donnan Laboratories, The University of Liverpool, England

I. INTRODUCTION

The development of a theory of the chemisorption of atoms by metals involves a search for a many-electron wave function which, although only an approximation to the true ground-state wave function, nevertheless leads to a systematization, and an understanding of a wide variety of experimental observations. For different systems, different approximations will be acceptable; not only at the fundamental level where the adequacy of, say, an orbital approximation without configuration interaction has to be decided, but also at a more superficial level when the details of the orbital approximation are being worked out. Examples of these two situations are numerous in molecular quantum mechanics. As an example of the former, I observe that the diatomic molecule F_2 is unstable in the ordinary Hartree-Fock molecular orbital approximation using g and u orbitals without configuration interaction (Wahl, 1964). As an example of the latter there is the success of Hückel's (1931) molecular orbital theory in treating planar unsaturated hydrocarbons (see for example Streitwieser, 1961) in spite of the fact, non known from all-valence-electron treatments, that σ-electron levels are embedded between π-electron levels in benzene.

For metal adsorbents where all the conduction electrons are likely to be involved in the atom-metal bond (the surface bond), the formalism of second quantization provides an economical notation, and now the development of the theory involves a search for an approximate Hamiltonian operator, not for an approximate wave function. But whichever formalism we use, a central problem is the treatment of the Coulomb interactions of the electrons. If we carry out second quantization in terms of a basis set of one-electron functions consisting of the valence electron orbitals of the adatom, and the orbitals describing the extended electron states in the energy bands of the semi-infinite metal, then because the former are localized, and the latter are not, it is natural to assume that the most important interactions are between electrons in the adatom orbitals. This simplification of a complex problem was made by Anderson (1961) when he put forward his model Hamiltonian to describe and extend Friedel's (1958) ideas on localized magnetic states in dilute alloys.

To see the importance of electron interactions in localized states, we refer to Mann's (1967) Hartree-Fock calculations for free atoms, and give in table 1 the average interaction energy $U(nl, nl)$ (Coulomb plus exchange) of a pair of electrons in the valence shell for several atoms. Now when an atom is near a metal surface, the chemical potential of its electrons is fixed (equal to the Fermi level of the metal), and to achieve this it will be necessary, in general, to transfer charge between the atom and the metal. In the Hartree-Fock approximation, this

Table 1. *Average interaction energy (Rydbergs*) of a pair of electrons.*

Na	$U(3s, 3s)$	= 0.4305
S	$U(3p, 3p)$	= 0.8432
Fe	$U(3d, 3d)$	= 1.6776
Ba	$U(6s, 6s)$	= 0.3436
W	$U(5d, 5d)$	= 0.9269

* Rydberg units of energy are used throughout; 1 Ry $\equiv$ 13.6 eV $\equiv$ 2.18 x 10^{-18} J.

charge transfer causes a self-consistent movement of the orbital energies of electrons in the atom, the extent of this movement being large if the pair interaction energy U is large. In this way it is possible for the chemical potential of electrons in the atom to achieve the required fixed value with only very small amounts of charge transfer. This is important. Without such a self-consistent movement of the orbital energies, multiply charged adsorbed ions would nearly always be formed.

Anderson's Hamiltonian explicitly contains terms to describe both the charge transfer, and the self-consistent movement of the chemical potential of electrons in the adatom, and this makes it a useful point from which to begin developing the theory of chemisorption.

II. THE HAMILTONIAN

For an atom with no low-lying excited states, and one valency electron occupying an orbital ϕ_A in the Hartree-Fock ground state, interacting with the surface of a semi-infinite metal whose one-electron states are ϕ_k, the Hamiltonian operator can be written (Grimley, 1970)

$$H = \sum_{rs\sigma} a_{r\sigma}^+ (r|\mathscr{H}_0|s) a_{s\sigma}$$

$$+ \tfrac{1}{2} \sum_{rstu\sigma\sigma'} a_{r\sigma}^+ a_{s\sigma'}^+ (rs|1/r_{12}|tu) a_{u\sigma'} a_{t\sigma}. \tag{1}$$

Here r, s, t and u stand for any member of the basis set $\{\phi_A, \phi_k\}$. $a_{r\sigma}^+$ and $a_{r\sigma}$ are creation and destruction operators for electrons in the orbital

$$\sum (S^{-1})_{sr} \phi_s \tag{2}$$

with spin σ which is either ↑ or ↓, S is the overlap matrix of the basis set $\{\phi_r\}$

$$S_{rs} = \int \phi_r^* \phi_s \, d\tau, \quad S_{rr} = 1. \tag{3}$$

$\mathscr{H}_0$ is the Hamiltonian operator for a single electron moving in the field of the ion cores, and

$$(rs|1/r_{12}|tu) = \int \phi_r^*(1) \phi_s^*(2)(1/r_{12}) \phi_t(1) \phi_u(2) \, d\tau_1 \, d\tau_2 \tag{4}$$

Because the set $\{\phi_r\}$ are not orthogonal, the anti-commutation rules of the operators in equation (1) are

$$\{a_{r\sigma}, a_{s\sigma'}^+\} = (S^{-1})_{rs} \delta_{\sigma\sigma'}$$

$$\{a_{r\sigma}, a_{s\sigma'}\} = \{a_{r\sigma}^+, a_{s\sigma'}^+\} = 0 \tag{5}$$

where $\{A, B\} = AB + BA$ denotes the anti-commutator of A and B.

The first step in the passage from equation (1) to Anderson's Hamiltonian is to discard all Coulomb terms (4) except the term J which refers to two electrons in the same orbital on the adatom

$$J = (AA|1/r_{12}|AA). \tag{6}$$

This means that we maintain the large difference between the ionization potential, and the electron affinity of the free atom in the theory, because it is this difference which prevents massive charge transfer from the metal to the adatom. But we do not include Coulomb interactions between the adatom and the metal, or between electrons in the metal, explicitly in the theory. These neglected Coulomb terms are responsible, amongst other things, for the classical image interaction between a charged atom and the metal, and although some of their effects can be incorporated in the Hamiltonian in an indirect way (see section III), it is better to accept now that systems where the adatom charge is known to be large lie outside the present discussion.

With the above assumption concerning the Coulomb terms, equation (1) reduces to

$$\left. \begin{array}{l} H = \sum_{\sigma} E_A \tilde{n}_{A\sigma} + J \tilde{n}_{A\sigma} \tilde{n}_{A-\sigma} + \sum_{k\sigma} \epsilon_k \tilde{n}_{k\sigma} + \sum_{kl\sigma} V_{kl} a_{k\sigma}^+ a_{l\sigma} + H_{\mathrm{c.t.}} \\[2mm] H_{\mathrm{c.t.}} = \sum_{k\sigma} (V_{Ak} a_{A\sigma}^+ a_{k\sigma} + V_{kA} a_{k\sigma}^+ a_{A\sigma}) \end{array} \right\} \tag{7}$$

Here $\bar{n}_{r\sigma} = a^+_{r\sigma}a_{r\sigma}$, E_A and ϵ_k are the diagonal elements of $\mathcal{H}_0$, and V_{kl} and V_{kA} non-diagonal ones. $H_{\text{c.t.}}$ describes the charge transfer, and the electron sharing between the atom and the metal: the fourth term in equation (7) is k-state scattering.

We recall that the operators $a^+_{r\sigma}$ and $a_{r\sigma}$ do not create and destroy electrons in the basis state ϕ_r when these are not orthogonal. These rôles are performed by different operators $b^+_{r\sigma}$ and $b_{r\sigma}$ defined by

$$b^+_{r\sigma} = \sum_t a^+_{t\sigma} S_{tr}, \qquad b_{r\sigma} = \sum_t S_{rt} a_{t\sigma}, \tag{8}$$

with different anti-commutation rules

$$\{b_{r\sigma}, b^+_{s\sigma'}\} = S_{rs}\delta_{\sigma\sigma'} \tag{9}$$

$$\{b_{r\sigma}, b_{s\sigma'}\} = \{b^+_{r\sigma}, b^+_{s\sigma'}\} = 0.$$

We can write the Hamiltonian (7) in terms of these operators if we wish, but some new terms appear because

$$H = \sum b^+_{r\sigma} X_{rs}(s|\mathcal{H}_0|t) X_{tu} b_{u\sigma}$$
$$+ \tfrac{1}{2}J \sum b^+_{r\sigma} X_{rA} b^+_{s\sigma'} X_{sA} X_{At} b_{t\sigma'} X_{Au} b_{u\sigma}, \tag{10}$$

with $X = S^{-1}$. Consequently by going over to the operators $b^+_{r\sigma}$ and $b_{r\sigma}$, effective Coulomb interactions between electrons in the metal, and between electrons in the adatom and those in the metal, have been introduced into H. In addition, there are some extra charge-transfer and k-state scattering terms, as well as A- and k-state level shifts. Because the anti-commutation rules (9) are hardly any simpler than those of equation (5), and because the Hamiltonian (10) is more complicated than (7), nothing is gained by using the operators $b^+_{r\sigma}$ and $b_{r\sigma}$. It is however clear that some care must be exercised in writing down the Hamiltonian operator in a non-orthogonal representation; in spite of the extra terms, equation (10) is entirely equivalent to equation (7).

Now let us introduce a set of operators $c^+_{r\sigma}$ and $c_{r\sigma}$ which obey the usual Fermion anti-commutation rules,

$$\{c_{r\sigma}, c^+_{s\sigma'}\} = \delta_{rs}\delta_{\sigma\sigma'}$$

$$\{c_{r\sigma}, c_{s\sigma'}\} = \{c^+_{r\sigma}, c^+_{s\sigma'}\} = 0. \tag{11}$$

These, like the operators $b^+_{r\sigma}$, $b_{r\sigma}$, are linear combinations of the original operators $a^+_{r\sigma}$ and $a_{r\sigma}$.

$$c^+_{r\sigma} = \sum a^+_{s\sigma} D_{sr}, \qquad c_{r\sigma} = \sum D_{rs} a_{s\sigma}, \tag{12}$$

with inverses

$$a^+_{r\sigma} = \sum c^+_{s\sigma} Z_{sr}, \qquad a_{r\sigma} = \sum Z_{rs} c_{s\sigma}, \qquad Z = D^{-1}$$

Consequently

$$H = \sum c_{r\sigma}^+ Z_{rs}(s|\mathscr{H}_0|t) Z_{tu} c_{u\sigma}$$
$$+ \tfrac{1}{2} J \sum c_{r\sigma}^+ Z_{rA} c_{s\sigma'}^+ Z_{sA} Z_{At} c_{t\sigma'} Z_{Au} c_{u\sigma}. \qquad (13)$$

This is the Hamiltonian (7) in an orthogonal representation specified by the matrix D in equation (12), and we can now say that non-orthogonality is handled by using an effective Hamiltonian operator with effective Coulomb interactions like those already encountered in equation (10), effective A- and k-state energies, effective k-state scattering, and effective charge-transfer terms. To recover the form of Anderson's original Hamiltonian, we drop the k-state scattering terms, and all Coulomb terms except $JZ_{AA}^4 \, n_{A\sigma} \, n_{A-\sigma}$ (here $n_{A\sigma} = c_{A\sigma}^+ c_{A\sigma}$) and then make the identifications

$$\sum Z_{Ar}(r|\mathscr{H}_0|s) Z_{sA} \longrightarrow E_A$$

$$\sum Z_{kr}(r|\mathscr{H}_0|s) Z_{sk} \longrightarrow \epsilon_k$$

$$\sum Z_{Ar}(r|\mathscr{H}_0|s) Z_{sk} \longrightarrow V_{Ak}$$

$$JZ_{AA}^4 \longrightarrow J.$$

In this way we arrive at the effective Hamiltonian operator

$$\left.
\begin{aligned}
H &= \sum_\sigma E_A n_{A\sigma} + J n_{A\sigma} n_{A-\sigma} + \sum_{k\sigma} \epsilon_k n_{k\sigma} + H_{\text{c.t.}} \\
H_{\text{c.t.}} &= \sum_{k\sigma} (V_{Ak} c_{A\sigma}^+ c_{k\sigma} + V_{kA} c_{k\sigma}^+ c_{A\sigma})
\end{aligned}
\right\} \qquad (14)$$

which has formed the basis of several papers on the theory of chemisorption (Edwards and Newns, 1967; Newns, 1969; Grimley, 1971a; for example).

III. SOME COMMENTS ON THE HAMILTONIAN

There are two important points to be discussed in connection with the Hamiltonian (14). First, the dropping of the k-state scattering terms, and second the neglect of all Coulomb terms save $J n_{A\sigma} n_{A-\sigma}$.

Dropping the k-state scattering terms is justified on the grounds that $H_{\text{c.t.}}$ itself leads to k-state scattering in second order, so that no new effects are encountered if the explicit scattering terms are retained. But of course, the *effective* k-state scattering provided by $H_{\text{c.t.}}$ is not quantitatively the same as the explicit scattering; the latter have a part to play in screening any adatom charge generated by $H_{\text{c.t.}}$.

This missing Coulomb terms comprise not only the effective interactions in equation (13), but also those discarded in going from equation (1) to equation

(7). One effect of the missing Coulomb terms is to alter once more, the meanings of V_{Ak} and V_{kA} in equation (14). Of course there are other effects, but I do not have space to discuss this problem here.

I now show that $H_{c.t.}$ leads to k-state scattering in second order. One way to do this is to use the canonical transformation method (Bailyn, 1966), but I shall use a different approach. The Hamiltonian (14) has the form

$$H = H_0 + H_{c.t.} \tag{15}$$

so we expand the exact many-electron ground state of H in terms of a basis set comprising all many-electron eigenfunctions of H_0. Starting from the ground-state eigenfunction of H_0, which we denote $|N\sigma\rangle$ (N for neutral, σ for the unpaired spin on the free atom), we can build up the required basis set by successive applications of the operators in $H_{c.t.}$. Thus

$$c_{A-\sigma}^+ c_{k-\sigma}|N\sigma\rangle, \quad c_{k\sigma}^+ c_{A\sigma}|N\sigma\rangle \tag{16}$$

are charge-transfer states which correspond respectively to the anion and the cation of the adatom. I denote these states $|C\sigma\rangle$. Two applications of the operators in $H_{c.t.}$ produce states like

$$c_{l\sigma'}^+ c_{A\sigma'} c_{A-\sigma}^+ c_{k-\sigma}|N\sigma\rangle, \quad c_{A\sigma'}^+ c_{l\sigma'} c_{k\sigma}^+ c_{A\sigma}|N\sigma\rangle. \tag{17}$$

I denote these states $|P\sigma\rangle$. Putting $\sigma' = -\sigma$ in the first, and $\sigma' = \sigma$ in the second yields the states

$$c_{k-\sigma}^+ c_{l-\sigma}|N\sigma\rangle, \quad c_{k\sigma}^+ c_{l\sigma}|N\sigma\rangle, \tag{18}$$

which are produced from $|N\sigma\rangle$ simply by k-state scattering. Since two applications of the operators of $H_{c.t.}$ are needed to produce the states (18), it is clear that $H_{c.t.}$ leads to k-state scattering in second order.

The other states in (17) are

$$c_{k\sigma}^+ c_{A-\sigma}^+ c_{l-\sigma} c_{A\sigma}|N\sigma\rangle, \tag{19}$$

(obtained with $\sigma' = \sigma$ in the first, and $\sigma' = -\sigma$ in the second state in (17)), and these are produced from $|N\sigma\rangle$ by k-state scattering with simultaneous spin flips in both A- and k-states. Consequently, if we take the theory to second order in $H_{c.t.}$ we find k-state scattering both with, and without spin flips. The existence of the spin flips induced by $H_{c.t.}$ is easily overlooked in the canonical transformation method (Bailyn, 1966), but along with the Coulomb-induced spin flip terms dropped in reaching equation (14),

$$(kA|1/r_{12}|Al) c_{k\sigma}^+ c_{A-\sigma}^+ c_{l-\sigma} c_{A\sigma},$$

they are essential to Schrieffer's induced covalent bonding theory of chemisorption (Schrieffer and Gomer, 1971). I shall mention the importance of the set of states $|P\sigma\rangle$ in hydrogen chemisorption later (section VI).

IV. HARTREE-FOCK THEORY OF ANDERSON'S HAMILTONIAN

When overlap is retained in the theory, the simplest starting point is the Hamiltonian (7) with, of course, the k-state scattering terms omitted. We form the equations of motion of the operators $a_{r\sigma}$ by commuting them with H, linearize the equations by replacing $n_{A-\sigma}$ by its ground-state expectation value $\langle n_{A-\sigma} \rangle$, and then take Fourier transforms. In this way we arrive at the familiar secular equation of the LCAOSCF (linear combination of atomic orbitals self-consistent field) theory (Roothaan, 1960).

$$\det |\epsilon S - F^\sigma| = 0. \tag{20}$$

Here F^σ is the Hartree-Fock matrix for spin σ electrons, with non-vanishing elements

$$F_{AA}^\sigma = E_A + J\langle n_{A-\sigma} \rangle = \epsilon_{A\sigma},$$

$$F_{kk}^\sigma = \epsilon_k,$$

$$F_{Ak}^\sigma = V_{Ak}, \quad F_{kA}^\sigma = V_{kA}.$$

The overlap matrix S has diagonal elements unity, and S_{kA} and S_{Ak} are its only finite non-diagonal elements.

To solve equation (20) we require $\langle n_{A-\sigma} \rangle$, and this is unknown until equation (20) has been solved. The self-consistency conditions are here just a pair of equations involving $\langle n_{A\sigma} \rangle$ and $\langle n_{A-\sigma} \rangle$ (Grimley 1970)

$$\left. \begin{array}{l} \langle n_{A\sigma} \rangle = -\dfrac{1}{\pi} \, \mathrm{Im} \displaystyle\int_{-\infty}^{\epsilon_F} d\epsilon \, \dfrac{1}{\epsilon + i0 - \epsilon_{A\sigma} - q_A} \\[4mm] q_A(\epsilon) = \displaystyle\sum_k \dfrac{|\epsilon S_{Ak} - V_{Ak}|^2}{\epsilon + i0 - \epsilon_k} = \alpha(\epsilon) - i\Gamma(\epsilon). \end{array} \right\} \tag{21}$$

If we introduce the function

$$w(\epsilon, \epsilon') = \pi \sum_k |\epsilon S_{Ak} - V_{Ak}|^2 \, \delta(\epsilon' - \epsilon_k) \tag{22}$$

then

$$\alpha(\epsilon) = \frac{1}{\pi} \int_{-\infty}^{+\infty} d\epsilon' \frac{w(\epsilon, \epsilon')}{\epsilon - \epsilon'}, \quad \Gamma(\epsilon) = w(\epsilon, \epsilon). \tag{23}$$

We observe that $\langle n_{A\sigma} \rangle$ is obtained by integrating to ϵ_F a level density function

$$\rho_{A\sigma} = -\frac{1}{\pi}\,\text{Im}\,\frac{1}{\epsilon + i0 - \epsilon_{A\sigma} - q_A}. \tag{24}$$

For the uncoupled system (free atom, free metal), $\rho_{A\sigma}$ is a δ-function at E_A, $\rho_{A-\sigma}$ one at $E_A + J$. The zeros of the real part of the denominator in equation (24), i.e., the roots of the equation

$$\epsilon - \epsilon_{A\sigma} - \alpha = 0 \tag{25}$$

indicate resonances associated with spin σ electrons in the surface bond.

V. HYDROGEN ON W(100)

To illustrate the theory, consider the adsorption of hydrogen on W(100) in the W_2H linear surface geometry (figure 1). As a matter of fact, the experimental evidence (Estrup and Anderson, 1966; Tamm and Schmidt, 1969)

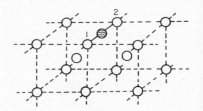

Fig. 1. The linear W_2H structure on W(100).

could support a planar, or inverted pyramidal W_4H local geometry. No new principles are involved in treating this situation, and nothing in the section on surface molecule formation needs to be altered. Only equation (26) below becomes more complicated.

We suppose that only the d orbitals on the tungsten atoms closest to the hydrogen atom, with lobes pointing towards the hydrogen atom, are involved in the surface bond, so that the hydrogen 1s orbital couples to the group orbital

$$\psi_d = \frac{1}{\sqrt{2}}\{d_{x^2-y^2}(1) + d_{x^2-y^2}(2)\}. \tag{26}$$

Consequently

$$V_{Ak} = \sqrt{2}V\langle\psi_d|\phi_k\rangle, \quad S_{Ak} = \sqrt{2}S\langle\psi_d|\phi_k\rangle,$$

where V is the coupling, and S the overlap between the hydrogen $1s$ orbital, and $d_{x^2-y^2}$ (1). Equation (22) can now be written

$$
\begin{aligned}
w(\epsilon, \epsilon') &= (\lambda\epsilon + 1)^2\, w(0, \epsilon'), \quad \lambda = -S/V, \\
w(0, \epsilon') &= 2\pi |V|^2 \sum_k |\langle \psi_d | \phi_k \rangle|^2\, \delta(\epsilon' - \epsilon_k).
\end{aligned}
\tag{27}
$$

Thus an application of the theory hinges on a knowledge of the function $w(0, \epsilon')$, and so ultimately on a level density function $\rho_d(\epsilon)$ defined by

$$
\rho_d(\epsilon) = \sum_k |\langle \psi_d | \phi_k \rangle|^2\, \delta(\epsilon - \epsilon_k).
\tag{28}
$$

This is the resonance in the d band of tungsten of the group orbital (26).

THE FORMATION OF A SURFACE MOLECULE

If $\rho_d(\epsilon)$ is a sharp resonance, the whole theory simplifies to the calculation of the orbital energy levels, and the binding energy of a pseudomolecule W_2H. To demonstrate this we put

$$
\rho_d(\epsilon) = \delta(\epsilon - \epsilon_d),
\tag{29}
$$

which means that ψ_d has a perfectly sharp resonance in the d band at ϵ_d. Then

$$
\alpha(\epsilon) = (\lambda\epsilon + 1)^2\, 2|V|^2/(\epsilon - \epsilon_d),
$$

$$
\Gamma(\epsilon) = (\lambda\epsilon + 1)^2\, 2\pi |V|^2\, \delta(\epsilon - \epsilon_d),
$$

and equation (25) for the orbital resonances is

$$
(\epsilon - \epsilon_{A\sigma})(\epsilon - \epsilon_d) = 2|V|^2(\lambda\epsilon + 1)^2.
$$

The latter is nothing more than the usual secular equation of molecular orbital theory for the energies $E1^\sigma$ and $E2^\sigma$ say, of the bonding, and anti-bonding molecular orbitals formed from the hydrogen $1s$ orbital at $\epsilon_{A\sigma}$, and the group orbital Ψ_d at ϵ_d. The level density $\rho_{A\sigma}$ of equation (24) has δ-functions of strengths

$$
\eta_{A1}^\sigma = (E1^\sigma - \epsilon_d)/(E1^\sigma - E2^\sigma), \quad \eta_{A2}^\sigma = (E2^\sigma - \epsilon_d)/(E2^\sigma - E1^\sigma)
$$

at $E1^\sigma$ and $E2^\sigma$,

$$
\rho_{A\sigma} = \eta_{A1}^\sigma\, \delta(\epsilon - E1^\sigma) + \eta_{A2}^\sigma\, \delta(\epsilon - E2^\sigma),
$$

so for example, if $E1^\sigma < \epsilon_F < E2^\sigma$ so that $E1^\sigma$ is occupied, and $E2^\sigma$ is vacant, then $\langle n_{A\sigma} \rangle = \eta_{A1}^\sigma$. These equations show that, when equation (29) holds, hydrogen is chemisorbed to form a surface molecule, W_2H in our case, with discrete molecular orbital energies like an ordinary molecule; the metal substrate is represented in it by the group orbital ψ_d at ϵ_d.

The results of a calculation assuming (see below) that ϵ_d coincides with the Fermi level of tungsten may be of interest. The work function of $W(100)$ is

0.354 Ry (Gomer, 1966), and this fixes the Fermi level, which we now take as our energy zero. Then $E_A = -0.646$ Ry if we ignore the difference (perhaps as much as 0.1 Ry) between E_A in Anderson's Hamiltonian, and the $1s$ level of hydrogen. For the Coulomb energy J, we take 0.912 Ry. This is the difference in the $1s$ orbital energies of H^- and H. The interaction parameters

$$|V|^2 = 0.0178, \quad \lambda = -0.5, \tag{30}$$

have been chosen solely to illustrate the theory. They reproduce the experimental binding energy, 0.22 Ry. (Tamm and Schmidt, 1969), but they cannot be fixed individually at present.

For the interaction parameters (30), the ground-state of the surface molecule has different molecular orbitals for different spins (DMODS). The results are

$$\langle \tilde{n}_A \uparrow \rangle = 0.8015 \qquad \langle \tilde{n}_A \downarrow \rangle = 0.4355$$
$$\epsilon_A \uparrow = -0.249\,\text{Ry} \qquad \epsilon_A \downarrow = 0.085\,\text{Ry}$$
$$E1 \uparrow = -0.382\,\text{Ry} \qquad E1 \downarrow = -0.167\,\text{Ry}$$

Binding Energy D = 0.221 Ry/atom,

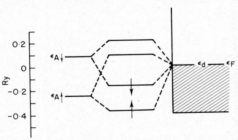

Fig. 2. Energy level diagram for the surface molecule formed by H on W(100).

and the energy level diagram is shown in figure 2. As a matter of fact, although the ground state is DMODS, the ordinary state with doubly occupied molecular orbitals,

$$\langle \tilde{n}_A \uparrow \rangle = \langle \tilde{n}_A \downarrow \rangle = 0.6332, \qquad E1 \uparrow = E1 \downarrow = -0.250\,\text{Ry}$$

gives only 0.001 Ry less binding energy.

When we allow for the fact that the concept of a surface molecule is only an approximation, the discrete levels $E1^\sigma$ and $E2^\sigma$ acquire certain widths, but still the levels $E1^\sigma$ should give the positions of the resonances associated with electrons in the surface bond. These resonances are being investigated experimentally by ion neutralization (Hagstrum and Becker, 1971), field emission (Plummer and Young, 1970; Gadzuk, 1970) and photoelectron spectroscopy (E. W. Plummer, 1970, private communication), but no results for hydrogen on tungsten are available at present. These methods give information on the total contribution $\Delta\rho_\sigma$ which the adsorbed atom makes to the density of states for spin σ electrons in the combined system. For our present model

$$\Delta\rho_\sigma = \delta(\epsilon - E1^\sigma) + \delta(\epsilon - E2^\sigma) - \delta(\epsilon - \epsilon_d),$$

and for example, in field emission spectroscopy, any enhanced tunnelling exists by virtue of the connection between the metal, and the vacuum, through the energy levels $E1^\sigma$ and $E2^\sigma$ of the surface molecule.

GROUP ORBITAL RESONANCES

The resonance (28) has not yet been properly investigated, although preliminary work indicates that it may be quite narrow, and located near the Fermi level. Preliminary investigations of other group orbital resonances have been made (Grimley, 1971a; 1971b). For an alkali atom interacting with the (100) surface of either a face-centred, or a body-centred cubic metal in the M_4A pyramidal local geometry, the appropriate resonance again appears to be a narrow one, so the concept of a surface molecule should be useful here. On the other hand, for a chalcogen on Ni(100) in the Ni_4A pyramidal local geometry, although the p_z orbital on the chalcogen couples to the same group orbital as an alkali atom, the p_x and p_y orbitals couple to group orbitals with quite broad resonances in the nickel d band. It appears therefore that the concept of a surface molecule may not be so useful for these systems.

VI. CONCLUSION

In Anderson's Hamiltonian, the electrons in the metal are described by non-localized wave functions, but even so the model shows how, in chemisorption, electrons can be well-localized in surface bonds, so that a simple "molecular" picture is possible. However, if this localization occurs, it raises a problem with Anderson's model. Is it sensible to neglect the Coulomb interactions of these electrons in the metal which are now well-localized on the metal atoms of the surface molecule? This question needs further study using a Hamiltonian in which these Coulomb interactions are retained (Grimley, 1969; Newns, 1970).

It is important also to investigate the validity of the Hartree-Fock approximation and in this connection the following result is interesting. When the concept of a surface molecule is used for hydrogen on W(100), the ground state of the Hamiltonian (14) can be expressed exactly in terms of the limited basis set consisting of $|N\sigma\rangle$, two states $|C\sigma\rangle$, and one state $|P\sigma\rangle$, as defined in section III, and an exact formula for the binding energy is easily obtained. When the interaction parameters are chosen to give $D = 0.221$ Ry, it turns out that the

Hartree-Fock approximation (DMODS) then gives $D = 0.169$ Ry. The approximation is clearly a useful one. We note also that, in the exact theory, the state $|P\sigma\rangle$ contributes 0.090 Ry to D. Since this state describes the k-state scattering with simultaneous spin flips in A- and k-states, these effects play an important, though not a decisive, rôle in the theory of hydrogen chemisorption as described by Anderson's Hamiltonian.

REFERENCES

Anderson, P. W. (1961). *Phys. Rev.* **124**, 41.
Bailyn, M. (1966). *Advan. Phys.* **15**, 179.
Edwards, D. M. and Newns, D. M. (1967). *Phys. Lett.* *A* **24**, 236.
Estrup, P. J. and Anderson, J. (1966). *J. Chem. Phys.* **45**, 2254.
Friedel, J. (1958). *Nuovo Cimento, Suppl.* **7**, 287.
Gadzuk, J. W. (1970). *Phys. Rev.* *B* **1**, 2110.
Gomer, R. (1966). *Discuss. Faraday Soc.* **41**, 14.
Grimley, T. B. (1969). *In* "Molecular Processes on Solid Surfaces" (E. Drauglis, R. D. Gretz and R. J. Jaffee, eds) p. 299. McGraw-Hill, New York.
Grimley, T. B. (1970). *J. Phys. C.* **3**, 1934.
Grimley, T. B. (1971a). *J. Vac. Sci. Technol.* **8**, 31.
Grimley, T. B. (1971b). To be published by the Honda Memorial Foundation, Japan.
Hagstrum, H. D. and Becker, G. E. (1971). *J. Chem. Phys.* **54**, 1015.
Hückel, F. (1931). *Z. Phys.* **70**, 204.
Mann, J. B. (1967). Los Alamos Sci. Lab. Rep. LA-3690.
Newns, D. M. (1969). *Phys. Rev.* **178**, 1123.
Newns, D. M. (1970). *Phys. Rev. Lett.* **25**, 1575.
Plummer, E. W. and Young, R. D. (1970). *Phys. Rev.* *B* **1**, 2088.
Roothaan, C. C. J. (1960). *Rev. Mod. Phys.* **32**, 179.
Schrieffer, J. R. and Gomer, R. (1971). *Surface Sci.* **25**, 315.
Streitwieser, A. (1961). "Molecular Orbital Theory for Organic Chemists". Wiley, New York.
Tamm, P. W. and Schmidt, L. D. (1969). *J. Chem. Phys.* **51**, 5352.
Wahl, A. C. (1964). *J. Chem. Phys.* **41**, 2600.

THE INTERPRETATION OF SLOW DESORPTION KINETICS

L. A. PÉTERMANN

Battelle Institute, Geneva Research Centre, Switzerland

I. INTRODUCTION

The main objective of thermal desorption experiments is to increase understanding of the desorption kinetics of a gas from a solid. In particular, the dependence of the desorption rate on temperature and on surface coverage can be used to test the concept of activation energy, to derive the order of the desorption reaction and to determine the number of different phases of the adsorbate co-existing on the solid surface (Hickmott and Ehrlich, 1958; Redhead, 1962; Carter, 1962; Ehrlich, 1961 and 1963).

Practical difficulties mainly arise from the limited size of the surface area of a well-defined solid sample, and consequently from the small amount of reactants available. In the case of sub-monolayer coverage of an active gas (such as hydrogen) on a single crystal of metal (such as nickel), the rate of desorption induced by raising the temperature of the substrate cannot be accurately derived from a recording of partial pressures in the experimental chamber, because of uncontrolled adsorption, desorption and other effects on other surfaces in the chamber ("wall effects", hot filaments, etc.). Furthermore, the temperature being a function of time during the experiment, the desorption rate itself becomes a very complicated function of time, from which it is usually quite difficult to derive the parameters of the reaction unambiguously, the order of the reaction for instance (McCarroll, 1969). It is more difficult, if not impossible, to test the applicability of a theoretical rate expression to the desorption reaction.

For these reasons, we have been studying desorption kinetics at constant temperature and under very low pressure, while avoiding all perturbations due to wall effects by measuring the gas coverage, on the surface of the sample, with the technique of electron probe surface mass spectrometry (Lichtman, 1965; Lichtman *et al.*, 1968). The interpretation of our experimental results so far has been mainly aimed towards testing the degree of applicability of the absolute rate theory to desorption reactions.

227

II. APPLICABILITY OF ABSOLUTE RATE THEORY

The quantum mechanical theory of reaction rates leads to a specific rate expression of great generality (Eyring *et al.*, 1944). However a less general expression, suitable for direct comparison with experiments, was obtained at the

Table 1. *General form and "classical limit" of the absolute rate, as proposed by Eyring, Walter and Kimball* (1944)

$$\kappa = \frac{(A_1)(A_2)\ldots}{F_1 F_2 \ldots} \int_0^\infty \sum_s \sum_n r_{ns}(p) \cdot \omega_n \exp\left(-\frac{E_n}{kT}\right) \exp\left(-\frac{p^2}{2mkT}\right) \frac{p \cdot dp}{mh}$$

Classical limit: (1°)
$$r_{ns}(p) = 0 \text{ if } \frac{p^2}{2m} \leqslant \Delta H \frac{k}{R}$$

$$\int_0^\infty \sum_s \sum_n r_{ns}(p) \ldots = \bar{r} \int_0^\infty \ldots \text{if} \frac{p^2}{2m} > \Delta H \frac{k}{R}$$

(2°)
$$\sum_n \omega_n \exp\left(-\frac{E_n}{kT}\right) = F\ddagger$$

Whence *absolute specific rate*:

$$\kappa' = \bar{r} \cdot \frac{kT}{h} \cdot \frac{F^{\ddagger}}{F_1 F_2} \cdot \exp\left(-\frac{\Delta H}{RT}\right)$$

(in general $\bar{r}$ is a function of T)

κ = rate of reaction
κ' = specific rate (for unity concentration of reactants)
A_i = concentration
F_i = partition function } for reactant i

$F\ddagger$ = partition function for the activated complex
r = transmission coefficient
m = mass of reactant particle
p = momentum of reactant particle
ω_n = statistical weight of configuration with energy E_n
k = Boltzmann constant
T = absolute temperature
h = Planck's constant
ΔH = activation energy
R = gas constant

expense of a number of limiting assumptions referred to as the "classical limit". The three main simplifying assumptions are:

(a) Equilibrium distribution of momentum of the reactants.
(b) Existence of a special partition function $F^{\ddagger}$, corresponding to an activated complex (see table 1).
(c) Existence of an average transmission coefficient $\bar{r}$ (usually considered as a constant, frequently even as equal to unity!) (see table 1).

Curtiss (1948), Prigogine and Xhrouet (1949), Prigogine and Mahieu (1950), Takayanagi (1951), Zwolinski and Eyring (1947), and Kramers (1940), have shown that assumption (a) is fully justified for all practical purposes when the ratio of activation energy ΔH to kT is larger than about 10. More recently, Montroll and Shuler (1958) have reviewed this question and confirmed the above conclusion. Therefore, we should not expect any difficulty on this account in the interpretation of the great majority of thermal desorption experiments.

Assumption (b) may be considered as the definition of the "activated complex", an unstable atomic arrangement or molecule, also referred to as the "transition state". In the case of desorption, this complex might be made up of atoms from both the adsorbate and the substrate. However, if the ratio of partition functions for the activated complex and the reactants, derived from rate measurements, turns out to depart from unity by more than a few powers of ten, the required size of the reactant molecule or activated complex may become unreasonable (Glasstone *et al.*, 1941 p. 18.).

Assumption (c) certainly is the most difficult to appraise *a priori*, since it deals with quantum mechanical wave propagation, for which many parameters play a very sensitive role. As defined in the "classical limit", $\bar{r}$ cannot in general be a constant for any particular reaction, but usually should be dependent on temperature. Furthermore, $\bar{r}$ may be very much smaller than unity (Glasstone *et al.*, 1941, p. 213; Daudel, 1967, p. 37) for several types of reactions, in particular for reactions involving 2 atoms and for many unimolecular reactions.

Table 2 shows that the theoretical expression for the absolute reaction rate may be considered as the product of three factors: the first, the average transmission coefficient (quantum mechanics and kinetics); the second, resulting for example from a random walk, or from a harmonic oscillator model; and the third, resulting from the assumptions of equilibrium thermodynamics. The frequency factor A is the product of three terms

$$A = \bar{r} \cdot \frac{kT}{h} \cdot \frac{F^{\ddagger}}{F_{\text{init}}}$$

When A is much smaller than kT/h, the particular reaction is called a "slow reaction". If, in addition, the average transmission coefficient $\bar{r}$ is assumed to be constant and equal to unity (therefore frequently left out of the rate expression and even forgotten!) the slowness of the reaction is entirely attributed to a very small value of the ratio of partition functions (or to a large and negative value of the activation entropy), sometimes so small as to lead to very serious problems of interpretation (Degras, 1967; Lapujoulade, 1967; Pétermann, 1967).

Table 2. *The three main factors of the specific absolute rate, at the classical limit, and the definition of slow reactions.*

$$K = \bar{r} \cdot \frac{kT}{h} \cdot \frac{F^{\ddagger}}{F_{\text{init}}} \cdot \exp\left(-\frac{\Delta H}{RT}\right)$$

"Equilibrium" thermodynamics

Random walk, or harmonic oscillator

Average transmission coefficient (kinetics)

Frequency factor $A = \bar{r} \cdot \dfrac{kT}{h} \cdot \dfrac{F^{\ddagger}}{F_{\text{init}}}$

Slow reactions: $A \ll \dfrac{kT}{h}$ or $\bar{r} \cdot \dfrac{F^{\ddagger}}{F_{\text{init}}} \ll 1$

Table 3. *The possible range of values of the frequency factor (without taking into account any effect due to implicit temperature dependence)*

	$\bar{r}$	$\dfrac{F^{\ddagger}}{F_{\text{init}}} = \exp\left(\dfrac{\Delta S}{R}\right)$	$A, [S^{-1}]$
Normal reactions	~ 1	~ 1	$\sim 10^{13}$
Fast reactions	~ 1	max ~ 100	max $\sim 10^{15}$
Slow reactions	min $\sim 10^{-13}$	min $\sim 10^{-5}$	min $\sim 10^{-5}$

Table 3 shows that the possible values of the frequency factor cover a very wide range, even with reasonable values of the ratio of partition functions, if the average transmission coefficient $\bar{r}$ is not always considered as equal to unity, but is allowed to span the range of values from 10^{-13} to unity. A test of applicability of the absolute rate expression is obtained by plotting the quantity

Ln $\kappa h/kT$ against $1/T$ for each pair of values κ, T determined experimentally. (This could be called the "Eyring plot", by analogy with the "Arrhenius plot".) If there exist well-defined and constant values of the activation energy ΔH and of the activation entropy $\Delta S[\cong R \ \mathrm{Ln}(F^{\ddagger}/F_{\mathrm{init}})]$, and if $\bar{r}$ is truly constant, the Eyring plot will yield a straight line. The slope will be equal to $-\Delta H/R$, and the intercept with the ordinate axis will yield the quantity Ln $\bar{r} + (\Delta S/R)$.

III. APPARENT VALUES OF ACTIVATION ENERGY AND ENTROPY

However, if $\bar{r}$ is really a function of temperature over the experimental range, the Eyring plot will yield apparent values ΔH_{app} and ΔS_{app}, conditioned by the extent of the temperature range covered and by the accuracy of rate measurements, though without simple physical meaning. Of course, a similar effect would result from some temperature dependence of ΔH or ΔS, but we assume here that such temperature dependence is very weak (Glasstone et al., 1941, p. 21, 194). Figure 1 shows the Eyring plot for a hypothetical reaction, in which the average transmission coefficient $\bar{r}$ is a constant over a certain temperature range (left-hand side of figure 1) and is a function of temperature over some other range (right-hand side of figure 1). With measurements covering a limited temperature range (solid lines), it is conventionally assumed that a straight line must fit the experimental points. And this is usually not difficult to achieve with large enough error rectangles. Figure 1 shows that the apparent value ΔH_{app}, due to a temperature dependence of $\bar{r}$, necessarily leads to an apparent value ΔS_{app}, and that there is a simple relationship between them (change in ordinate of intercept due to change of slope), as given in table 4. Drawing a straight line through the experimental points is equivalent to approximating the true temperature dependence of $\bar{r}$ by an exponential function over the limited experimental range, of the form $\bar{r}(T) = r_0 \ \exp - (B/T)$. The quantity r_0 obviously has no physical meaning, since it would be the value of the average transmission coefficient at infinite temperature, assuming that the exponential behaviour could be extrapolated that far! Nevertheless, figure 2 shows how this meaningless value of r_0 may be directly responsible for the entire difference between ΔS and ΔS_{app}. It is also apparent that a modest change of $\bar{r}$ over the experimental range may lead to a large negative value of ΔS_{app}. In the example shown in figure 2, the frequency factor appears to be extremely small, of the order of $10 \ \mathrm{s}^{-1}$, even though neither $\bar{r}$, nor $F^{\ddagger}/F_{\mathrm{init}}$ are particularly small within the experimental temperature range.

As shown in figures 1, 2 and table 4, the apparent values ΔH_{app} and ΔS_{app} are related to the true ones by a compensation relationship. Isokinetism is obtained for any value of the temperature T_m within the experimental range,

232   L. A. PÉTERMANN

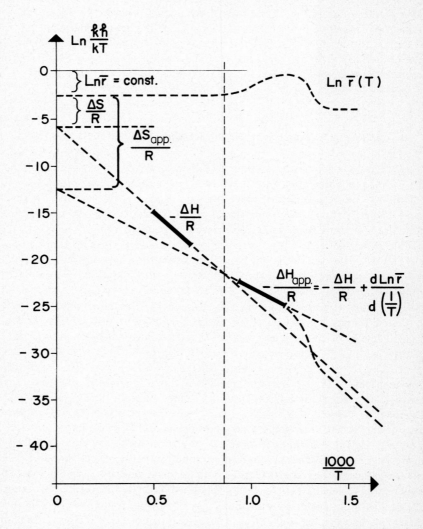

Fig. 1. Eyring plot of a hypothetical reaction, showing: (left-hand side) true values of ΔH and ΔS deduced from experiments where $\bar{r}$ = constant; (right-hand side) apparent values of these parameters where $\bar{r}$ is a function of temperature.

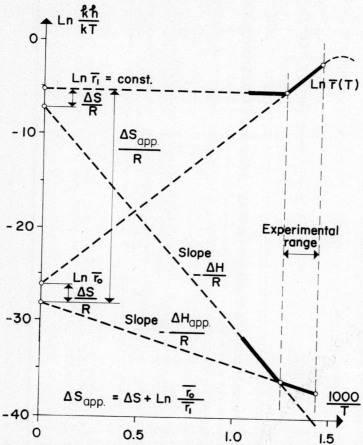

Fig. 2. Eyring plot of a hypothetical reaction, showing that fitting the experimental data with a straight line is equivalent to approximating $\bar{r}(T)$ by an exponential function over a narrow range of T. The apparent value of the activation entropy is simply related to $\bar{r}_0 = \bar{r}(T)$ for $T \to \infty$.

which is a direct consequence of drawing a straight line through the experimental points in the Eyring plot. This means that the rate equations

$$\kappa = \bar{r}(T) \cdot \frac{kT}{h} \cdot \exp\frac{\Delta S}{R} \exp-\frac{\Delta H}{RT}$$

and

$$\kappa' = \bar{r}(T_m) \cdot \frac{kT_m}{h} \exp\frac{\Delta S_{app}}{R} \exp-\frac{\Delta H_{app}}{RT_m}$$

are identical for any T_m within the experimental range, since the compensation relationship of table 4 can be re-written as

$$\frac{\Delta S}{R} - \frac{\Delta H}{RT_m} = \frac{\Delta S_{app}}{R} - \frac{\Delta H_{app}}{RT_m}$$

Table 4. *Compensation relationship between apparent and true values of* ΔH *and* ΔS, *as resulting from temperature dependence of the average transmission coefficient* $\bar{r}(T)$.

$$\Delta H_{app} = \Delta H - R\,\frac{d\,\mathrm{Ln}\,\bar{r}}{d\left(\dfrac{1}{T}\right)}$$

$$\Delta S_{app} = \Delta S - \frac{R}{T_m}\cdot\frac{d\,\mathrm{Ln}\,\bar{r}}{d\left(\dfrac{1}{T}\right)}\ (T = T_m)$$

whence

$$\Delta H_{app} - \Delta H = T_m\,(\Delta S_{app} - \Delta S)$$

(Compensation relationship, with isokinetic temperature T_m within experimental range.)

IV. TEMPERATURE DEPENDENCE OF THE AVERAGE TRANSMISSION COEFFICIENT $\bar{r}$

The behaviour of the transmission coefficient was already the subject of detailed theoretical work as far back as 1939 (Hirschfelder and Wigner, 1939; Eyring *et al.*, 1944). In the extreme case of atomic hydrogen recombination in vacuum, the excess energy of the newly formed hydrogen molecule can only be carried away by photon emission, and the transmission coefficient is of the order of 10^{-14} (Glasstone *et al.*, 1941, p. 213). In the presence of a third body, other mechanisms can take care of this excess energy, thereby increasing the value of the transmission coefficient far above the lower limit of 10^{-14}. Such a situation would be expected to occur for desorption of hydrogen from a solid surface when recombination takes place, the third body being the solid surface itself.

A temperature dependence of the transmission coefficient could result if the efficiency of energy transfer mechanisms were dependent on temperature.

In this respect, Suhl and his collaborators (1970) have been studying the possible role of spin fluctuations in the desorption of hydrogen from paramagnetic metals. A large change of the average transmission coefficient with temperature is expected in the vicinity of a paramagnetic-ferromagnetic transition.

Other mechanisms, not necessarily related to atomic recombinations or to spin fluctuations, could result in a temperature dependence of the average transmission coefficient for desorption reactions. Therefore, Suhl's proposal does not in any way imply that slow desorption kinetics could be restricted to cases involving a magnetic transition in the solid surface and/or a recombination of atomic adsorbates.

V. EXPERIMENTAL

Thermal desorption of hydrogen from a single crystal of nickel [100] was studied under conditions designed to overcome some of the difficulties associated with conventional thermal "flash" desorption techniques. The temperature of the sample was kept constant during each measurement, and the rate of desorption was obtained by measuring the hydrogen coverage, as a

SYSTEM FOR DESORPTION EXPERIMENTS

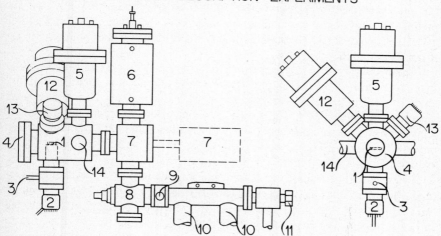

Fig. 3. System for desorption experiments at constant temperature (see text). 1. Single crystal. 2. Grid-controlled electron gun. 3. To diffusion pump. 4. Viewing port. 5. Quadrupole RGA. 6. Orbitron pump. 7. High speed UHV valve. 8. Isolation valve. 9. TC gauge. 10. Cryo-sorption pumps. 11. Isolation valve. 12. Quadrupole surface ion detector. 13. Surface probe electron gun. 14. To leak-valve.

function of time, with the technique of electron probe surface mass spectrometry (Lichtman, 1965 and Lichtman *et al.*, 1968). The experimental system is
shown schematically on figure 3. The single crystal (1) rests on top of a
vacuum-tight enclosure (2) containing a grid-controlled electron gun used to
regulate the crystal temperature. This enclosure is fitted with a separate pumping
line (3): (4) is a viewing port. The residual gas in the experimental chamber, and
the pure gas during adsorption periods, are analysed with an EA1 QUAD 250
quadrupole mass spectrometer (5); the special ionization chamber is fitted with a
low temperature oxide cathode, operating down to 1 μA ionizing current. An
orbitron pump (6) is connected to the experimental chamber via a high-speed,
high-conductance valve (7) operated by a pneumatic cylinder. This design allows
a very rapid pressure drop at the end of an adsorption period (for hydrogen: 1 x
10^{-7} to 1 x 10^{-10} Torr in 0.8 second). A leak valve (not shown) on port (14)
controls the flow of the pure test gas, together with a fine adjustment of the
residual conductance of the high-speed valve in its closed position. The surface
probe consists of an electron gun (13) with focussing and scanning facilities, and
of a second quadrupole head (12) fitted with a special 3-electrode ion lens
instead of the normal ionization chamber. The overall sensitivity of the surface
proble is sufficient to detect the release from the sample of about 50 H^{+} ions per
second. Isolation valve (8) separates the experimental chamber from the
auxiliary pumping systems: cryosorption pumps (10) and diffusion pump
(alumina/zeolite baffle) behind valve (11). A total pressure gauge is only used for
preliminary tests, and is thereafter isolated from the system behind a closed
valve to avoid large perturbations, especially bothersome with hydrogen
(Pétermann and Baker, 1965).

Figure 4 shows a typical recording of surface ion current as a function of
time, as measured at the output of the secondary electron multiplier of the
second quadrupole head. In this series of measurements, the background
pressure was below 10^{-10} Torr at the beginning of the recording, and the initial
temperature of the sample was about room temperature ($\sim$300°K). The surface
probe was operated at an electron energy of 300 eV and a current density of
about 10^{-7} A/cm^2 on the sample. The electron beam was focussed on the
sample, with a spot diameter of 2 mm, and scanned to form a rectangular raster
of three adjacent lines, 6 x 7 mm, at a frame frequency of about 100 s^{-1}. Under
these conditions, electron impact desorption had a negligible effect on the
surface coverage. The position of the bombarded rectangle was so chosen as to
avoid direct line of sight from the first dynode of the electron multiplier in the
quadrupole head, thus suppressing the large background signal associated with
"Bremsstrahlung" radiation from the sample. The sample was heated by means
of servo-controlled electron bombardment of the stainless steel base on which it
sits. The base reached a very stable temperature within a few minutes, but it
took about 20 min for the sample temperature to reach equilibrium. From 20 to

50 min, on figure 4, the surface ion current decreased exponentially with time, with a time constant of 400 ± 10 s at this particular temperature. Two series of measurements were taken on the same nickel crystal, separated by about 90

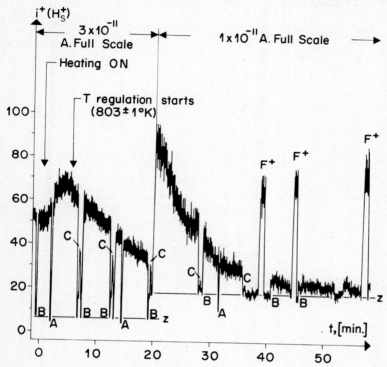

Fig. 4. Thermal desorption at constant temperature ($T = 803°$K). Surface ion current (H_s^+) as a function of time.

A: electron current check, 4.0×10^{-8} A (sample no longer bombarded during check)

B: zero check (quadrupole detuned to $M = 1.5$)

A-B: residual X-ray photocurrent (Bremsstrahlung) $\sim 8 \times 10^{-13}$ A at the output of the secondary electron multiplier.

C: D_s^+ ion current (from previous experiments with deuterium)

F$^+$: surface contamination (fluorine)

Electron energy: 300 eV; current density $\sim 10^{-7}$ A/cm^2.

days. In all cases, both tests for first and second order kinetics were performed, with a very clear-cut result: surprisingly, all desorption reactions observed were of first order, as shown for example on figure 5 obtained by re-plotting direct recordings (such as on figure 4) with semi-log coordinates.

The only phase of adsorbed hydrogen, from which desorption could be followed by this method, required temperatures in the range between about 700 and 800°K for convenient values of the desorption rate. This unexpected result

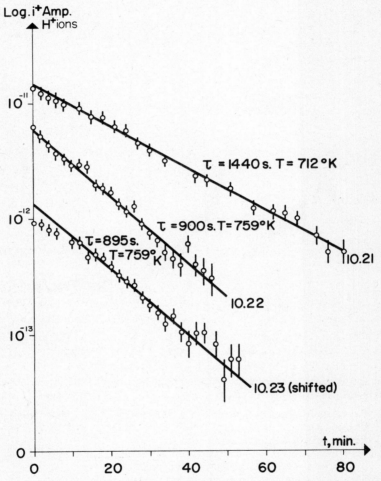

Fig. 5. Hydrogen desorption at constant temperature. The logarithm of surface ion current decreased linearly with time, except for the first few points (equilibrium temperature not yet reached), showing first order kinetics.

is probably due to some residual surface contamination of the nickel crystal, most likely an oxide layer. Nevertheless, the temperature dependence of the rate of this particular desorption reaction could be investigated, even though the initial conditions of the reactants were not known.

VI. INTERPRETATION

Plotting the first series of results, as described earlier, suggests that the experimental points were closer to an S-shaped rather than a straight line. The average slope and the intercept of the Eyring plot (figure 6) gave the following apparent values:

$\Delta H_{app} \cong 12.6$ kcal/mole
$\Delta S_{app} \cong -56.4$ cal/mole degree
　　　(assuming $\bar{r} = 1$)

These values are clearly unusual for hydrogen desorption, and the large and negative activation entropy can hardly be given a credible physical meaning. Therefore, we interpreted our experimental data according to Suhl's suggestion (see section IV). An almost perfect fitting of our data can be obtained with the following values of the reaction parameters.

$\Delta H = 45.8$ kcal/mole
$\Delta S = -4$ cal/mole degree
$\bar{r}$ decreasing (as shown on figure 6) from 0.1 to 0.02 in the temperature range 690-830°K.

These values are not unique, and therefore somewhat arbitrary. But they prove the possibility of explaining the observed slow kinetics of desorption with a temperature-dependent transmission coefficient $\bar{r}$, without having to lend physical significance to the anomalous values of ΔH_{app} and ΔS_{app}. Clearly, an extension of the rate measurements to higher and lower temperatures would be extremely valuable to check the validity of this interpretation.

Figure 7 shows an enlarged portion of figure 6, where the first and second series of experimental values appear with an indication of the accuracy of the rate and temperature measurements. The second series of measurements (triangles) yields a much straighter experimental line, with an even lower apparent activation energy

$\Delta H_{app} = 6.8$

and a large, negative apparent activation entropy (assuming $\bar{r} = 1$)

$\Delta S_{app} = -65$

again quite anomalous and typical of very slow kinetics. Only one of the new experimental points (lower right-hand corner of figure 7) falls quite close to the extrapolated curved line fitted to the first series of measurements. These two series show that the reproducibility of measurements over a long period of time (~ 90 days) is not good and so more detailed interpretation is not justified. New measurements are under way, with a greatly improved crystal processing

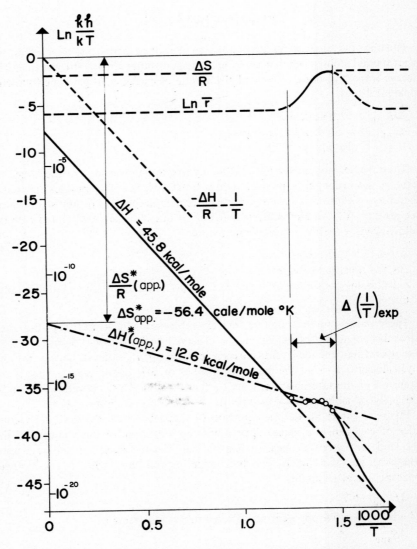

Fig. 6. Interpretation of desorption rate measurements. The six experimental points shown as opened circles fall close to an S-shaped line. The assumed behavior of Ln $\bar{r}$ with temperature within the experimental range is just a plausible one among an infinite number of possibilities.

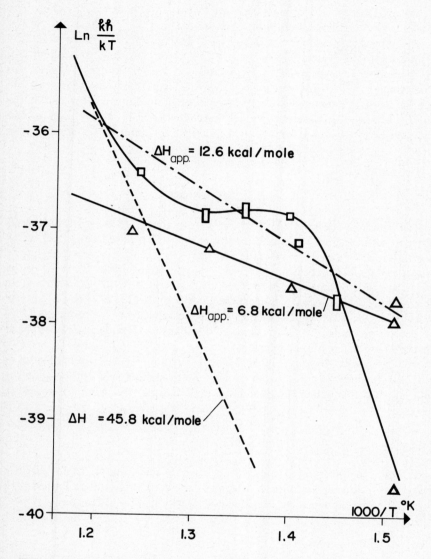

Fig. 7. Thermal desorption at constant temperature (enlarged portion of Eyring plot). First series of measurements: rectangles. Second series of measurements: triangles.

technique. Desorption rate measurements are expected to become possible over a much wider temperature range than was covered so far.

Preliminary calculations by Suhl (1970) have predicted a sharp peak of $\bar{r}$ around the Curie temperature of nickel (631°K). At this time, it would clearly be premature to attempt an interpretation of the differences between this prediction and the approximate behavior of $\bar{r}$ deduced from our experiments. On the one hand, our experimental reproducibility is not yet sufficient, and on the other, the nickel surface, possibly being contaminated by nickel oxide, would not necessarily have the same magnetic transition temperature as pure bulk nickel.

VII. CONCLUSION

The frequent assumption that the average transmission coefficient $\bar{r}$ is constant and equal to 1, and the usually narrow temperature range over which desorption rates can be measured, may lead to apparent values ΔH_{app} and ΔS_{app} very different from the true values, and related to them by a compensation relationship. It is possible to ascribe such effects to a temperature dependence of the average transmission coefficient $\bar{r}(T)$; as might occur, for instance, if the adsorbent undergoes some transition in that temperature range or, more generally, if energy-transfer mechanisms between adsorbent and adsorbate are temperature-dependent.

The experimental results reported here, concerning the desorption kinetics of a peculiar phase of hydrogen from nickel, tend to support Suhl's proposal on the role of spin fluctuations in the desorption of hydrogen from paramagnetic metals.

ACKNOWLEDGEMENT

It is a great pleasure to acknowledge the financial support of the Battelle Institute for this entire research program. I am deeply grateful to Professor H. Suhl and to my colleague Dr. E. Bergmann for very fruitful discussions.

REFERENCES

Carter, G. (1962). *Vacuum,* 12, 245.
Curtiss, C. F. (1948). Report CM-476, University of Wisconsin.
Daudel, R. (1967). "Théorie Quantique de la Réactivité Chimique". Gauthier-Villars, Paris.
Degras, D. A. (1967). *Nuovo Cimento, Suppl.* 5, (2), 420.
Ehrlich, G. (1961). *J. Appl. Phys.* 32, 1, 4.
Ehrlich, G. (1963). *Advan. Catal. Relat. Subj.* 14, 255.

Eyring, H., Walter, J. and Kimball, G. E. (1944). "Quantum Chemistry". Chap. 16. John Wiley, New York.

Glasstone, S., Laidler, K. J. and Eyring, H. (1941). "The Theory of Rate Processes". McGraw Hill, New York.

Hickmott, T. W., Ehrlich, G. (1958). *J. Phys. Chem. Solids,* **5**, 47.

Hirschfelder, J. O. and Wigner, E. (1939). *J. Chem. Phys.* **7**, 616.

Kramers, H. A. (1940). *Physica* **7**, 284.

Lapujoulade, J. (1967). *Nuovo Cimento, Suppl.* **5**, (2), 433.

Lichtman, D. (1965). *J. Vac. Sci. Technol.* **2**, 70.

Lichtman, D., Simon, F. N. and Kirst, T. R. (1968). *Surface Sci.* **9**, 325.

McCarroll, B. (1969). *J. Appl. Phys.* **40**, 1.

Montroll, E. W. and Shuler, K. E. (1958). *Advan. Chem. Phys.* **1**, 361.

Pétermann, L. A. and Baker, F. A. (1965). *Brit. J. Appl. Phys.* **16**, 487.

Pétermann, L. A. (1967). *Nuovo Cimento, Suppl.* **5**, (2), 364.

Prigogine, I. and Xhrouet, E. (1949). *Physica,* **15**, 913.

Prigogine, I. and Mahieu, M. (1950). *Physica,* **16**, 51.

Redhead, P. A. (1962). *Vacuum,* **12**, 203.

Suhl, H., Smith, J. H. and Kumar, P. (1970). *Phys. Rev. Lett.* **25**, 1442.

Suhl, H. (1970). Private communication.

Takayanagi, K. (1951). *Progr. Theor. Phys.* **6**, 486.

Zwolinski, B. J. and Eyring, H. (1947). *J. Amer. Chem. Soc.* **69**, 2702.

FIELD EMISSION AND FLASH DESORPTION STUDY OF ADSORPTION AND THERMAL DECOMPOSITION OF AMMONIA ON MOLYBDENUM

M. ABON, B. TARDY and S. J. TEICHNER

Department de Chimie-Physique, Institut de Recherches sur la Catalyse, C.N.R.S., Villeurbanne, France

I. INTRODUCTION

The studies of the interaction between ammonia and metallic surfaces, and in particular the kinetic study of the catalytic decomposition of this gas, have usually been performed under pressures higher than 0.1 Torr (Bond, 1962; Tamaru, 1964). Lacking an agreement on the reaction mechanism, a few authors have recently undertaken the study of the interaction between ammonia at low pressures and the ultra-high vacuum decontaminated surface of tungsten; in some cases crystallographically oriented. Many of the results still do not agree.

In the second series of experiments Dawson and Hansen (1968) found by field-emission microscopy that chemisorption of ammonia at 200°K proceeds without dissociation and that it decreases the work function Φ of tungsten. However between 200 and 400°K a progressive dissociation into hydrogen and nitrogen was recorded. The rate-determining step above 1200°K would be the decomposition of a surface complex of the type W_2NNH_2. When the definite crystallographic planes (100) (Estrup and Anderson, 1968) and (211) (May *et al.*, 1969) are considered the adsorption of ammonia at room temperature would be non-dissociative. These experiments, performed by the LEED technique, confirmed the decrease (−1 eV) of Φ of tungsten due to the chemisorption of undissociated ammonia.

Now concerning thermal desorption of ammonia from a polycrystalline filament of tungsten, Matsushita and Hansen (1969) found on their desorption curves one peak at low temperature due to hydrogen and a second peak at higher temperature due to nitrogen. More recently Matsushita and Hansen (1970) have found a second peak of x-nitrogen at a temperature slightly lower than that recorded previously for $\beta-N_2$ peak. The rate determining step in the decomposition of ammonia would be the desorption of $x-N_2$. An extended interaction between ammonia and tungsten at 300°K leads to the ratio of 0.65

245

$NH_{1.5}$ adsorbed species per every tungsten atom on the surface. This result is explained, in agreement with previous experiments (Dawson and Hansen, 1968), by a partial decomposition of ammonia during its adsorption at $300°K$.

The thermal desorption work on a single crystal of tungsten, done on the face (100) by Estrup and Anderson (1968), on which there is no dissociation of ammonia at room temperature, shows two peaks of partial pressure of hydrogen; termed A for low temperature and B for high temperature. In the same way two peaks of partial pressure of nitrogen are recorded (mass 14). The first small peak, at the beginning of the thermal desorption, is attributed to the desorption of undecomposed NH_3. The second much higher peak is recorded for the same temperature that the peak B of hydrogen. These results are interpreted by the following scheme:

Intermediate temperatures:

$$NH_{3(ads)} \longrightarrow NH_{3(g)}$$
$$NH_{3(ads)} \longrightarrow NH_{2(ads)} + H_{(ads)}$$
$$H_{(ads)} \longrightarrow \tfrac{1}{2}H_{2(g)} \quad \text{(peak A)}$$

High temperature:

$$NH_{2(ads)} \longrightarrow \tfrac{1}{2}N_{2(g)} + H_{2(g)} \text{(peak B)}$$

May et al. (1969) studied the decomposition of ammonia on face (211) of tungsten and their results were in agreement with the previous scheme, established for face (100).

Adsorption of ammonia on molybdenum was much less studied. Field-emission microscopy work of Ishizuka (1969) tends to show that ammonia is dissociated into hydrogen and nitrogen already at room temperature. Hydrogen is desorbed between 400 and $700°K$ whereas nitrogen forms with molybdenum a nitride which is decomposed above $900°K$.

The results described below on adsorption and desorption (or decomposition) of ammonia on molybdenum were performed by field-emission microscopy and flash filament desorption technique. These two methods produce complementary results.

II. FIELD EMISSION MICROSCOPY

The apparatus, made out of "Pyrex" has been already described (Abon and Teichner, 1966). Its essential parts are: the microscope itself, a vacuum rotary pump and oil diffusion pump together with a Biondi trap and a gas introduction device (metallic variable leak). A pressure of 10^{-10} Torr, measured with an ionization gauge, is obtained into the microscope after baking at $620°K$. The molybdenum tip is then decontaminated by heating at high temperature (of the order of $2100°K$). The purity of ammonia is of 99.9%. In order to avoid the difficulties connected with the use of the ionization gauge (decomposition of ammonia into hydrogen and nitrogen on the tungsten filament, formation of carbon monoxide) the gauge was not used during the experiments. The variations of the mean work function Φ of molybdenum, due to adsorption of ammonia, were calculated from the Fowler-Nordheim equation (Gomer, 1961). All the emission patterns were recorded on a Polaroid film (3000 ASA) with the same emission current (1.5×10^{-7} A) the tip being maintained at room temperature.

ADSORPTION OF AMMONIA

The adsorption of ammonia on the decontaminated molybdenum tip was followed as a function of contact time with this gas, maintained during this process at a constant pressure (5×10^{-8} Torr) at room temperature. Figure 1, on

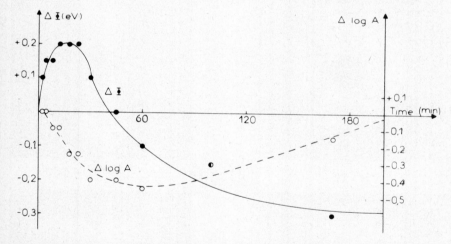

Fig. 1. Modification of Φ and of $\log A$ with time by adsorption of ammonia on molybdenum (p_{NH_3} =5 x 10^{-8} Torr, T = $300°K$).

which $\Delta\Phi$ is plotted as a function of time (in minutes), shows that $\Delta\Phi$ exhibits first a positive maximum (+ 0.2 eV), then decreases and remains negative. At the same time $\log A$ (the parameter of Fowler-Nordheim equation) varies only very slightly. Figure 2 shows the parallel modifications of the emission pattern during this adsorption. The pattern A, which was taken 1 minute after decontamination, is characteristic of a still clean surface of molybdenum.

The decrease of Φ observed in the second part of adsorption (figure 1), which tends to a value of $\Delta\Phi = -0.4$ eV after 24 hours, cannot be explained by a

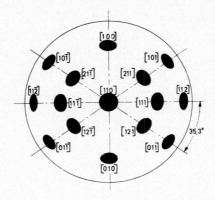

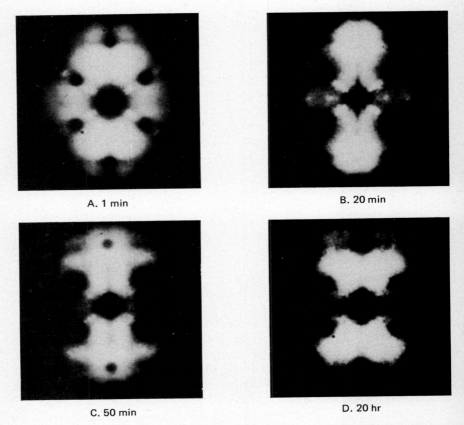

A. 1 min

B. 20 min

C. 50 min

D. 20 hr

Fig. 2. Electron-field emission patterns during the adsorption of ammonia.

complete dissociation (into adsorbed H_2 and N_2) of adsorbed ammonia, because adsorption of hydrogen as well as of nitrogen, at saturation (18 hours), increases the work function of molybdenum (Abon and Teichner, 1967).

In the case of tungsten, when non-dissociative adsorption of ammonia was postulated (Estrup and Anderson, 1968; May et al., 1969) the decrease of the work function was higher than that recorded here for molybdenum. However Dawson and Hansen (1968) and Dawson and Peng (1970) consider that on tungsten at room-temperature adsorption of ammonia, may result in a decomposition of a small fraction of it, mainly on regions of low value of Φ. The mean work function of molybdenum is smaller by 0.3 eV than the mean work function of tungsten. The hypothesis of previous authors assumes the possibility for ammonia to partially decompose, depending on the value of Φ of the corresponding region. The decomposition of ammonia on molybdenum then should be more pronounced than on tungsten. This process would be favoured during the first minutes of adsorption, when the tip, after decontamination at high temperature, has not yet cooled to the ambient temperature. This hypothesis would explain (i) the relatively small decrease of Φ at saturation ($\Delta\Phi = -0.4$ eV), (ii) the positive maximum of Φ in the first stage of adsorption (figure 1). This maximum would be due to the presence of adsorbed hydrogen. Indeed, if nitrogen was the only species adsorbed during the first stage the work function would decrease, as it was shown previously (Abon and Teichner, 1967). In the second stage, the decrease of Φ would be due to the adsorption of undecomposed ammonia which, according to Estrup and Anderson (1968) as well as May et al. (1969) is competitively adsorbed with hydrogen and displaces this species.

The modification of the pattern of figure 2 suggests that non-dissociative adsorption of ammonia mainly occurs on regions surrounding planes (100). Indeed, these regions which include in particular planes (210), (310), (610), (311) and (611) are highly emissive on the pattern D (saturation by ammonia) which is normal behaviour when a decrease of the work function is registered. The initial increase of the emission (pattern B) on planes (100) would be due to the complete lack of dissociation of ammonia on these planes (Dawson and Peng, 1970). If these planes look, in the next steps (patterns C and D), less emissive it is because the work function decreases more, by adsorption of ammonia, on surrounding planes than on planes (100) themselves.

The log A parameter of the Fowler-Nordheim equation is proportional to the emitting area of the tip. Emission patterns of figure 2 show that this area varies only slightly. Indeed, the minimum of log A of figure 1 seems to correlate with the minimum of the emitting area, recorded for pattern B of figure 2.

THERMAL DESORPTION AND DECOMPOSITION OF AMMONIA

After saturation of the tip under 5×10^{-8} Torr of ammonia (pattern D) the variable leak is closed and the tip is heated during one minute at successively increasing temperatures. The tip is then cooled down to room temperature in order to measure $\Delta\Phi$ and observe the pattern. Figure 3 represents $\Delta\Phi$ and $\Delta \log A$ as a function of the temperature of the tip during heating. The original value of Φ is gradually recovered between 620° and 800°K. However $\Delta \log A$ shows a minimum around 800°K, afterwards recovering its initial value (at 1300°K).

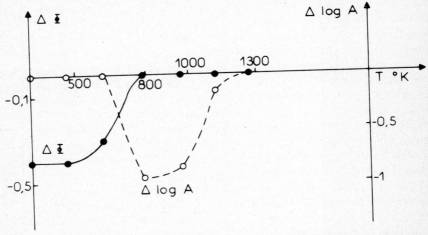

Fig. 3. Modification of Φ and of $\log A$ as the function of the temperature of the tip precovered by ammonia.

Despite that $\Delta\Phi = 0$ already at 800°K, the corresponding pattern (B, figure 4) shows that the tip is still covered by some adsorbed species. This behaviour must be correlated with the corresponding behaviour of tungsten saturated by ammonia at room temperature (Dawson and Hansen, 1968; May et al., 1969). In this case also $\Delta\Phi = 0$ after thermal desorption at 800°K. It should be also pointed out that $\Delta\Phi$ does not exhibit any positive maximum like that, observed on figure 1, during the adsorption. We previously suggested that this maximum is due to some decomposition of ammonia into hydrogen which remains adsorbed at room temperature. On the other hand, flash desorption experiments, described below, show that adsorbed hydrogen is entirely desorbed around 800°K. The departure of ammonia and of hydrogen during thermal desorption up to 800°K (figure 3) may explain the increase of Φ (figure 3).

Finally, despite the recovery of the initial value of Φ at 800°K, the pattern B of figure 4 and $\log A$ of figure 3 are not those of an entirely decontaminated tip. If the minimum value of $\log A$ at 800°K (figure 3) is connected with a minimum

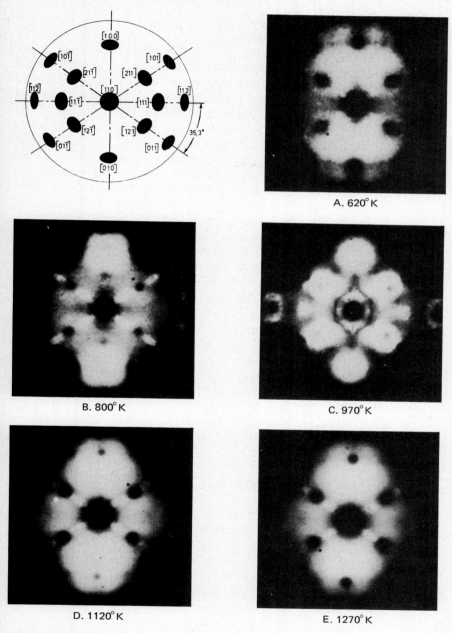

A. 620° K

B. 800° K

C. 970° K

D. 1120° K

E. 1270° K

Fig. 4. Electron-field emission patterns after the heating of the tip at increasing temperatures.

value of the emissive area, this should appear on patterns of figure 4, which is not the case. However, it was suggested by Holscher (1967) that a decrease of log A may also result from a rearrangement of metal surface atoms. It was, indeed, observed by Estrup and Anderson (1968) by electron diffraction, that after adsorption of ammonia at room temperature on plane (100) of tungsten and heating thereafter at 800°K, the pattern is modified. The same observation was also made by May *et al.* (1969) on planes (211) of tungsten. These modifications were interpreted as resulting from a peculiar bi-dimensional periodic structure of remaining adsorbed NH_2 radicals. But this modification could perhaps also result from some rearrangement of metal surface atoms, because of the interaction (at temperatures around 800°K) between metal and adsorbate. The minimum of log A (figure 3) would then be correlated with a surface modification of metal atoms and the emission patterns B and C with the presence of some adsorbed species. Indeed, it is shown below by flash desorption experiments, that nitrogen and hydrogen are desorbed (not as ammonia) at temperatures between 800° and 1300°K. The initial value of Φ, which is restored after heating of the tip at 800°K ($\Delta\Phi$ = o) and does not change any more with further heating, could therefore result from the formation on the surface during thermal desorption up to 800°K of some species of the type NH_x ($x < 3$) decreasing the value of Φ and eventually of the remaining nitrogen (not combined with hydrogen) increasing the value of Φ. But flash desorption experiments give evidence that nitrogen which is desorbed at 1300°K comes from the decomposition of some intermediate NH_x species and not from eventually adsorbed nitrogen atoms or molecules. It is therefore necessary to assume that any NH_x-adsorbed species (with $x < 3$, because these species cannot be ammonia) do not influence the value of the work function. It is only at 1300°K that all initial conditions of a decontaminated tip are restored (figure 3 and pattern E of figure 4).

III. FLASH DESORPTION EXPERIMENTS

The molybdenum filament, of 17 cm length and 0.11 mm diameter is soldered to two tungsten leads in a pyrex flask of 0.5 l capacity. An ionization gauge and essentially a quadrupole gas analyser monitor the pressure changes during flashing. Ammonia is introduced through a metallic variable leak after an initial baking and evacuation with an ionic pump down to 4×10^{-9} Torr. The residual gases are then H_2 (95%), CO (3%) and CH_4 (1%). This composition is in relation with the presence of parts made out of stainless steel (Redhead *et al.*, 1968). After outgassing of the filament in vacuum, an ammonia flow at constant pressure at room temperature is established in order to realize well-defined conditions of adsorption according to the method already described (Tardy and

Teichner, 1970). The filament is then heated at 1500°K during 30 sec before the adsorption experiments start. Then after a given contact time t_a with ammonia, the filament is flashed and the variations of the partial pressure are recorded with quadrupole analyser for a given mass number. These experiments are repeated for all required mass numbers. The temperature of the filament is determined from the voltage and intensity curves of the filament current during the temperature increase which takes 3 seconds. Three minutes are then required to cool the filament to room temperature. The use of the ionization gauge, which favours the dissociation of NH_3, is excluded during desorption experiments.

DESORPTION CURVES

The partial pressures monitored during flashing are p_{NH_3} (M = 17), p_{H_2} (M = 2) and p_{N_2} (M = 14). For p_{N_2} the mass 14 is preferred to the mass 28 because of the presence of some CO. For a contact time with ammonia, t_a = 3 min (the gas atmosphere has then a composition of 30% of hydrogen for a total pressure of 2.7 x 10^{-7} Torr) the flash desorption curves, recorded all with the same

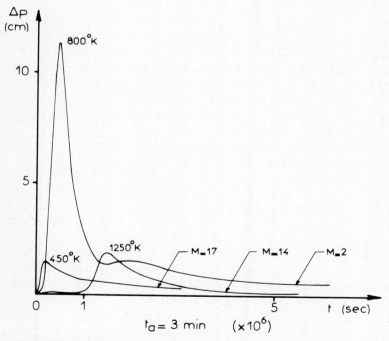

Fig. 5. Variation of p_{H_2} p_{NH_3} and p_{N_2} during the flash desorption (table 1, experiment 2).

sensitivity, are given on figure 5. The partial pressure of hydrogen (M = 2) gives an important peak at 800°K. A much smaller peak follows at 1400°K. It will be shown below that this peak may be considerably increased by increasing the pressure of ammonia for the same contact time. Only one desorption peak of ammonia (M = 17) is recorded at low temperature (450°K), below that of the first hydrogen peak. Finally nitrogen (M = 14) gives a very noticeable peak only at 1250°K.

These results lead to the conclusion that ammonia is desorbed at very low temperatures (450°K) and that hydrogen previously adsorbed from the gas phase and perhaps also coming from some decomposition of ammonia, is desorbed at 800°K. Therefore all the adsorbed species which still remain after this temperature may be no more in the form of ammonia and of hydrogen, atomic or molecular. The problem is therefore to assign the nitrogen peak at 1250°K and the second peak of hydrogen at 1400° to the decomposition of some nitrogen- and hydrogen-containing species of the type NH_x. The previous hypothesis was checked in the following way.

DESORPTION CURVES FOR VARIABLE PRESSURE OF AMMONIA

In table 1 are given the results of the flash desorption experiments which follow the adsorption of ammonia during t_a = 3 min, at increasing pressures. Five experiments were performed in this way (numbered from 0 to 4, number 0 corresponding to the experiment made with the residual gas pressure, which is mainly hydrogen, in the absence of ammonia). For the three masses (M = 17, 2 and 14) are given in arbitrary units the heights ΔP of the main peaks (recorded respectively at 450°K, 800°K and 1250°K) and the composition in the same units of the gas phase, previous to the flashing.

The amount of ammonia (M = 17) desorbed by flashing at 450°K increases with its pressure during adsorption. The desorption curves have the same shape as the curve on figure 5. In the same way, the amount of nitrogen (M = 14), desorbed at 1250°K, increases with ammonia pressure and again the shapes of the desorption curves are the same as in figure 5. But the hydrogen peak (M = 2) registered at 800°K is almost constant for all pressures of ammonia during adsorption. This means that this peak is mainly due to adsorption of hydrogen initially present in the gas phase. However, the slight second peak of hydrogen (~ 1400°K) in figure 5 increases with ammonia pressure (experiments 1 to 4, figure 6) whereas this peak is not recorded at all, in the absence of ammonia in the gas phase during adsorption (experiment 0).

It can therefore be assumed that the second desorption peak of hydrogen is not connected with the previous presence of molecular hydrogen in the gas phase, but only to the presence of ammonia. In this case it is possible to increase the height of the second peak of hydrogen and also of nitrogen (1250°K) by

Table 1

Mass number and ($T°K$)	0 $p_{H_2} =$ 5.2 x 10^{-8} Torr		1 P total = 1.6 x 10^{-7} Torr p_{H_2}/p_{NH_3} = 1.1		2 P total = 2.7 x 10^{-7} Torr p_{H_2}/p_{NH_3} = 0.52		3 P total = 1 x 10^{-6} Torr p_{H_2}/p_{NH_3} = 0.91		4 P total = 1.3 x 10^{-6} Torr p_{H_2}/p_{NH_3} = 1.1	
	Comp.	Δp	Comp.	Δp	Comp.	Δp	Comp.	Δp	Comp.	Δp
17 (450°)	0.004	0.004	1.9	0.8	6.95	1.45	16.5	2.5	18	2.5
2 (800°)	0.45	8.65	0.6	11.6	1.05	11.35	4.35	11.5	5.85	10
14 (1250°)	0.003	0.02	0.05	1.15	0.25	1.85	0.55	2.1	0.75	2.15

enrichment of the surface into adsorbed species, which by decomposition at high temperature give out N_2 and H_2. These species are not adsorbed ammonia because this compound is desorbed at fairly low temperatures (450°K). For this reason the flash desorption experiments were performed in the cumulative way in order to increase the concentration on the surface of molybdenum of species giving out nitrogen and hydrogen at high temperatures. After a contact time t_a with ammonia at room temperature the filament was flashed to $\sim 800°K$ in order to desorb hydrogen corresponding to the first peak, and also ammonia. But hydrogen corresponding to the second peak, as well as nitrogen remain adsorbed. A second adsorption of ammonia at the same pressure during t_a at

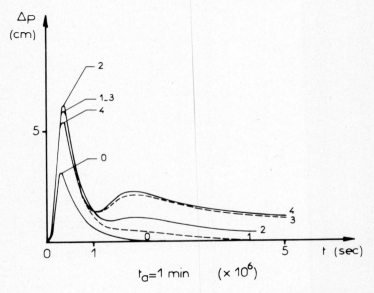

Fig. 6. Variation of p_{H_2} for increasing pressures of presorbed ammonia.

room temperature and a second flash desorption can only increase the amount of adsorbed species giving high temperature hydrogen and nitrogen peaks. The cycle is repeated a few times but the last flashing is carried out in the normal way, up to the high temperature ($\sim 1500°K$). The cumulative t_a is equal to 8 min 45 sec. On the other hand, for comparison purposes, an ordinary flashing experiment is made, but t_a of adsorption of ammonia is also made equal to 8 min 45 sec. This experiment, without intermediate heating at 800°K would give the picture of the presence, without enrichment, of species with hydrogen and nitrogen peaks at high temperature. Figure 7 describes flashings corresponding to masses 17, 2 and 14, with and without intermediate heatings.

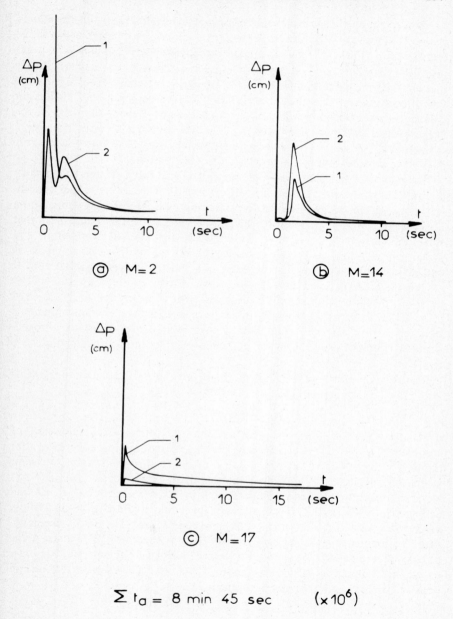

$\Sigma \, t_a = 8 \; \text{min} \; 45 \; \text{sec} \qquad (\times 10^6)$

Fig. 7. Variation of p_{H_2} (curves a), p_{N_2} (curves b) and p_{NH_3} (curves c) (1) without intermediate flashing, (2) after intermediate flashings at 800°K.

Then for mass 2 (figure 7a), after intermediate heatings (curve 2), the second peak of hydrogen is very much increased in comparison to the peak observed for direct flashing (curve 1) after the same contact time with ammonia. The first peak of hydrogen for curve 2 really represents a contact time with H_2 $t_a = 1$ min, whereas the same peak for curve 1 represents $t_a = 8$ min 45 sec. In the same way a very definite increase is observed for nitrogen desorbed at $1250°K$ after cumulative intermediate heating at $800°K$ (figure 7b). But for ammonia (figure 7c), which is desorbed after each intermediate heating, the peak of curve 1 represents $t_a = 8$ min 45 sec whereas the peak of curve 2 corresponds to $t_a = 1$ min of the last adsorption of ammonia before final flashing.

These experiments show that the cumulative adsorption of ammonia, and its desorption mainly without decomposition ($800°K$), increase the surface concentration of species which gives out nitrogen and hydrogen at high temperature. This species would be of the form NH_x ($x \leqslant 3$). If $x = 3$, this would mean that some ammonia remains adsorbed after $800°K$ and then decomposes at still higher temperatures, giving out H_2 and N_2. But the presence of ammonia at $800°K$ was ruled out by field-emission experiments. Ammonia after $800°K$ remains in a decomposed state, in the form of hydrogen and nitrogen species desorbed at higher temperatures. Indeed, no peak of desorption of hydrogen at $T > 800°K$ is recorded when hydrogen is adsorbed alone, without ammonia (figure 6, curve 0). This rules out by symmetry the presence of nitrogen-adsorbed species coming from the complete decomposition of ammonia above $800°K$.

The decomposition of ammonia during heating molybdenum up to $1500°K$, is not a one-step process of the type

$$2NH_{3(ads)} \longrightarrow N_{2(g)} + 3H_{2(g)}$$

This decomposition must proceed through the intermediate formation of adsorbed NH_x ($x < 3$) species. Because the first peak of hydrogen increases after the first introduction of ammonia (table 1) and then remains almost constant when the pressure of ammonia is increased, it can be concluded that this increased desorption of hydrogen is due to partial decomposition of adsorbed ammonia into NH_x and H: this first species remaining adsorbed, according to the model of Estrup and Anderson (1968) given previously. The partial decomposition of ammonia is liable to occur at room temperature, at least on some planes, according to Dawson nd Hansen (1968). The species NH_x are decomposed finally at high temperature ($1200-1400°K$) giving out the second peak of hydrogen and the peak of nitrogen.

IV. CONCLUSIONS

The interaction of ammonia with metallic surfaces and in particular with the tungsten surface, has been studied in order to elucidate the mechanism of the adsorption and of the thermal decomposition of ammonia. These are the necessary steps before a study of the synthesis of ammonia may be undertaken.

However, even a relatively simple process of adsorption and thermal decomposition of ammonia has been described differently by some authors. Thus, ammonia may be adsorbed and desorbed on molybdenum at moderate temperatures without decomposition. This behaviour has been envisaged though not directly demonstrated as in this work. The most conflicting results are related to the mechanism of the decomposition itself. According to some authors,ammonia may be adsorbed at least at room temperature, mainly without decomposition. However the thermal desorption produces the departure of a fraction of unchanged ammonia, as particularly shown in this work, and of fragments which are the result of its decomposition. For some authors decomposition occurs in one step, with the formation of elementary N and H adsorbed species, while for others the process involves the intermediate formation of hydrogen deficient species of the type NH_x $(x < 3)$. This work shows that the second mechanism is that found on molybdenum. Indeed, the thermal desorption of ammonia as well as the field-emission measurements show that below $800°K$ the ammonia may be desorbed without decomposition, but the fragments which remain above this temperature must be of the type NH_x and not in the form of elementary adsorbed N and H species which finally desorb at 1200-$1400°K$.

REFERENCES

Abon, M. and Teichner, S. J. (1966). *J. Chim. Phys.* 2, 272.
Abon, M. and Teichner, S. J. (1967). *Nuovo Cimento, Suppl.* 5, 521.
Bond, G. C. (1962). "Catalysis by Metals", p. 371 Academic Press, New York and London.
Dawson, P. T. and Hansen, R. S. (1968). *J. Chem. Phys.* 48, 623.
Dawson, P. T. and Peng, Y. K. (1970). *J. Chem. Phys.* 52, 1014.
Estrup, P. J. and Anderson, J. (1968). *J. Chem. Phys.* 49, 523.
Gomer, R. (1961). "Field Emission and Field Ionisation", Harvard University Press, Cambridge, Massachusetts.
Holscher, A. A. (1967). Thesis (Leiden).
Ishizuka, K. (1969). *Shokubai.* 11, 212.
Matsushita, K. and Hansen, R. S. (1969). *J. Chem. Phys.* 51, 472.
Matsushita, K. and Hansen, R. S. (1970). *J. Chem. Phys.* 52, 4877.
May, J. W., Szostak, R. J. and Germer, L. H. (1969). *Surface Sci.* 15, 37.
Redhead, P. A., Hobson, J. P. and Kornelsen, E. V. (1968). "The Physical Basis of Ultrahigh Vacuum", Chapman and Hall, London.
Tamaru, K. (1964). *Advan. Cataly Relat. Subi.,* 15, 65.
Tardy, B. and Teichner, S. J. (1970). *J. Chim. Phys.* 67, 1962.

ADSORPTION OF H₂, CO, N₂, AND O₂ ON TUNGSTEN FIELD EMITTERS: IMAGING OF ADSORPTION LAYERS BY CHANNEL PLATE FIELD EMISSION MICROSCOPES

W. A. SCHMIDT and O. FRANK

Fritz-Haber-Institut der
Max-Planck-Gesellschaft, Berlin-Dahlem, Germany

I. INTRODUCTION

Much work has been done to observe adsorbed gases in the field-ion microscope (Müller, 1960; Müller and Tsong, 1969; Ehrlich and Hudda, 1960; 1962; Ehrlich, 1963; Holscher, 1967). Normally in these studies a field of the order of 4-6 V/Å was required with helium as the image gas. However, under these conditions, the adsorbed entities often became unstable during image formation, so that it is questionable whether the spots in the resulting field-ion micrographs represent the adsorbed species themselves, or the perturbed surface atoms after the adsorbates had been field desorbed.

The situation is complicated by the fact that in field-ion microscopy, even with inert gases, an adsorption-free surface is non-existent (Tsong and Müller, 1970). As they point out each of the more protruding surface atoms, having a field high enough to cause ionization, will be covered by a gas atom.

In the present work a channel plate has been used to intensify field-electron and field-ion images. The high gain of this plate makes it possible to use considerably reduced fields and low gas pressure. Also, surface imaging is possible with any gas, because the conversion from ions into electrons in the channel plate, greatly enhances the efficiency of light emission at the fluorescent screen. Therefore, gases with low ionization energies can now be used which gives rise to a further field reduction. Such gases were unsuitable in the FIM because of their heavy masses. These facts point to a more successful study of adsorption layers on field emitter surfaces.

The work further deals with the ionization probability at gas-covered metal surfaces. It will be shown that adsorption layers can either enhance or suppress field ionization probabilities of the image gas.

261

II. EXPERIMENTAL

Images were intensified by a Mullard one inch diameter plate with 40 μm channels. The adsorption of H_2, CO, N_2, and O_2 was established either at 77°K or at 300°K on thermally annealed clean tungsten emitters. Adsorption was stable after gas exposures at 10^{-6} Torr pressure lasting 1 to 10 min. At 10^{-6} Torr pressure and 77°K tip temperature field-ion images of these layers were taken using either krypton or the adsorbed gases themselves.

Field emission of electrons was used to monitor the adsorbed layer both before and after imaging surfaces by ions. These patterns and also the measured Fowler-Nordheim work function remained unchanged, provided that a certain limit of the ionization field strength had not been exceeded. Resulting ion emission currents were in the region of 10^{-12} A.

Mass spectra of flash-desorbed adsorption layers indicated contamination by reactive residual gases, mainly like H_2, CO, and CO_2. A representative mass spectrum yields the following partial pressures: p (H_2) = 6 x 10^{-10} Torr, p (CO) = 2 x 10^{-10} Torr, and p (CO_2) = 1 x 10^{-11} Torr. Evidently, the hydrogen layers are contaminated by CO and the carbon monoxide layers by H_2. The nitrogen layers contain H_2 and CO. The highest degree of cleanliness is supposed in the case of oxygen adsorption. Flash desorption data from oxygen layers are uncertain owing to the use of hot filaments.

III. EXPERIMENTAL RESULTS AND DISCUSSION

FIM PATTERNS OF GAS-COVERED SURFACES

Figures 1-4 show field-electron and field-ion micrographs for H_2, CO, N_2, and O_2 on tungsten. The electron emission patterns before and after ionization are identical except for the minor deviations in figures 3(d) and 3(f) which are probably due to residual gas co-adsorption. This proves that neither the ionization field nor the imaging gas have had an influence on the established adsorption structure.

These ion images will be compared with those of the clean surface, though such a comparison is difficult because the clean surface is covered with a field-induced image inert gas layer as stated by Tsong and Müller (1970). This statement is born out in figure 5. The field-electron pattern figure 5(c), taken after krypton ion imaging of the initially clean surface, is that of an inert-gas-covered surface (Ehrlich and Hudda, 1959). Despite this, it is assumed that the ion image of figure 5(b) still represents the tungsten atom surface structure. The ion images of figures 1-4 demonstrate that changes in surface structure are owing to adsorption.

Ion micrographs of gas-covered surfaces (figures 1-4) are characterized by a number of bright spots more or less statistically concentrated on special surface areas. Generally these spots exceed those of the tungsten atoms in size and brightness. There are no significant differences between ion patterns obtained with krypton or with the adsorbed gases themselves.

IONIZATION PROBABILITY AT GAS-COVERED SURFACES

Following the simplest model for field ionization, the ionization probability D according to WKB calculations is proportional to $\exp\left[-A(I - \Phi)I^{1/2} F^{-1} \times f(F, I)\right]$, where I is ionization energy, Φ the work function, and F the field strength.

This theoretical model requires that surface areas with low work function (bright in the FEM) are dark in the FIM and those with high work function (dark in the FEM) are bright in the FIM. Some deviations from the expected behaviour occur in most cases. To decide whether the work function alone is an adequate physical parameter to describe the field ionization phenomenon the following considerations are made: $(I - \Phi)I^{1/2}$ versus field ionization voltage V (proportional to F) should be a straight line for constant ion current (in our case 9×10^{-12} A). The I values are well-known and the Φ values of the clean and of the gas-covered tungsten surfaces were calculated from Fowler-Nordheim plots. Figure 6 shows the results obtained.

The values for the noble gases Xe, Kr, and Ar, ionized at the clean surface, satisfy the expected relation only if the Φ value of the clean tungsten surface (4.5 eV) is used. Some arguments can be given to support and to justify the assumption of a clean surface Φ value. First, the various values of Φ for noble gas adsorption from the work of Ehrlich and Hudda (1959) although different do not contradict the assumed 4.5 eV value used here. The emitting areas in the case of the Fowler-Nordheim evaluation (areas around the {100} poles) are different to the areas emitting ions (around {110} and {111}) as seen in figure 5(b) and 5(c). Second, the same authors stated that adsorption is weakest for the ion-emitting regions. Therefore the Φ value valid during ionization must be closest to that of the clean surface.

Probing the covered surfaces with krypton, H_2, CO, N_2, and O_2 shows promotion, suppression or the expected behaviour depending on the ion efficiencies of the ionized gases.

Typical examples are

(1) Nitrogen adsorbed at 77°K promotes Kr^+ formation, and even stronger N_2^+ formation.

(2) Hydrogen adsorbed at 77°K suppresses H_2^+ formation heavily, and when adsorbed at 300°K it promotes Kr^+ formation.

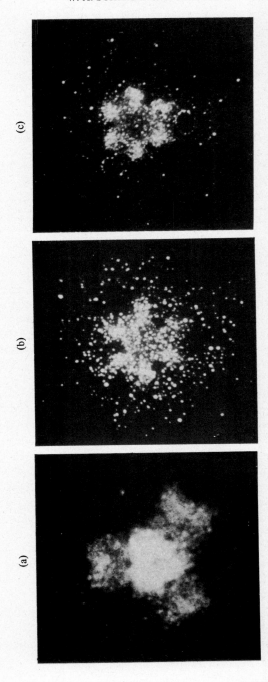

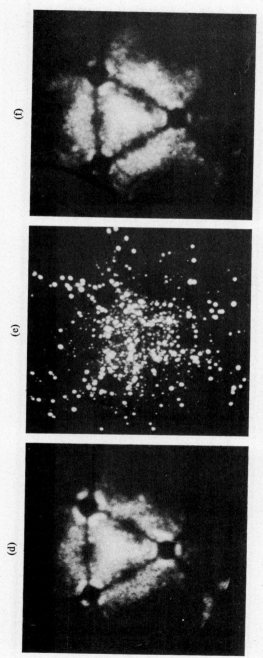

Fig. 1. Hydrogen adsorbed on tungsten, 3×10^{-6} Torr, 10 min, images taken at 77° K. (a) FEM: H_2 adsorbed at $T = 77^{\circ}$ K, (b) FIM: state (a) imaged with Kr, (c) FIM: state (a) imaged with H_2, (d) FEM: H_2 adsorbed at $T = 300^{\circ}$ K, (e) FIM: state (d) imaged with Kr, (f) FEM: state after (e).

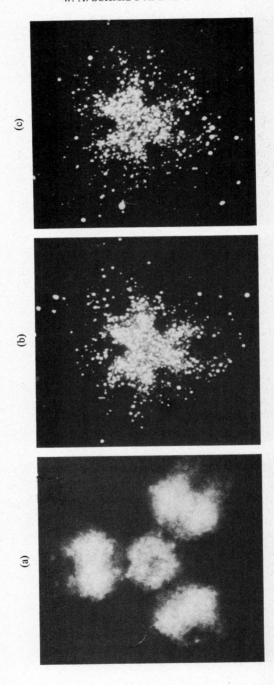

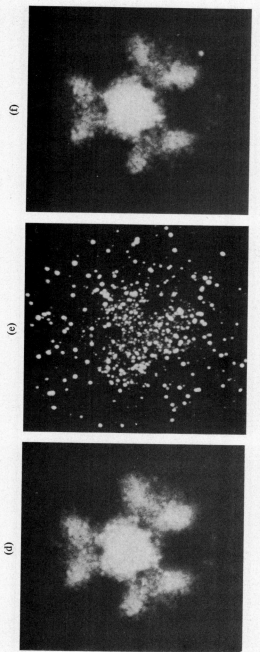

Fig. 2. Carbon monoxide adsorbed on tungsten, 3×10^{-6} Torr, 10 min, images taken at $77°K$. (a) FEM: CO adsorbed at $T = 77°K$, (b) FIM: state (a) imaged with Kr, (c) FIM: state (a) imaged with CO, (d) FEM: CO adsorbed at $T = 300°K$, (e) FIM: state (d) imaged with Kr, (f) FEM: state after (e).

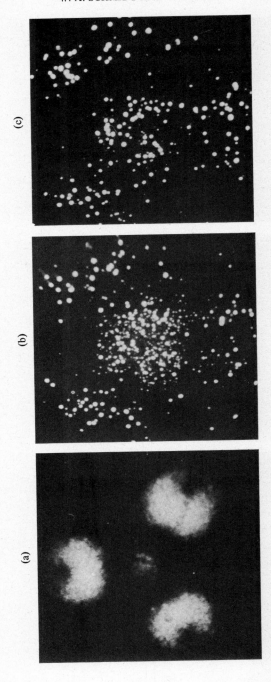

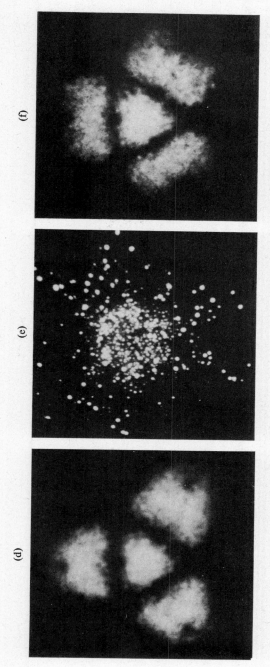

Fig. 3. Nitrogen adsorbed on tungsten, 3×10^{-6} Torr, 10 min, images taken at $77°$K. (a) FEM: adsorbed at $T = 77°$K, (b) FIM: state (a) imaged with Kr, (c) FIM: state (a) imaged with N_2, (d) FEM: N_2 adsorbed at $T = 300°$K, (e) FIM: state (d) imaged with Kr, (f) FEM: state after (e).

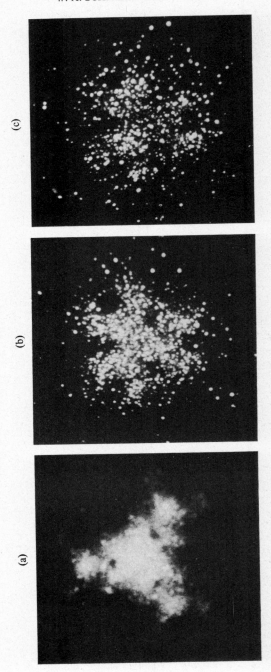

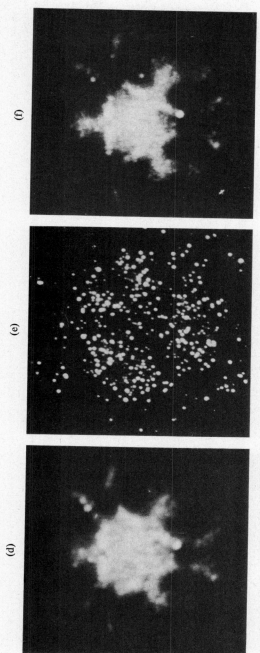

Fig. 4. Oxygen adsorbed on tungsten, 3×10^{-6} Torr, 10 min, images taken at $77°K$. (a) FEM: O_2 adsorbed at $T = 77°K$, (b) FIM: state (a) imaged with Kr, (c) FIM: state (a) imaged with O_2, (d) FEM: O_2 adsorbed at $T = 300°K$, (e) FIM: state (d) imaged with Kr, (f) FEM: state after (e).

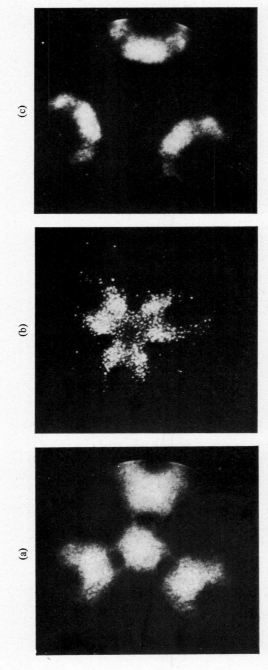

Fig. 5. Tungsten imaged at 77°K. (a) FEM: clean surface, (b) FIM: clean surface imaged with Kr, (c) FEM: state after (b).

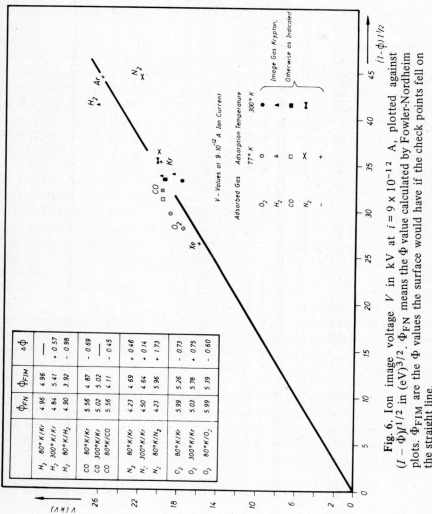

Fig. 6. Ion image voltage V in kV at $i = 9 \times 10^{-12}$ A, plotted against $(I - \Phi)^{1/2}$ in $(eV)^{3/2}$. Φ_{FN} means the Φ value calculated by Fowler-Nordheim plots. Φ_{FIM} are the Φ values the surface would have if the check points fell on the straight line.

	Φ_{FN}	Φ_{FIM}	$\Delta\Phi$
H_2 80° K/Kr	4.96	4.96	—
H_2 300° K/Kr	4.84	5.41	+ 0.57
H_2 80° K/H_2	4.90	3.92	− 0.98
CO 80° K/Kr	5.56	4.87	− 0.69
CO 300° K/Kr	5.02	5.02	—
CO 80° K/CO	5.56	4.11	− 0.45
N_2 80° K/Kr	4.23	4.69	+ 0.46
N_2 300° K/Kr	4.50	4.64	+ 0.14
N_2 80° K/N_2	4.23	5.96	+ 1.73
O_2 80° K/Kr	5.99	5.26	− 0.73
O_2 300° K/Kr	5.03	5.78	+ 0.75
O_2 80° K/O_2	5.99	5.39	− 0.60

(3) Carbon monoxide adsorbed at $77°K$ suppresses Kr^+ formation as well as CO^+ formation.

(4) Oxygen adsorbed at $77°K$ suppresses Kr^+ formation as well as O_2^+ formation, and at $300°K$ it promotes Kr^+ formation.

(5) Kr^+ formation is as expected for $W-N_2-300°K$, $W-H_2-77°K$, and $W-CO-300°K$.

An estimation of the experimental errors proves that the deviations from the expected straight line in figure 6 must be ascribed to real effects. Due to adsorption, different crystal planes show large and diverging variations in work function. Even within those planes with highest work function, where ionization should predominate, selected sites occur in the FIM. Consequently we have to postulate virtual $\Delta\Phi$ values with large local variations and which are even different in sign from the field electron Φ values. The mean value of these locally different Φ values determines the position of the check points in figure 6. Obviously during promoted field ionization, large and widespread spots are observed. If field ionization is suppressed the spots are more dense, smaller, and the patterns resemble more those of the clean tungsten. There is evidence that local spots are created by peculiar adsorption states which cause particular local charge densities and local field enhancement of locally diminished fields. A high polarizability of a strongly adsorbed molecule will enhance both negative and positive fields. A rigid, negative surface charge (permanent dipole due to surface bond) will only diminish the local ionization field. Then suppressed field ionization is observed, although Φ increases. A comparable model can be derived for permanent positive surface dipoles.

From the introduced model, the results may be summarized in two rules. First, with the exception of nitrogen, permanent negative dipoles occur for adsorption layers established at $77°K$. Second, adsorption layers established at $300°K$ show the influences of permanent positive dipoles and/or those of the induced dipoles.

The permanent positive dipoles found for $300°K$ adsorption are supposed to be a re-arranged surface and are in agreement with Holscher and Sachtler (1966). Any explanation of the permanent negative dipoles as concluded from low temperature layers is not yet possible. More detailed statements can only be expected if measurements are carried out on single crystal planes.

ACKNOWLEDGEMENT

The authors gratefully acknowledge valuable discussions with J. H. Block, they are indebted to the Deutsche Forschungsgemeinschaft for generous financial support of this work.

REFERENCES

Ehrlich, G. and Hudda, F. G. (1959). *J. Chem. Phys.* **30**, 493.
Ehrlich, G. and Hudda, F. G. (1960). *J. Chem. Phys.* **33**, 1253.
Ehrlich, G. and Hudda, F. G. (1962). *J. Chem. Phys.* **36**, 1233.
Ehrlich, G. (1963). *In* "Advances in Catalysis" (D. D. Eley, H. Pines and P. B. Weisz, eds.) Vol. **14**, p. 255. Academic Press, New York and London.
Holscher, A. A. and Sachtler, W. M. H. (1966). *Discuss. Faraday Soc.* **41**, 29.
Holscher, A. A. (1967). Thesis, Rijksuniversiteit, Leiden.
Müller, E. W. (1960). *Advan. Electron. Electron Phys.* **13**, 82.
Müller, E. W. and Tsong, T. T. (1969). "Field Ion Microscopy". American Elsevier Publishing Co., New York.
Tsong, T. T. and Müller, E. W. (1970). *Phys. Rev. Lett.* **25**, 911.

INFRARED REFLECTION SPECTRA AND SURFACE POTENTIALS OF CARBON MONOXIDE CHEMISORBED ON COPPER, SILVER AND GOLD

M. A. CHESTERS, J. PRITCHARD and M. L. SIMS

Queen Mary College, London, England

I. INTRODUCTION

Infrared spectra of carbon monoxide adsorbed on copper, silver, and gold have been observed by transmission through very thin evaporated films (Bradshaw and Pritchard, 1970). Anomalous dispersion distorted the sharp intense bands and obscured the coverage dependence of the spectra, but there were clear indications of two adsorption stages on copper and of a shift to lower frequency with increasing coverage on silver. A similar frequency shift has been found with CO on supported gold (Yates, 1969). The high frequency (2160 cm^{-1}) of the band on silver films could have been due to impurities such as oxygen. Incomplete reduction of supported gold yields a much higher frequency band than on the fully reduced surface (Yates, 1969), and in a recent study of CO adsorption on supported silver Keulks and Ravi (1970) found a band at 2180 cm^{-1} on a partially reduced surface but no band at room temperature on a fully reduced surface. Co-adsorption of CO and oxygen gave a band at 2162 cm^{-1}.

Good spectra of CO on copper films, undistorted by anomalous dispersion, have been obtained by multiple reflections at high angles of incidence (Bradshaw *et al.*, 1968). The application of this technique to the weak adsorptions on silver and gold at low temperatures involves the complication of providing surfaces which can be moved and cooled in vacuum. A fuller investigation of the multiple reflection system (Pritchard and Sims, 1970) indicated that useful spectra may be obtained from a single reflection in such favourable cases as CO on copper. As the infrared optical properties of silver and gold are similar to those of copper we have tried a simple single reflection method with polycrystalline films of all three metals.

The interpretation of infrared spectra must be related to the results of other measurements, such as surface potentials, adsorption heats, LEED. A broad correlation of available data for polycrystalline copper, silver and gold has been

277

made (Bradshaw and Pritchard, 1970), but there is a clear need for data on individual crystal planes. We have combined the single reflection method with surface potential measurements to study CO adsorption on Cu (100) and Cu (111) surfaces. A preliminary report of results on Cu (100) has appeared already (Chesters *et al.*, 1970).

II. EXPERIMENTAL

EVAPORATED FILMS

The infrared cell (figure 1a) was constructed from pyrex glass with sapphire windows sealed to the body with Araldite AT 1 epoxy resin. The glass re-entrant had a flat polished face, 20 mm x 30 mm, on to which the films (> 100 nm

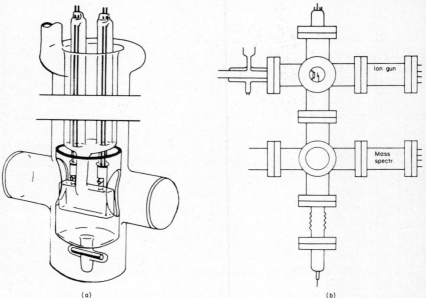

(a)   (b)

Fig. 1. (a) glass cell for evaporated films. (b) stainless steel cell for single crystals.

thick) were deposited from beads on two tungsten filaments. A rotatable glass shield screened the windows during evaporation. Tantalum foil shields on the evaporation sources prevented films being deposited on the upper part of the re-entrant where they would have interfered with the measurement of adsorbed amounts when the re-entrant was cooled. The vacuum system, spectrometer and recording system have been described (Pritchard and Sims, 1970). After baking overnight with the cell at $470°K$ the background pressure was 10^{-9} Torr and during film deposition the pressure did not exceed 5×10^{-8} Torr. The noise level of the computed spectra was ±0.1% of the reflected intensity. As several digital readings were taken within the resolution of the system further smoothing was carried out by eye.

Surface potentials and isosteric heats of adsorption were measured by the combined diode and vibrating capacitor method (Alexander et al., 1968).

SINGLE CRYSTALS

The stainless-steel chamber (figure 1b) consisted of two 38 mm six-way cross-pieces. The upper cross-piece carried two sapphire windows, an ion gauge, the crystal holder and an ion-bombardment gun. The crystal holder was a flange-mounted pyrex glass re-entrant with a flat stainless-steel end attached through a glass-Kovar seal. The copper crystals ($\sim$5 mm thick) were clipped to the stainless-steel plate with tungsten pins. A copper-constantan thermocouple monitored the crystal temperature. The crystal was cooled by filling the re-entrant with refrigerant, or heated by inserting a small electric heater into the re-entrant. The temperature range was $77°K$ to $\sim 600°K$.

The lower cross-piece connected the chamber via metal valves to an ion pump (AE1 P8), a getter bulb, and to a glass vacuum line which served both for the initial evacuation and for gas admissions. It also carried a small mass spectrometer (V.G. Micromass 1) and a bellows-mounted reference electrode reaching to the crystal for surface potential measurements by the vibrating capacitor method. The 5 mm square reference electrode, of raw tantalum foil, was inert to the gases being used. It could be moved away from the crystal during ion-bombardment, gas admission, and infrared measurements, or it could be adjusted close to the crystal to allow surface potential measurements to better than ±1 mV.

Copper single crystals were obtained from Metals Research Ltd., cut by spark erosion to within $1°$ of the desired orientation, smoothed by polishing with a saturated solution of cupric chloride in concentrated hydrochloric acid, and finally electropolished in 50% orthophosphoric acid. The (100) crystal was 10 mm square and the (111) crystal had an elliptical face 16 mm x 12 mm.

After baking at $570°K$ pressures of 5×10^{-10} Torr were reached. The crystals were cleaned by bombardment with xenon ions (10^{-6} A, 350 to 500 V) for several 30 minute periods until the contact potential difference with respect to the reference electrode remained constant. A constant surface potential for a xenon monolayer at $77°K$ was used as the final criterion of cleanness after annealing at $570°K$.

Infrared reflection measurements were made by focussing radiation from the monochromator on the leading edge of the crystal face to give one reflection at angles of incidence between 79 and 87°. Higher angles, as with the evaporated films, were not possible without sacrificing much of the radiation, because of the small size of the crystals. The crystal face could be given timed exposures to gas by controlling the metal valve to the gas admission line. The increase in CO coverage was monitored by both the infrared spectra and the more sensitive surface potential measurement in the same experiment.

Further studies of CO adsorption on the same Cu (100) crystal have been carried out by LEED and Auger electron spectroscopy using a Vacuum Generators L2A LEED system with 3-grid optics. The same cleaning procedures were employed, and surface potential measurements by the vibrating capacitor method were used as the common element between the two systems.

III. RESULTS

COPPER FILMS

The adsorption of CO at 77°K on a copper film deposited at room temperature led to the spectra in figure 2(a). As in the transmission spectra with thin films there were two distinct stages of adsorption. During the first stage the band at 2103 cm^{-1}, and the associated shoulder band, grew in intensity without any perceptible frequency shift. The peak absorbance was directly proportional to the amount of CO adsorbed. In the second stage the peak frequency moved to higher values by up to 6 cm^{-1}. These results are in good agreement with the earlier transmission and multiple reflection spectra, apart from the absence of the shoulder band in the transmission spectra. The shoulder band was also absent in the reflection spectrum from a *thin* copper film deposited and maintained at 77°K on a thick film previously deposited at room temperature. As figure 2(b) shows, the initial adsorption gave a sharp band at 2113 cm^{-1} which shifted to 2119 cm^{-1} during the second stage. After desorbing and annealing at room temperature the re-adsorption of CO at 77°K gave an ill-defined broad band centred near 2090 cm^{-1} (figure 2c), resembling the shoulder band observed with the other films.

The variability of the spectra, depending on the conditions of film deposition, suggests that the copper surfaces are heterogeneous and that structural effects are important. Even when the deposition conditions are similar the spectra may show significant differences. Figure 2(d) shows spectra at 77°K and room temperature for a film deposited at room temperature. Although the sharp peak attained nearly its maximum intensity at room temperature and 1 Torr the shoulder band was much less well developed, suggesting that it corresponds to a low heat of adsorption.

A similar variability in the surfaces of copper films has been revealed by surface potential measurements on polycrystalline films deposited on glass under ultrahigh vacuum conditions. Measurements by the retarding field diode or vibrating capacitor showed the two stages of adsorption (Pritchard, 1963) in which the surface potential first increases to a maximum positive value and then decreases. However, the positive maximum for different films varied between 0.29 and 0.45 V in a random manner. At 195°K only the positive first stage

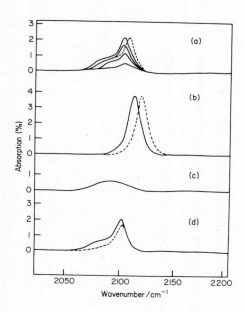

Fig. 2. Spectra of CO adsorbed on copper films: (a) at 77°K on film deposited at room temperature; (b) at 77°K on thin film deposited at 77°K; (c) at 77°K after annealing the thin film at room temperature; (d) at 77°K and at room temperature (broken line) on a film deposited at room temperature.

occurred and it was noted that higher pressures were required to reach the maximum surface potentials on those films which gave high values, suggesting that a high surface potential corresponds to a lower heat of adsorption. Figure 3 shows some results for a film giving a value of 0.38 V at 195°K. Isotherms of surface potential versus pressure were recorded with the vibrating capacitor at temperatures between 252.5 and 295.5°K. The corresponding isosteres at surface potentials between 50 and 300 mV are shown together with the heats of adsorption. There is a marked fall in the heat of adsorption from 67 to 50 kJ mol^{-1}.

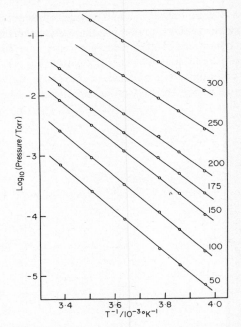

Fig. 3. CO adsorption isosteres on a copper film. The surface potential corresponding to each isostere is given in mV. Heats of adsorption varied from 67 kJ mol^{-1} at 50 mV to 50 kJ mol^{-1} at 300 mV.

SILVER FILMS

Silver films deposited at room temperature, or annealed at room temperature after deposition at 77°K, gave similar spectra after the adsorption of CO at 77°K. A weak broad band centred near 2155 cm^{-1} was observed (figure 4a) at pressures from 10^{-5} to 10^{-3} Torr. A much sharper and more intense band appeared on films deposited and maintained at 77°K (figure 4b). The high pressures reflect the weakness of the adsorption. The frequency shift to lower values with increasing coverage is quite striking, and extrapolation of the peak frequency to zero coverage (figure 5a) gives a value of about 2155 cm^{-1} for an isolated CO molecule, close to the peak frequency at saturation after annealing (figure 4b, broken line).

GOLD FILMS

The adsorption of CO at 77°K on a gold film deposited at room temperature also led to spectra (figure 6a) showing a frequency shifting to lower values with increasing coverage. Two more films, deposited on top of the original one, gave

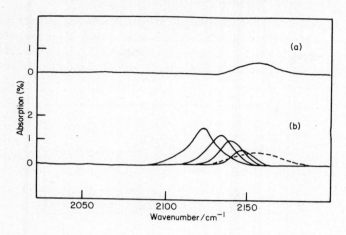

Fig. 4. Spectra of CO adsorbed on silver films at 77°K: (a) film deposited at room temperature, pressure $\sim 10^{-3}$ Torr; (b) film deposited and maintained at 77°K (full lines), pressures $< 5 \times 10^{-6}$, 5×10^{-5}, 2×10^{-4} and 1.5 Torr, and (broken line) at 10^3 Torr after annealing.

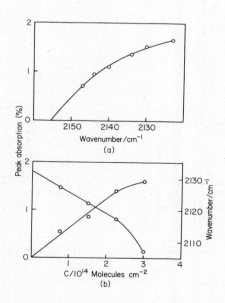

Fig. 5. Dependence of CO spectra on coverage: (a) silver film: (b) gold film.

spectra (figures 6b, c) differing slightly from the first set, but the frequency shift was common to them all. The coverage dependences of the peak adsorption and frequency are shown for the first film in figure 5(b). The other results were similar except for the rather marked changes on this film during the final stages

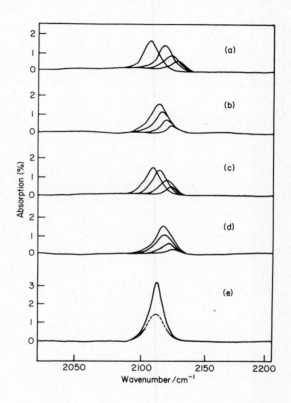

Fig. 6. Spectra of CO adsorbed on gold films: (a) (b) (c) at 77°K on three successive gold films deposited at room temperature; (d) at 195°K on the same film as (c) at pressures 1.0×10^{-4}, 1.2×10^{-3}, 2.5×10^{-2} and 1.5 Torr; (e) at 77°K on a thin film deposited and maintained at 77°K (full line) and after annealing (broken line).

of adsorption. At 195°K a similar but smaller frequency shift was observed (figure 6d).

A further *thin* gold film deposited and maintained at 77°K gave the sharp intense band shown in figure 6(e) after saturating with CO. Annealing at room temperature (broken line) simply reduced the intensity.

COPPER (100)

The xenon surface potential was consistently 0.45 V after several cycles of ion bombardment cleaning and annealing at 470°K. Exposing the crystal to CO at 77°K at a pressure of 2×10^{-8} Torr gave a positive surface potential initially, followed by a negative change (figure 7a). A rather higher maximum positive potential than that indicated in the figure could be attained by allowing the

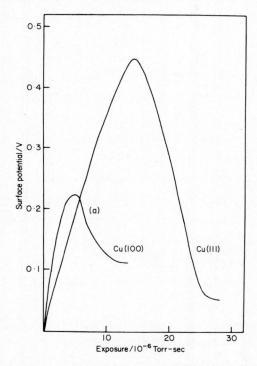

Fig. 7. Surface potential of CO on Cu (100) and Cu (111) single crystal surfaces.

crystal to warm slowly, in agreement with the observation that the second stage of weakly adsorbed CO is not taken up at 195°K. The highest value recorded was 0.25 V. Surface potential isotherms were measured between 250 and 300°K in order to determine the isosteric heat of adsorption. During these measurements some small irreversible changes of surface potential occurred which we suspect may be caused by reaction of CO with nickel in the chamber, leading to the transport of nickel as carbonyl to the copper surface. The effect was not eliminated by allowing the CO to stand in contact with freshly evaporated copper films before admission to the chamber, so it is unlikely to have been due to an impurity in the admitted gas. As a result the isosteric heat determination was not very accurate, a value of 59 ± 5 kJ mol^{-1} being derived at a surface potential of 0.05 V.

CO reflection spectra at 77°K from the surface produced by annealing at 470°K are shown in figure 8(a) for the first stage of adsorption and in 8(b) for the final saturated surface. There is an asymmetry in the bands which was much reduced when the crystal was annealed at 570°K (figure 8c and d). Although

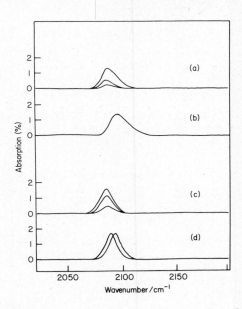

Fig. 8. Spectra of CO on Cu (100) surface at 77°K; (a) growth of first stage, and (b) the second stage, on crystal annealed at 470°K; (c) after annealing at 570°K, at surface potentials of 116, 185 and 236 mV; (d) as (c) but during the second stage at surface potentials of 175 and 120 mV.

the surface potential and peak frequencies were practically unchanged by the higher annealing temperature the half-widths of the bands were appreciably reduced from about 18 cm^{-1} to 13 cm^{-1}. The spectra in figure 8(c) correspond to increasingly positive surface potentials, and the growth in intensity clearly occurred at fixed frequency as on polycrystalline films. The frequency shifts (figure 8d) after higher exposures to CO occurred only as the surface potential decreased. These results confirm the suggested connection between the infrared and surface potential data on polycrystalline films (Bradshaw and Pritchard, 1970).

The crystal surface may not have been entirely clean after ion bombardment as appreciable partial pressures of methane appeared during the bombardment. (This effect was much reduced during the subsequent measurement on a copper (111) surface, probably because the chamber received an unintentional bake at atmospheric pressure after the glass vacuum line cracked!) Surface potentials measured on the same crystal in the LEED system, where methane levels were much lower, were only slightly different, e.g. 0.47 V for Xe, 0.23 V for CO at the positive maximum, and final values of 0.07 V for CO. The agreement is quite reasonable, particularly as the temperature was probably slightly lower in the LEED system because the refrigerant was drawn through the crystal holder under reduced pressure. Auger spectra established that carbon contamination was removed by the ion bombardment.

LEED studies of CO adsorption on copper (100) at the nominal temperature of 77°K led to results contrary to the conclusions drawn from results on polycrystalline films (Pritchard, 1963). There was no evidence whatsoever for a structure giving $\frac{1}{4}$ coverage. As the positive surface potential grew a $c(2 \times 2)$ LEED pattern appeared, corresponding to $\frac{1}{2}$ coverage, and during desorption, as the crystal was warmed, this pattern slowly faded without any additional features appearing. At higher coverages, as the surface potential fell to between 0.15 and 0.10 V a complex pattern appeared which suggested that the initial array was being compressed. If this interpretation is supported by further experiments the distinction between two adsorbed states may disappear.

COPPER (111)

After ion bombardment cleaning and annealing at 570°K xenon adsorption at 77°K gave a surface potential of 0.48 V. The behaviour of CO is shown in figure 7(b). The maximum surface potential was much higher (0.44 V) than on the (100) surface, and it fell to 0.05 V during the second stage. As the crystal was warmed the highest surface potential reached was 0.47 V. Preliminary infrared measurements have shown a sharp band at 2075 cm^{-1} (half-width 11 cm^{-1}) with an intensity similar to that on the (100) plane. This band grew during the development of the positive surface potential and then appeared to diminish markedly during the second stage of adsorption.

The chemisorption of CO on the (111) face appeared to be considerably weaker than on the (100). Whereas at 195°K and 10^{-5} Torr a surface potential of 0.23 V was recorded on the (100) face, approaching the maximum value, the surface potential under similar conditions on the (111) face was only 0.15 V compared with a maximum of 0.47 V.

IV. DISCUSSION

The reflection spectra from polycrystalline films have broadly confirmed the features of the transmission spectra. A single sharp band is observed on films deposited at 77°K, and the high frequency of the band on silver is a reproducible feature of adsorption on both annealed and unannealed thick films deposited under good vacuum conditions, making it very unlikely to be a consequence of impurities.

However, for any one metal the spectra are influenced by the conditions of deposition and annealing. Annealing films deposited at 77°K does not simply reduce the intensity by decreasing the surface roughness and adsorbate concentration. It also alters the shape and frequency of the band. Gold is a possible exception (figure 6e) but the variability in the spectra on room temperature gold and copper films is quite appreciable. It seems reasonable to associate the variability of the spectra with varying proportions of particular crystal facets in the film surfaces. The surface potentials and relative heats of adsorption on Cu (100) and Cu (111) are consistent with the surface potential results on films and the lower frequency of the band on Cu (111) is in qualitative agreement with the relative weakness of adsorption of the CO giving the low frequency shoulder band on room temperature deposited films. Nevertheless, the results obtained so far on the two crystal faces are inadequate to account for the spectra from films. Indeed, sharp features at 2085 or 2075 cm^{-1} were conspicuously absent from the film spectra although one might have expected (100) and (111) facets to predominate after annealing. Such single crystal studies must be extended to include several more low index planes.

A model of linearly adsorbed CO, bound mainly by σ-bonds, has been used (Bradshaw and Pritchard, 1970) to account for the main features of CO adsorption on the I_B metals. The most striking additional features revealed here are the frequency shifts with coverage. Shifts to higher frequency with increasing coverage on platinum have been explained (Blyholder, 1964) in terms of competition for d-electrons available for π-back bonding leading to a decrease in π-bonding with increasing coverage. This may be generally true for CO on the transition metals, but it seems improbable that the reverse shift on silver and gold is due to an increase in π-bonding, and coupling between oscillators may be expected to cause a frequency increase.

The vibrational frequency may be affected in other ways. In the absence of any direct influence on the C–O bond any bonding between the carbon atom and the metal would raise the frequency relative to that of a free CO molecule. A decrease in the strength of the metal-carbon bond could cause a frequency decrease. If the bond to silver is predominantly σ in character the charge transfer to the surface metal atoms could cause a charge accumulation which progressively hinders σ-bonding with increasing coverage. Where π-bonding is important, as on the transition metals, charge accumulation is reduced by back donation. Thus we might expect a reversal of the direction of frequency shift in going from a transition metal to silver. When the mutual charge displacements due to σ- and π-bonding are in balance there would be little influence of coverage. Although this simple argument is very incomplete (for example, charge transfer from the carbon atom must influence the C—O bond, as is clearly shown by the higher vibrational frequency of CO^+ relative to CO, a factor which may well contribute to the high frequency on silver) it shows that the observed frequency shifts are consistent with the earlier model of CO chemisorption on copper, gold and silver. As the π-bonding component diminishes the coverage frequency shift increases. The constant frequency on copper may be compared with that of nitrogen on nickel (Van Hardeveld and Van Montfoort, 1969) where the bonding is probably very similar.

The frequency shift in the second adsorption stage on copper is quite different. It has been proposed previously that two distinct adsorbate states co-existed, the second being unobserved in the infrared spectrum, and that the shift to higher frequency was due to their interaction (Chesters *et al.*, 1970). The LEED results show that the first adsorption stage leads to a coverage of $\frac{1}{2}$ on the (100) face. The intermolecular distance is then less than in solid CO, leaving very little space for a second distinct set of adsorption sites. What further adsorption takes place appears to be accompanied by compression of the initial array so that the CO molecules are no longer all in optimum bonding positions. The frequency shift may well arise from an overall diminution of bonding, leading to frequencies closer to the gas phase frequency, and with an overall reduction in the dipole moment.

ACKNOWLEDGEMENTS

We wish to thank the Royal Society and the Central Research Funds Committee of London University for equipment grants, and the Science Research Council for equipment and for studentships.

REFERENCES

Alexander, C. S., Ford, R. R. and Pritchard, J. (1968). Fourth International Congress of Catalysis. Moscow.

Blyholder, G. (1964). *J. Phys. Chem.* **68**, 2772.

Bradshaw, A. M. and Pritchard, J. (1970). *Proc. Roy. Soc. London, Ser. A* **316**, 169.

Bradshaw, A. M., Pritchard, J. and Sims, M. L. (1968). *Chem. Comm.* 1519.

Chesters, M. A., Pritchard, J. and Sims, M. L. (1970). *Chem. Comm.* 1454.

Keulks, G. W. and Ravi, A. (1970). *J. Phys. Chem.* **74**, 783.

Pritchard, J. (1963). *Trans. Faraday Soc.* **59**, 437.

Pritchard, J. and Sims, M. L. (1970). *Trans. Faraday Soc.* **66**, 427.

Van Hardeveld, R. and Van Montfoort, A. (1969). *Surface Sci.* **17**, 90.

Yates, D. J. C. (1969). *J. Colloid Interface Sci.* **29**, 194.

THE CHEMISORPTION OF HYDRAZINE ON METALS AND INSULATORS

R. C. A. CONTAMINARD, R. C. COSSER
and
F. C. TOMPKINS

Chemistry Department,
Imperial College of Science and Technology,
London, England

The heterogeneous decomposition of hydrazine on various surfaces of solids has been the subject of several investigations (see reference list given by Cosser and Tompkins (1971)), but there has been little, or no, detailed study of the chemisorption of this compound. However, two recent papers by Cosser and Tompkins (1971) and Contaminard and Tompkins (1971), concerned with the mechanism of decomposition of this compound on polycrystalline films of tungsten and molybdenum, provide data from which information about the nature of the bonding to, and stability of, hydrazine on various surfaces can be extracted. In general, chemisorbed hydrazine on weak adsorbents, such as Pyrex and silica, decomposes at temperatures of *ca.* $600°K$ according to the equation

$$3N_2H_4 \rightarrow 4NH_3 + N_2, \tag{1}$$

but on the more active surfaces of transition metals the overall process conforms approximately to equation (2)

$$2N_2H_4 \rightarrow 2NH_3 + N_2 + H_2. \tag{2}$$

Szwarc (1949) has postulated that reactions (1) and (2) occur by decomposition of two activated surface complexes the mechanism of breakdown being as follows

(I)   (II)

The mechanisms involve the production of N_2 without rupture of the $> N-N <$ bond and are consistent with the production of non-randomized $^{15}N^{14}N$ from labelled $H_2^{15}N-^{14}NH_2$. We believe, however, that mechanism (II) is not the main source of hydrogen evolution, but that mechanism (I) correctly predicts the formation of N_2 and NH_3.

On tungsten and molybdenum, we found* that the amount of H_2 produced during the decomposition of N_2H_4, chemisorbed on evaporated films of these metals at 195°K, with rising temperature, was always directly proportional to the film weight: i.e. to the surface area of the film, and indeed plots of the amount of hydrogen as a function of temperature from unit area of molybdenum and tungsten films were virtually coincidental; this fact suggests that the adsorption potentials of the two metals for N_2H_4 must be very similar. However, the amount of N_2 and NH_3 evolved could not be normalized with respect to surface area alone, but was more largely controlled by the extent of "overdosing" the film at 195°K with N_2H_4. The "normal" dose procedure was to terminate the addition of N_2H_4 to the film as soon as the presence of permanent gas (N_2 and/or H_2) could be detected on a Pirani gauge; it was found, however, that by further addition of N_2H_4 higher adsorbed amounts were obtained and in the subsequent decomposition larger amounts of both N_2 and NH_3 were evolved, although there was only a slight increase of H_2 production. Moreover, although the ratios of the amounts of each gas produced, $[H_2]/[N_2]$ and $[H_2]/[NH_3]$, varied appreciably during a run and also in different runs, the $[N_2]/[NH_3]$ ratio was always approximately 1/4. These facts indicate the occurrence of two independent concurrent modes of decomposition, the one giving only H_2 and the other being responsible for the simultaneous production of N_2 and NH_3 according to reaction (1). Now, the production of N_2 is unlikely to involve the rupture of the strong metal-nitrogen bond so that a comparatively weak associative chemisorption of hydrazine molecules undoubtedly proceeds as suggested by Szwarc on a quartz surface. However, dissociative chemisorption would be expected on the surface of a transition metal with unfilled *d*-orbitals. Thus, hydrogen is dissociately chemisorbed at 185°K on both W and Mo, although this process involves the cleavage of the strong H–H bond with a dissociation energy of 104 kcal mol^{-1}, whereas the N–N bond energy in N_2H_4 is only 60 kcal mole^{-1} (Szwarc, 1949). Dissociative chemisorption to form $-NH_2$ adradicals strongly bonded to the surface should take place, particularly in view of the formation of these radicals (together with H-adatoms) when NH_3 is chemisorbed on these metals and here the $=N-H$ bond energy is much higher (103 kcal mol^{-1}). Indeed, since the heat of chemisorption of NH_3 varies from 73 to 40 kcal mole^{-1} with increasing coverage (Wahba and Kemball, 1953) and the W–H bond energy similarly decreases from 74 to 59 kcal mole^{-1} (brennan

* Cosser and Tompkins, 1971; Contaminard and Tompkins, 1971.

and Hayes, 1964) then using the above $=N-H$ bond energy, we estimate that $W-NH_2$ bond energy decreases from 103 to 85 kcal mole with the increasing coverage. Thus, the heat of dissociative chemisorption of N_2H_4 on W is estimated to vary from 146 down to approximately 110 kcal mole^{-1} with increasing coverage. Surface dissociation of $-NH_2$-adradicals to $=NH$-adradicals and mobile H adatoms probably follows and hydrogen is evolved as a result of surface recombination of these adatoms.

We therefore believe that N_2H_4 is dissociatively chemisorbed to form the primary surface layer on both tungsten and molybdenum, and we must now consider the various possibilities for simultaneous associative chemisorption of molecular hydrazine to occur. Possible suggestions are (i) by H-bonding of N_2H_4 molecules to $-NH_2$-adradicals in the primary layer; (ii) bonding via one H atom of the N_2H_4 molecule to single W sites not occupied by the primary $-NH_2$ (a) layer to form an α state, since there are many such free sites available (see later); (iii) molecular chemisorption on an inactive crystal plane, probably the (110) plane, on which dissociative chemisorption cannot occur. The (110) plane is selected since for W it is the only plane which does not dissociatively chemisorb $N_2(g)$ at these temperatures (Delchar and Ehrlich, 1965).

We adduce considerable evidence in favour of the third hypothesis, viz., (a) if N_2H_4 is molecularly adsorbed on a specific plane and reaction (1) proceeds to completion, then on raising the temperature to around 400°K, this plane should be completely freed of adsorbate. Consequently on readsorption of N_2H_4 at 195°K, subsequent to this decomposition, the same kinetic curves for the production of NH_3 and N_2 with increasing temperature should be obtained and indeed this was confirmed experimentally; (b) presaturation with H_2 which populates the (110) plane with H adatoms should largely prevent subsequent adsorption of N_2H_4 so that little or no evolution of N_2 and NH_3 should proceed on raising the temperature, whereas presaturation with $N_2(g)$ (which is not chemisorbed on the (110) plane) should have no effect on the $N_2 + NH_3$ evolution. Again, experimental confirmation was obtained.

Similarly, on the active planes where dissociative chemisorption of N_2H_4 does occur, only hydrogen is involved during the decomposition leaving N adatoms and some surface hydrogen either as $-NH(a)$ or $H(a)$; chemisorption of N_2H_4 on this surface would be largely inhibited and indeed a greatly reduced H_2 evolution occurred. It was also confirmed that by presaturation of the active planes of the clean film with $N(a)$ adatoms by exposure to $N_2(g)$, hydrogen evolution was also greatly inhibited. In the light of these facts, the alternative hypothesis (i) and (ii) can no longer be valid, since it would be unreasonable to expect that planes saturated with N adatoms and other contaminants would function as an adsorbent and catalyst with the same properties as the clean surface.

A further conclusion also emerges; since NH_3 is evolved at low temperature from the (110) plane then this gas can not be chemisorbed on such a surface. This conclusion has recently been reported by May *et al.* (1969) and there is no previous evidence in the literature that contradicts this statement. Advantage can therefore be taken of this result, by separate chemisorption of NH_3 and H_2 on a series of similarly prepared films of both tungsten and molybdenum under the same conditions of 10^{-3} Torr pressure and $195°K$, to evaluate the areas of the (110) plane and of the remaining active planes. If monolayer dissociate chemisorption of NH_3 (except on the (110) plane) and H_2 are assumed, the area of the (110) plane and that of the other active planes can be both evaluated. Since in any decomposition the total amount of NH_3 (and N_2) has been measured, then the amount of N_2H_4 originally chemisorbed as molecules on the (110) plane can be calculated. For normal dosing, the number of N_2H_4 per W surface atom on this plane was about 0.40 but approaches unity for heavy over-dosing. Hydrogen bonding to form the surface complex seems probable since a structure of this type has been identified in some hydrazine derivatives (Audrieth and Ogg, 1951).

However, the lattice spacing of metal atoms on the (110) plane appears to be too short to accommodate one N_2H_4 molecule per W site if the van der Walls' radius of N_2H_4 is considered. Hence, we suggest that co-operative adsorption takes place as is implicit in equation (1), where simultaneous decomposition of three N_2H_4 molecules is required to conform with Szwarc's suggested mode of reaction. On a Mo surface, the coverage is always less than a monolayer even with large overdosing, although the lattice constants of this plane are very close to those of tungsten.

The absence of hydrogen evolution from weakly adsorbent surfaces of insulators (Pyrex, silica), semi-conductors and some less active metals (such as copper) suggests that associative molecular chemisorption is the main process and that little dissociative adsorption occurs on such adsorbents.

However, on W, Mo, Pt, Re, etc. there is considerable production of $-NH_2$-adradicals from all the active planes (excluding, for example, (110) plane for Mo and W). Specific information is not available for the presence of similar inactive planes for Pt and Re.

As has been done for the (110) plane of W and Mo, the number of sites occupied by chemisorbed hydrazine on the active planes can be evaluated. The number of adsorption sites available is evaluated as before from the separate adsorptions of NH_3 and H_2. And if we now confine attention to Mo, evolution of H_2 is complete at $630°K$ and only N adatoms (which are not expected to be desorbed much below $1000°K$) remain on the active planes. From the total H_2 evolution, the number of N_2H_4 molecules originally chemisorbed on these planes may be calculated and then compared with the total number of sites available; it is found that each dissociatively chemisorbed N_2H_4 molecule occupies four (H-adatom) adsorption sites, so that the surface structure is probably

Furthermore, if it is assumed that $E(W\equiv N) : E(W=NH) : E(W-NH_2)$ vary roughly in the same ratio as $D(N\equiv N, 255) : D(N=N, 99) : D(N-N)$, 38 kcal mole^{-1}, then on a fairly full surface, $E(W-NH_2)$ would be about 85, $E(W=NH)$ about 108, and $E(W\equiv N)$ is known to be 155 kcal mol^{-1}. Since the bond energy W–H is nearly 70 kcal mole^{-1} on a half-covered (H-adatom) surface, and since 103 kcal mole^{-1} is required for cleavage of the N–H bond of NH_2(a), dissociation of this adradical to W=NH(a) and W–H(a) would be endothermic to the extent of $(-103 + (108 - 85) + 70) = 10$ kcal mole^{-1}, consequently further surface dissociation is probable. We therefore believe that hydrogen evolution arises from bimolecular surface recombination of H-adatoms. The energetics are in fact more favourable for the subsequent surface dissociation

$$W - NH(a) \rightarrow W\equiv N(a) + W - H(a) \tag{3}$$

provided H-atom sites are available, since this reaction is exothermic to the extent of about 14 kcal mole^{-1} using our approximate thermal data given above. The surface composition would therefore be expected to comprise $-NH_2$(a), =NH(a) and H(a) in the initial stages of decomposition, and largely $\equiv$N(a) and H(a) in the later stages.

Finally, any metal such as W and Mo, which dissociatively chemisorbs nitrogen gas, and indeed probably some metals such as Ni, Pd and Pt, which dissociatively chemisorb hydrogen but not nitrogen, should be capable of chemisorbing N_2H_4 to form NH_2(a), because of the comparative weakness of the N–N bond in hydrazine. From such metals, H_2 arises from $-NH_2$(a) on the above planes, and N_2 and NH_3 (approximately in the ratio 1:4) from the inactive (110) plane.

REFERENCES

Audrieth, L. F. and Ogg, B. A. (1951). "The Chemistry of Hydrazine". J. Wiley, New York.
Brennan, D. and Hayes, F. H. (1964). *Trans. Faraday Soc.* **60**, 589.
Contaminard, R. C. A. and Tompkins, F. C., (1971). *Trans. Faraday Soc.* **67**, 545.
Cosser, R. C. and Tompkins, F. C., (1971). *Trans. Faraday Soc.* **67**, 526.
Delchar, R. A. and Ehrlich, G. (1956). *J. Chem. Phys.* **42**, 2686.
May, J. W., Szostak, R. J. and Germer, L. H. (1969). *Surface Sci.* **15**, 37.
Szwarc, M. (1949). *Proc. Royal. Soc. Ser. A*, **198**, 267.
Wahba, M. and Kemball, C. (1953). *Trans.Faraday Soc.* **49**, 1351.

THE ADSORPTION OF HYDROGEN ON NICKEL: A STUDY OF KINETIC AND EQUILIBRIUM PROPERTIES

N. TAYLOR and R. CREASEY

School of Chemistry, The University of Leeds, England

I. INTRODUCTION

The adsorption of hydrogen on nickel has been studied by a wide variety of techniques but, apart from two recent attempts using filament substrates (Gasser *et al.,* 1969; Eley and Norton, 1970), there has been little published kinetic data except that concerning the slow activated steps occurring near saturation (Gundry and Tompkins, 1956). It is our purpose to present for this system a whole body of rate data in terms of the sticking probability S and to relate the temperature and coverage dependencies of S with the structure and binding energies of the adsorbed state.

II. THE MEASUREMENT OF STICKING PROBABILITY

Most early determinations of sticking probability using film adsorbents were based on the assumption that a uniform pressure existed in the reaction cell during adsorption. At a high sticking probability with large area adsorbents, this is far from true as most molecules in the void of the cell are merely in transit between gas inlet and surface. Results may therefore be widely in error unless specially designed cells are used (Clausing, 1961; Hayward and Taylor, 1967, Harra and Hayward, 1967).

The cell used in this work (figure 1) is similar in concept to that used by Clausing. Gas is distributed symmetrically in all directions from the centre of the cell through the diffuser D, which is a small hollow glass sphere punctured randomly with a number of small holes. The adsorbing metal film is evaporated onto the inner wall of the cell from the filament F with the gas inlet system withdrawn into the vertical tube and protected from metal vapour by a magnetically held nickel disk ND. The filament F is positioned so that no film is deposited in the sidearm leading to the ionization gauge IG. After evaporation the diffuser is lowered into the cell and hydrogen admitted through the gas inlet I, from a constant pressure source, via solenoid-operated all-glass valves.

The collision rate of molecules with the film surface may be divided into n_A, because of molecules arriving direct from the gas inlet and, secondly, n_B, because of molecules which have already made at least one collision within the

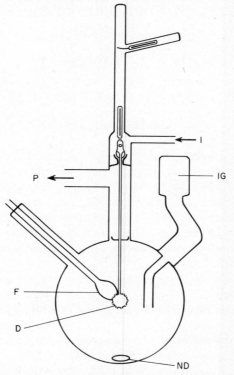

Fig. 1. Cell used for sticking probability measurements. F—filament; D—diffuser; ND—nickel disk; IG—ionization gauge; P—to pumps; I—gas inlet.

cell. When desorption is negligible n_A is sensibly equal to the total rate of adsorption and the sticking probability S is given by

$$S = \frac{n_A}{n_A + n_B} \tag{1}$$

The gauge IG is positioned so that it receives only molecules contributing to n_B and since these have approximately random motions we may use the Hertz-Knudsen expression and write

$$n_B = P.A. (2\pi m k T)^{-1/2} \tag{2}$$

where P is the pressure measured by IG and A is the geometric area of the film. It is sufficient in the above to use P measured with the gauge at room temperature irrespective of the temperature of the film provided that the collision factor is also calculated at room temperature. Hence we have

$$S = \frac{n_A}{n_A + PA(2\pi mkT)^{-1/2}} \tag{3}$$

This gives a particularly sensitive measure of S when $S \rightarrow 1$.

When desorption becomes appreciable S should be calculated from the total rate of adsorption; i.e. that including the contribution of the adsorbate as a source of gas. It is easily shown (see Hayward and Taylor, 1967) that in the presence of an equilibrium pressure P_{EQ} a suitably modified expression for S involves the substitution of $P' - P_{EQ}$ for P in equation 3. At the lowest temperatures studied drifts in cell pressure occur for long periods after interruption of the gas supply. In this case it is often convenient to continue to display results in terms of sticking probability calculated without correction from equation 3. Care must be taken to distinguish such values from the true S calculated from the sharp pressure changes at the cessation and restart of the supply of gas to the film.

III. EXPERIMENTAL DETAILS

Gas purity, construction and operation of ionization gauges etc. were as described previously (Hayward and Taylor, 1966; Hayward et al., 1966). The outgassing of the 0.5 mm nickel filament presented some problems because of the low evaporation temperature but it was found possible after a lengthy period (~50 hours) to reduce the pressure change at the end of deposition to less than 5 x 10^{-11} mm Hg; film weights varied between 10 and 50 mg. In regions of high equilibrium pressure gas could be pumped via a standard capillary interposed between cell and pumps; the amount removed was measured from the resulting pressure-time profiles.

IV. RESULTS AND DISCUSSION

THE VARIATION OF S WITH TEMPERATURE AND SURFACE COVERAGE

On sintered films the values of S are initially in the range 0.35 to 0.55 and, at a given temperature, vary by a maximum of about 15% for the range of film thickness. To determine the temperature dependence of S experiments were performed in which single films were continuously thermally cycled between

300 and 78°K and during each cycle gas was added to the film in increments of some $2.5 - 5 \times 10^{12}$ molecules per cm^2 of geometric surface. In this way about 700 values of S were obtained at 14 different temperatures within the range before rising equilibrium pressures rendered unobservable the sharp pressure changes at cessation and restart of the supply of gas. Each temperature cycle involved the total addition of only $3.5 - 7 \times 10^{13}$ molecules per cm^2 of surface (all coverages will refer to geometric areas); the values of S obtained, therefore, refer to adsorption onto a surface pre-equilibrated at 300°K within this coverage range. The combined results of one such experiment are displayed in figures 2 and 3. Initial values of S are approximately constant within the range of 300 to 195°K and below 195°K increase slightly with temperature. At a given temperature, S falls almost linearly over a wide range of coverage; above 195°K there is little variation of S with temperature at a given coverage; below 195°K, S falls progressively less rapidly with coverage.

Using films held at fixed temperatures enabled the above results to be confirmed and observations to be extended to regions of higher coverage by backing off the ion current due to the equilibrium pressure, thus enabling the $P - P_{EQ}$ components to be distinguished. At the highest temperature used $-360°K - S$ was measurable to only about 0.05 of the coverage at which the linearly extrapolated S values become zero (about $8 - 17 \times 10^{14}$ molecules per cm^2 depending on film thickness). At lower temperatures S could be studied over a more extensive range.

Within the temperature range 273 to 195°K there is an abrupt decrease in the variation of S with coverage after S has fallen to about $0.06 - 0.03$. At $T > 235°K$ the equilibrium pressure is already high ($> 10^{-8}$ mm Hg) in the "break" region and a change in the form of the isotherm is observed. Figure 4 illustrates the separate behaviour at 273, 208 and 195°K, and figure 5 the behaviour at 235°K including the form of the isotherm. Maximum coverages at 195°K (i.e. for $P_{EQ} \sim 10^{-4}$ mm Hg) are around 40% greater than for a projected fall of the early values of S to zero.

Below 195°K the sticking probability falls off initially less steeply with coverage than at high temperature but at coverages beyond the "break" described above S begins to fall more rapidly, with coverage to approach the high temperature values. This is illustrated in figure 3 in data drawn from the thermal cycling experiments and is substantiated in the results presented in figure 6 for an experiment performed at 175°K. Final confirmation was obtained by holding a film alternatively at 78 and 300°K and adding small increments of gas at each temperature. As shown in figure 7 the variation of S with coverage at 78°K is linear until the break region at the higher temperature is reached after which S begins to decrease more rapidly. In the latter region at 78°K parts of the surface saturate with respect to the ambient gas pressure after the addition of only some 3×10^{13} molecules per cm^2 since the previous

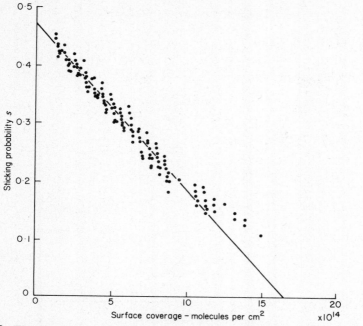

Fig. 2. The variation of sticking probability with surface coverage. Results, using a single film, for the temperature range 268 to 206°K.

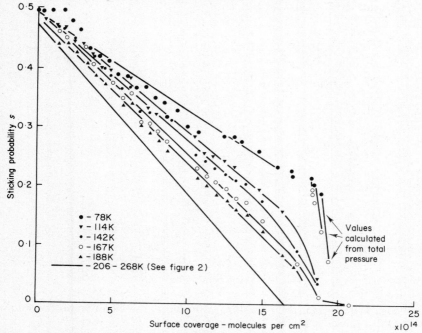

Fig. 3. The variation of sticking probability with surface coverage. Results, using same film as figure 2, for the temperature range 188 to 78°K.

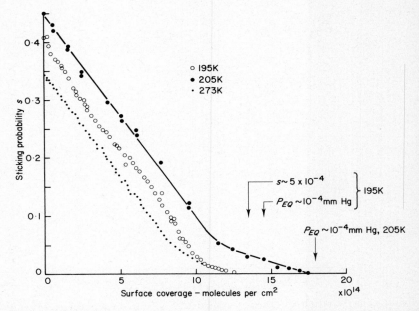

Fig. 4. The variation of sticking probability with surface coverage. Separate experiments at 273, 205 and 195°K.

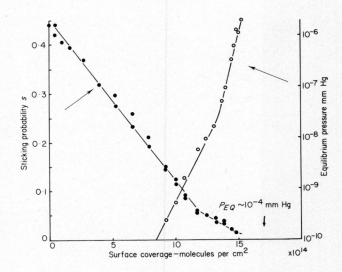

Fig. 5. The variation of sticking probability and Equilibrium Pressure with surface coverage at 235°K.

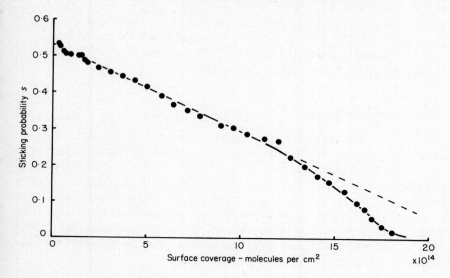

Fig. 6. The variation of sticking probability with surface coverage at 175°K.

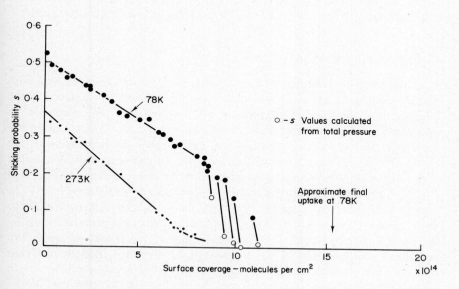

Fig. 7. The variation of sticking probability with surface coverage. Measurements made alternately at 78 and 273°K.

equilibration at the high temperature. It should be noted that the sticking probability curves at the various temperatures apparently never cross and this is true even for values of S calculated at $78°K$ from the total cell pressure. The significance of this feature will be discussed further.

DESORPTION SPECTRA

From the last experiment desorption spectra were constructed from the pressure variations during slow warming of the cell to $300°K$ after small additions $(3 - 7 \times 10^{13}$ molecules per cm^2) at $78°K$. Results are similar to those published recently by Wedler *et al.,* (1970) in showing good evidence for only one low temperature state and, at low and medium coverage, this being largely desorbed by about $130°K$. Above $200°K$ there is a slight shoulder on the rising pressure profile which is probably caused by a rearrangement in the primary chemisorbed layer as mobility improves. Wortman *et al.,* (1957) concluded that mobility set in above $240°K$ on nickel field emitters.

In other experiments attempts were made to saturate films at $78°K$ and then record the desorption spectrum on pumping on the cell during slow warming to $300°K$. Results indicated the persistence of the low temperature peak which became broadened; there was also considerably more evidence for rearrangements around $200°K$. On recooling to $78°K$ and again warming without further addition of gas the spectrum showed a greatly reduced low temperature peak.

HEATS OF ADSORPTION

Isotherms were constructed from equilibrium pressure data obtained both during addition of gas and its removal via the standard conductance in the pumping lead. In some experiments several temperatures were used and on occasions isosteres were constructed directly by variation of the temperature at a fixed coverage. Isosteric heats of adsorption obtained by application of the Clausius-Clapeyron equation to the various data are shown in figure 8 together with some sticking probability data obtained in the same experiment. (The filled circles on the heat curve refer to some calculated data to be discussed later.) The general features of the variation of q with the surface coverage are similar to those obtained isosterically by Rideal and Sweett (1960) and calorimetrically by Wahba and Kemball (1953) and Brennan and Hayes (1964) all using film adsorbents. The heat data of Eley and Norton (1970) using a filament adsorbent as a resistance calorimeter differ widely from other work and results appear to be inapplicable to films. The heat is seen in figure 8 to fall off after the "break" region—we will thus tentatively associate the high S region with the region of slowly falling heat and likewise the regions of low S and rapidly falling heat.

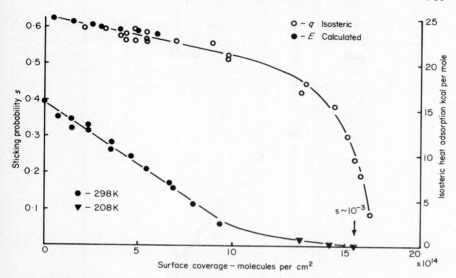

Fig. 8. The sticking probability and isosteric heat of adsorption versus surface coverage. The E values are calculated from the isotherm displayed in figure 11 (see text).

ADSORPTION AT 78°K

At 78°K parts of the surface become effectively saturated with respect to the ambient gas phase after a total addition of about 7×10^{14} molecules per cm^2. There is a slow redistribution of gas on closure of the supply—the decays of the "pseudo-equilibrium" pressure following the law reciprocal pressure linear with time as found for molybdenum at 78°K (Hayward *et al.*, 1966). It is easily shown that this behaviour is consistent with a redistribution via the gas phase from a layer obeying the Temkin type equation ($\ln P/P_0 = \alpha N$ where N is the number of molecules per cm^2 raising the pressure from P_0 to P and α is approximately a constant at a given temperature) to a surface adsorbing with an effectively constant sticking probability. We find

$$\frac{1}{P} - \left(\frac{1}{P}\right)_{t=0} = \alpha SZt \tag{4}$$

Within this derivation is the tacit assumption that the exchange processes attempting to maintain equilibrium are more rapid than the process giving net adsorption. Reference to figures 9 and 10 indicates that S is always high on restarting gas flow. From the lack of observable sharp pressure changes on cessation of gas supply we conclude that S is still high (> 0.1) at high ambient pressures. It was noted in the discussion of figure 7 that the values of S calculated at 78°K from the total cell pressure during addition of gas were always larger than at temperatures above 195°K; this is consistent with the general idea that the only regions which may have high S values under the prescribed circumstances for redistribution are those which are in pseudo-

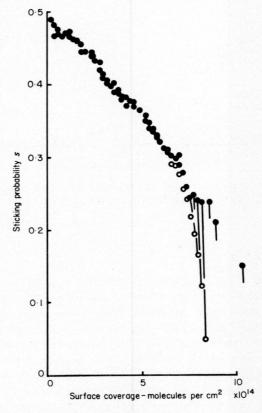

Fig. 9. The variation of sticking probability with surface coverage, film held solely at 78°K.

equilibrium with the gas phase. The slopes of the plots of reciprocal pressure versus time vary between 10^8 and 10^5 per mm Hg per second with increasing surface coverage. The constant α was estimated from attempts to construct isotherms at 78°K as about 7.5×10^{-14} per molecule per cm^2 of geometric surface; this indicates a variation of S for the net process from > 0.1 to around 10^{-4}. The high initial values are no doubt due to clean surface somewhat hidden from the gas phase.

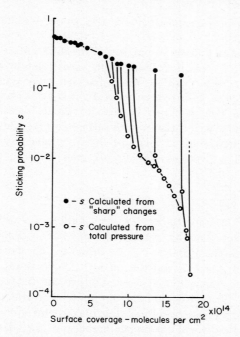

Fig. 10. The variation of sticking probability with surface coverage, film held solely at 78°K.

V. CONCLUSIONS

In the face-centred structure of nickel the lowest energy most close-packed planes are the 111 and 100 and these have 1.88 and 1.65 x 10^{15} surface atoms per cm^2 respectively. An estimate of the lower limit of the roughness factor of films sintered at 320°K as 2.0 thus indicates that at least 3.5 x 10^{15} Ni atoms per cm^2 are exposed to the gas phase. We shall assume that there is in the surface one site for primary chemisorption per nickel surface atom and also that hydrogen is, initially at least, adsorbed as atoms. The thinnest films are found to have the lowest uptakes of hydrogen and are judged the smoothest; at high temperature these adsorb about 10^{15} hydrogen molecules per cm^2 before S values of the initial coverage region extrapolate to zero. Thus, about one-half of the surface sites are contributing to S initially and if such sites may be grouped together without affecting their kinetic behaviour then S must be unity at zero coverage over such a surface. This now considerably simplifies a discussion of the temperature dependence of S at high temperature since we may postulate that it is unity on collision with vacant sites and, from the linear fall of S with coverage, zero on collision with occupied sites in this group. We shall designate such a surface as type A.

The above implies that about one-half of the total surface adsorbs with a low sticking probability even at collision with vacant sites. We shall again group together these sites as an "area" on the surface and designate this as type B.

The distribution of hydrogen between A and B demands some consideration. We could assume that it occurs because of kinetic factors i.e. the different magnitudes of S and their variation with surface coverage. It is easy to show that if $S_A{}^0$ and $S_B{}^0$ are the initial values of sticking probability and if each surface shows a linear fall of S with coverage θ_A, θ_B then

$$\frac{S_A{}^0}{S_B{}^0} = \frac{\ln(1 - \theta_A)}{\ln(1 - \theta_B)} \tag{5}$$

By assuming $S_A{}^0 \sim 1$ and $S_B{}^0 \sim 0.1$ the variation of S with total uptake over the whole surface can be calculated and has a shape similar to that experimentally observed above 195°K. However, the above implies a lack of mobility between A and B and this, at high temperatures, is in conflict with much evidence including direct observations in the field-emission microscope (Wortman et al., 1957). We will retain the above as an effect at low temperature but must construct our argument for high temperatures on the basis of thermodynamic factors and a complete mobility of the adatoms. The degree of overlap of the partial free energies is now the critical factor in determining the distribution of hydrogen between A and B. Since our discussion is to be only qualitative we shall equate the free energy factors with the binding energies at the surfaces and thus ignore the entropy differences of A and B. In our discussion of the heat data of figure 8 we considered that on surface A q fell only slowly whereas on B a rapid fall ensued. The binding energies of A and B apparently overlap somewhat since there is no very abrupt change in an isotherm traversing the intermediate region. It is a simple procedure to show that the change in heat variation associated with the filling of A occurs simultaneously with a change in the variation of S with total coverage. The value $S_B{}^0$ is again considered to be about 0.1 of $S_A{}^0$. There appears, in the results presented both here and elsewhere, a body of evidence indicating the homogeneity of the adsorption sites on single planes of nickel surfaces. The heat falls on nickel may therefore be usefully discussed in terms of an heterogeneity induced by interactions within the adsorbed layer. Only a small part of the observed falls may, however, be ascribed to the relatively long-range dipole-dipole forces or to the short-range repulsions of the charge clouds of neighbouring adatoms; Grimley (1957) and Culver et al. (1959) have suggested alternatives which would considerably increase the short-range interactions. On the basis of a pairwise nearest neighbour interaction Wang (1937) and others have deduced from statistical mechanics the effect of the interaction energy V on the equilibrium distribution of nearest neighbours. Results show an almost linear fall in heat with coverage at the limit of high temperature or low V (i.e. random distribution of adatoms) and increasingly sigmoid variations as T decreases or V increases. Reference to figure 8 indicates that V on surface A may be only 0.2 of that on B and the values of V_A, V_B estimated from the total falls in heat as θ_A, $\theta_B \rightarrow 1$ indicate an almost linear heat fall on A whereas on B the indicated variation is highly sigmoid. On these simple bases it is possible to reproduce the experimental heat and S variations with total coverage at high temperature.

A kinetic balance equation for equilibrium between surface A and the gas phase (i.e. before overlap of the binding energies) which includes our observations of the magnitude of $S_A{}^0$ and the linear fall of S with θ_A and also assumes a combination of energetic adatoms before desorption is of the form

$$S_A{}^0(1-\theta_A)ZP = 4n_S \cdot \theta_A{}^2 \cdot v \cdot \left\{\frac{E}{RT}+1\right\}e^{-E/RT} \tag{6}$$

where n_S is the number of sites assuming a 100 surface and v is a frequency factor taken as 10^{13} per second. The energy requirements are such that the bonds binding the pair of adatoms to the surface must have a combined energy in excess of E for desorption to occur. Retaining the θ^2 term may not be an approximation for non-random distributions since Laidler (1953) has shown that interactions influence both the distribution of nearest neighbours and their rate of reaction, the effects apparently cancelling exactly. To test equation 6 an isotherm was constructed from equilibrium pressure data obtained at 355.6°K, since at this temperature (see figure 11) the equilibrium pressures were

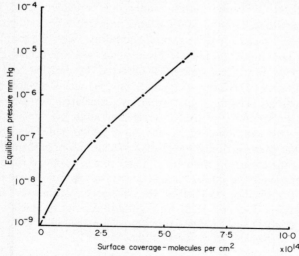

Fig. 11. Isotherm at 355.6°K. Data used to calculate E values.

measurable near zero coverage and calculations could thus be extended to a region in which the ratio of the θ terms was varying appreciably. The values of E obtained are shown in figure 8 and are in excellent agreement with the experimentally determined isosteric heats. These results are to be published in greater detail elsewhere.

Below 195°K there is evidence for a small temperature dependence of S. One possible cause is precoverage of the surface during deposition of the film; it is often forgotten that it is not the final pressure of "active" gases that determines contamination but rather a cumulative effect during the whole period of deposition. If precoverage is negligible then a part of the surface other than A must be adsorbing with an enhanced S at low temperatures. The change in behaviour observed below 195°K after A has become filled suggests that a mobile precursor state exists "on top of" the adatoms on A and in this state hydrogen may migrate to unoccupied sites in the surface. A relevant model in which all precursor finally chemisorbs has been used to explain results in the hydrogen-molybdenum system (Hayward and Taylor, 1968). The resulting expression for the variation of S is

$$S = S_A^{\,0}\left\{1 - \left(\frac{S_A^{\,0} - S_A^{\,1}}{S_A^{\,0}}\right)\theta_A\right\} \qquad (7)$$

where $S_A^{\,1}$, the sticking probability on top of a filled surface of A, is estimated as 0.5 at 78°K. This behaviour is in agreement with the observations of Wortman et al. (1957) who found that attempts to deposit hydrogen on a partially covered nickel field emitter held below 250°K resulted in further spreading of the primary layer indicating mobility only over occupied sites. The results presented earlier for adsorption at 78°K are consistent with an identification of the precursor with the adsorbate in pseudo-equilibrium with the gas phase during redistributions and this is tentatively assumed to be a molecular species. The positive transients in surface potential observed by Delchar and Tompkins (1968) on addition of hydrogen to partially covered nickel surfaces at 90°K are probably caused by a similar species. They calculated for this species an activation energy for mobility of 3.4 k cals per mole. At temperatures much above 120°K slow recoveries of S, due to a similar redistribution, are not observed and the precursor must have a low concentration, perhaps because of a combination of the increasing mobility and rate of desorption. A simple "random walk" treatment assuming similar frequency factors for mobility and desorption indicates that the distance travelled by the precursor molecules is "raised to the power" 2.5 between 195 and 78°K for a given difference of activation energies. On this basis at the higher temperature there is presumably little sampling of adjacent sites.

REFERENCES

Brennan, D. and Hayes, F. H., (1964). *Trans. Faraday Soc.* **60**, 589.

Clausing, R. E. (1961). *Trans. 7th A.V.S. Nat. Vac. Symp.* 1960 Pergamon, New York 345.

Culver, R. V., Pritchard, J. and Tompkins, F. C. (1959). *Z. Elektrochem.* **63**, 741.

Delchar, T. A. and Tompkins, F. C. (1968). *Trans. Faraday Soc.* **64**, 1915.

Eley, D. D. and Norton, P. R. (1970). *Proc. Royal Soc. Ser.* **A314**, 319.

Gasser, R. P. H., Roberts, K. and Stevens, A. J. (1969). *Trans. Faraday Soc.* **65**, 3105.

Grimley, T. B. (1957). *In* "Chemisorption" (Garner, W. E. ed.). Butterworths, London.

Gundry, P. M. and Tompkins, F. C. (1956). *Trans. Faraday Soc.* **52**, 1609.

Harra, D. J. and Hayward, W. H. (1967). *Nuovo Cimento, Suppl.* **5**, 56.

Hayward, D. O. and Taylor, N. (1966). *J. Sci. Instr.* **43**, 762.

Hayward, D. O. and Taylor, N. (1967). *J. Sci. Instr.* **44**, 327.

Hayward, D. O. and Taylor, N. (1968). *Trans. Faraday Soc.* **64**, 1904.

Hayward, D. O., Taylor, N. and Tompkins, F. C. (1966). *Discuss. Faraday Soc.* **41**, 75.

Laidler, K. J. (1953). *J. Phys. Chem.* **57**, 318.

Rideal, E. K. and Sweett, F. (1960). *Proc. Royal Soc. Ser.* **A257**, 291.

Wahba, M. and Kemball, C. (1953). *Trans. Faraday Soc.* **49**, 1351.

Wang, J. S. (1937). *Proc. Royal Soc. Ser. A* **161**, 127.

Wedler, G., Fisch, G. and Papp, H. (1970). *Z. Elektrochem.,* **74**, 186.

Wortman, R., Gomer, R. and Lundy, R. (1957). *J. Chem. Phys.* **27**, 1099.

PHOTODESORPTION OF
CARBON MONOXIDE FROM TUNGSTEN

P. KRONAUER and D. MENZEL

Lehrstuhl für Physikalische Chemie,
Technische Hochschule Darmstadt, West Germany;
Institut für Physikalische Chemie und Elektrochemie,
Technische Universität München, West Germany

I. INTRODUCTION

Photodesorption of gases from metals seems to be a controversial subject. Lange and Riemersma (1962) and Lange (1965) found a photodesorption effect for CO on Ni by recording total and partial pressure changes caused by irradiation. They reported a desorption probability of 1 to 2×10^{-8} for photons of about 350 nm, with a threshold at about 430 nm. For CO on W, Lange (1965) was not able to identify a photodesorption signal from a full adsorption layer, because of large thermal desorption effects caused by irradiation; from a layer containing strongly bound CO only, a desorption probability of about 2×10^{-8} was found between 230 and 320 nm. Adams and Donaldson (1965) did not detect photodesorption for CO_2, H_2, and N_2 on Ni, Fe, and Zr, and for CO on Mo and W; they did, however, find an effect for CO on Ni, Zr, and Fe. The relative energy dependence for CO/Ni given by them agrees roughly with that of Lange (1965), but their absolute values were higher by about a factor 100. Moesta and Breuer (1968) and (1969), and Moesta et al. (1969) found a large increase in the work function of a CO covered Ni surface when it was irradiated with light of wavelengths below 250 nm. From CO on W, irradiated with 254 nm light, they found desorption of CO, CO_2, and smaller amounts of more complicated particles (Moesta and Trappen, 1970). In both cases the quantum efficiency seemed to be very high (close to 1). On the other hand, Schram (1970) claims that the photodesorption cross section of CO from Ni is below 10^{-23} cm^2 and assumes that the higher values reported previously are due to thermal desorption.

Our interest in this subject arose in connection with our work on electron impact desorption. As shown by different authors (for a survey see Menzel (1970)), slow-electron (0 to a few hundred eV) impact on adsorbed layers can lead to desorption of ions and neutrals. These processes most probably proceed via excited or ionized states of the adsorbate. It seemed possible that similar processes could be induced by photon impact, and more information about excited adsorbate states be derived from such experiments. Measurements of the energy dependence close to the threshold are easier with photons than with electrons, and contain more information. CO on W was selected as the first system to be investigated, since a large amount of information is available on it (see for instance Ford, 1970) as well as from electron desorption experiments (Menzel and Gomer, 1964b; Redhead, 1964; Menzel, 1968), and the disagreement among the previous workers on photodesorption would have to be resolved, before any conclusions could be reached.

Measurements were also done on oxygen on tungsten.

II. EXPERIMENT AND RESULTS

As in the electron desorption experiments, two basic experimental approaches have been used, one of them employing two different techniques. Firstly, the amount of photodesorbed particles has been measured by recording total and partial pressure changes caused by irradiation of CO layers on polycrystalline tungsten foils. Secondly, the change induced in the surface layer has been followed by work-function measurements on a tungsten field emission tip. As much higher sensitivities were possible in the measurements of total rather than of partial pressure changes, the former have been employed for the study of functional dependences of the photosignal. The measurements of partial pressure changes complement these by showing the main constituent of the photosignals. The advantage of the field-emission microscope is that the light beam's effect can be integrated over a long time, and there is the chance of detecting changes other than desorption. In particular, such effects as the large increase of the work function, as reported by Moesta et al. (1969) for CO/Ni, should become obvious.

OPTICAL SYSTEM

For most measurements a 200 W high-pressure mercury discharge lamp has been used with quartz lenses and interference filters for 578 ± 20 (half-width) nm, 434 ± 20 nm, 366 ± 20 nm, 311 ± 20 nm, 280 ± 10 nm, and 250 ± 10 nm. The power density at the site of the target has been measured with a calibrated thermopile, and the number of photons per second and cm^2 of beam cross section has been calculated from it. The beam cross section was about 2×10^{17} photons/sec cm^2 at 578 nm and 1×10^{16} at 250 nm, and changed somewhat with ageing of the lamp.

MEASUREMENTS OF TOTAL PRESSURE CHANGES

A small glass uhv system pumped by two mercury diffusion pumps (during bakeout and gas inlet) or by the ionization gauge employed for pressure recording (during photodesorption experiments) was used. The base pressure after usual processing was about 1×10^{-10} Torr as measured by the ion gauge. However, the contamination rate of the field-emission microscope contained in the same system (see section II.D.) showed that most of this pressure was due to helium diffusing through the walls, and that the partial pressure of active (i.e. adsorbable) gases was below 1×10^{-12} Torr. The measuring cell is shown in figure 1. The target was a polycrystalline tungsten ribbon (30 x 3 x 0.025 mm), which was mounted to a tungsten rod press seal via platinum wires of 1 mm diameter. After bake-out of the system, the tungsten ribbon was cleaned either by repeated flashing to $2500°K$ in an ultrahigh vacuum, or by periodic heating (2 sec on, 20 sec off) to $2000°K$ for several hours in 10^{-6} Torr oxygen to remove carbon, followed by vacuum-flashing to $2500°K$. No difference in the photodesorption measurements were induced by these two pre-treatments. It was then exposed to 1×10^{-6} Torr CO for 5 min and photodesorption runs started after pumpdown.

For these experiments the center 10 mm of the ribbon were irradiated through a Suprasil I window at an angle of incidence of about 60° to the surface normal. Photodesorption as well as thermal desorption caused by the beam leads to an increase of the steady-state pressure. The analysis of this signal was the same as given by Lange (1962, 1965) and Adams and Donaldson (1965). The pressure change caused by a given additional influx of particles is inversely proportional to the pumping speed S. Therefore, the pumping action of the ion gauge only (about 0.1 litre/sec at 8 mA emission) was used during the photodesorption runs, in order to obtain high sensitivity. S was determined from the response time to a stepwise change in influx. The quantum efficiency for desorption, or desorption probability, γ (desorbed particles per incident photon), can then be calculated according to

$$\gamma = \frac{\Delta i \cdot S \cdot b \cdot h\nu}{a \cdot I \cdot F \cdot \cos \vartheta}$$

where Δi is the measured change of the ion current of the ionization gauge caused by irradiation with photons of energy $h\nu$, a is the gauge constant, S the pumping speed of the gauge, b the constant converting from vacuum units (Torr·litre/sec) to particles/sec, F the irradiated area and θ the angle between its normal and the beam, and I the measured power density of the beam. If the number N per cm^2 of the adsorbed species concerned is known, the desorption cross section q can be calculated from $q = \gamma/N$. By compensating off most of the background ion current, changes below 10^{-13} amps, corresponding to 10^{-12} Torr, could be measured, which corresponds to a minimum additional influx of about 3×10^6 particles/sec. The limit of sensitivity in γ is then about 10^{-10} for 578 nm and 2×10^{-9} at 250 nm. This analysis assumes that the only cause of

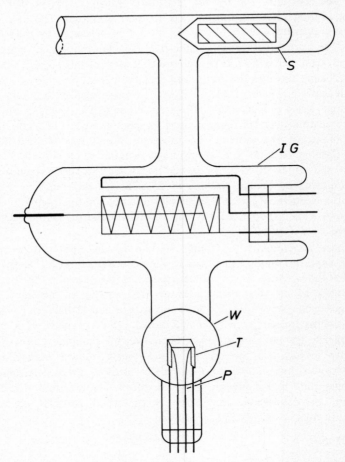

Fig. 1. Sketch of the experimental cell for measurements of total pressure changes by irradiation. W: quartz window (Suprasil I); T: target (W ribbon); P: potential leads; IG: ionization gauge; S: glass-encapsuled slug.

pressure rises accompanying irradiation is photodesorption from the target, i.e. that there are no contributions from thermal desorption from the target, or from thermal and/or photodesorption from the window or the walls. No desorption effects were found, when the ribbon was clean. This shows that thermal and/or photodesorption from the quartz window and the glass walls were below the detection limit, and the desorption effects to be reported must originate on the ribbon. As there are doubts about the contribution of thermal desorption effects in photodesorption measurements, and these have been reported to be especially severe in the system CO on W (see section I. above), great care was taken to avoid heating of the ribbon by the light beam. Without precautions, strong thermal desorption was indeed found to be caused by light of all wavelengths, as reported by Lange (1965). To avoid it, the ribbon was held at a temperature of about 15°K above ambient by a fast automatic temperature controller, which worked similar to that described by Hansen and Gardner (1968), using two potential leads (0.011 mm diameter tungsten filament) to sample the voltage drop across the center third of the ribbon. When the light beam is directed onto the ribbon, thus increasing the total power input, the controller decreases the electrical power to keep the temperature constant. Using this device, no desorption was detected at 578 nm even at 800 W/m², while at lower wavelengths definite effects were found at considerably lower power densities. The desorption probabilities calculated from the signals are shown in figure 2, with the power density at each wavelength. The inverse variation of these two quantities with wavelength is a strong argument against a thermal nature of the signals: the quotient between the desorption signal at 250 nm and the limit of detectability at 578 nm amounts to about 5×10^3.

The largest error connected with the calculation of absolute γ-values probably rests with the estimation of the irradiated area of the ribbon. In order to keep this error constant and thus make the relative error smaller than that of the absolute values of γ, the optical arrangement was left unchanged for each series of measurements except for the interchange of the interference filters. The same procedure was used when measuring the power density with the thermopile. The shape of the wavelength dependence is therefore considered to be quite accurate. The total reproducibility of the absolute values of γ for different sets of measurements was in the order of 30%; the absolute error may be somewhat larger.

In order to examine from which adsorbed state the signal originates, the dependence of the photosignal on temperature was measured at 250 nm. The results are shown in figure 3. As the background pressure was increased considerably by this heating, making the measurements difficult, the following procedure was adopted. With the temperature controller, the ribbon was heated to the temperature given by the abscissa for 1 min, starting at the lowest temperature. The initial pressure burst had largely subsided by the end of this period. The heating current was then switched off and the photosignal measured immediately afterwards. As even then the background pressure was considerably higher in these measurements than in those of figure 2, the values are less accurate, especially those above 500°K. The residual signal at 600°K may be due to re-adsorption occurring during the measurement. It can be concluded,

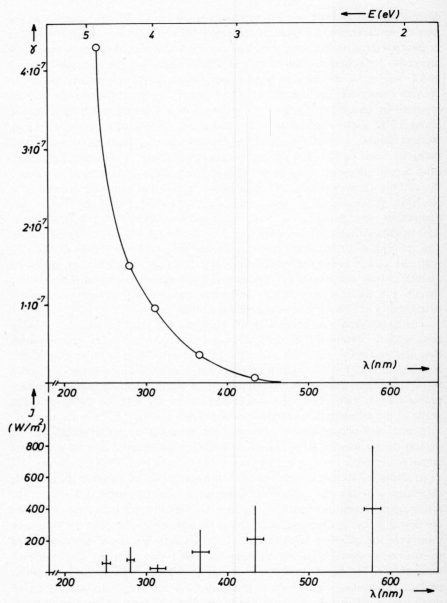

Fig. 2. Upper part: Photodesorption probability of CO from W as a function of wavelength or energy. Lower part: Intensity of the irradiation used in these experiments (the horizontal bars indicate the half-widths of the interference filters used).

however, that the main photodesorption signal from CO on W stems from an adsorbed species which is thermally desorbed below 600°K.

No photodesorption signal could be detected at any wavelength for an adlayer of oxygen on W.

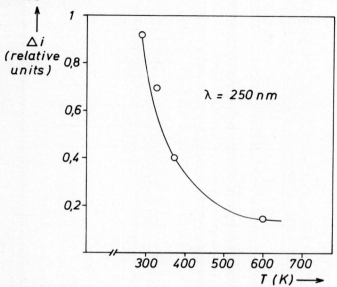

Fig. 3. Photodesorption signal from CO on W for 250 nm light as a function of the preheating temperature (in relative units).

PARTIAL PRESSURE CHANGES

The measurements of photoinduced partial pressure changes used the same principles as those of total pressure changes. They were performed with a quadrupole mass spectrometer in a stainless-steel uhv-system (see figure 4) of about 2 litres volume, which was pumped with an ion pump (nominal pumping speed 15 litres/sec). The light beam entered the system through a sapphire window and hit the tungsten ribbon at an angle of 60° to the normal to the surface. Because of the inevitably much larger residual influx in this system, the base pressure was only about 1×10^{-9} Torr after baking and somewhat higher after several gas dosings. This could have been improved by increasing the pumping speed. However, it was found preferable to work at these conditions rather than to go to lower residual pressures, as an increase of the pumping speed makes the desorption signal, as measured for a certain number of desorbed particles, smaller by the same factor. As the residual gas mainly consisted of CO, no contamination problems of the CO layer (prepared as in section II.B.) were encountered, except after long standing at the residual pressure.

As strong signals were found to originate from the light beam hitting the stainless-steel walls, an "exponential horn" made of pyrex was placed behind the ribbon to absorb the uv radiation passing by and being reflected from the ribbon. This was indeed found to remove most of the signal from the walls. The

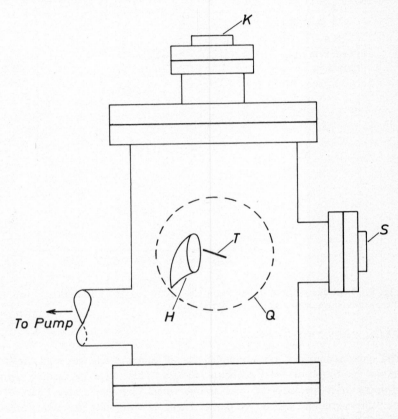

Fig. 4. Sketch of the system used for measurements of partial pressure changes induced by irradiation. S: sapphire window; T: target (W ribbon; mounting not shown); H: exponential horn; Q: flange with quadrupole mass spectrometer; K: window (kovar glass).

remaining contribution was eliminated by determining the apparent desorption signal with clean W ribbon (held at a temperature of $800°K$ to avoid adsorption from the residual gas), and subtracting it from the signal from the CO-covered ribbon. The necessity of this procedure and the higher pumping speed led to a decrease of the accuracy and sensitivity by more than a factor 100, as compared to section II.B. Therefore these measurements were only used to determine the nature of the majority of desorbing particles. A refinement of the system which would allow functional dependences to be measured was not necessary on account of the results: When a CO layer, which had been formed by an exposure of about 3×10^{-4} Torr·sec CO immediately after flashing to 2500°K, was irradiated with light of 250 or 280 nm, the desorbing particles were found to be CO only (i.e. to at least 95%). However, when the ribbon had been covered by standing in the residual gas for several hours, or when the pure CO layer was allowed to interact with the residual gas for more than an hour, up to 10% of the photodesorbed particles were CO_2 (the partial pressure of CO_2 was about 1×10^{-10} Torr), when light of 250 nm was used.

This suggests that the photodesorbed CO_2 stems from adsorbed CO_2. No CO_2 was found with light of 280 nm, which could be due to different thresholds for the photodesorption of CO and CO_2. Large amounts of CO_2 were also observed to be desorbed from the walls by the light beam. This was especially pronounced, when a low-pressure mercury lamp similar to that used by Moesta and Trappen (1970) (with about 90% of its emission at 253.7 nm) was placed in front of the sapphire window: under these conditions the photosignals of CO and CO_2 were about equal. The low-pressure Hg lamp was not used normally, as it is difficult to focus it properly and to measure its intensity.

EXPERIMENTS WITH THE FIELD-EMISSION MICROSCOPE

The small glass system described in section II.B. also contained a field-emission microscope (FEM) similar to those used by Menzel and Gomer (1964), with an attached Suprasil I window to allow the light beam to be directed onto the tungsten FEM tip from the side (see figure 5). Work function changes caused by irradiation could be measured by taking Fowler-Nordheim characteristics (see for example Gomer (1961)). Visual observation of the emission pattern was also possible. When the tungsten tip was covered with CO at room temperature and irradiated with light of 250 nm, the pattern was found to become spotty after about 60 min. The work function was found to increase slowly with time of irradiation (see figure 6); because of the spots, the scatter in these measurements is somewhat larger than normal.

The qualitative variation of the work function suggests that irradiation leads to desorption of an adsorbed species with a positive surface dipole, i.e. a species which decreases the work function. On the other hand, the appearance of spots in the emission pattern, which are probably caused by molecules adsorbed in exposed sites, seems to indicate that irradiation induces other changes besides desorption. However, upon irradiation of the "clean" tip the appearance of spots is also observed; at the same time the work function also rises, though much less than with a CO-covered tip (see curve b of figure 6). Under both conditions, the spots are seen distributed over the entire pattern, not only over the irradiated area. It seems likely, then, that during the very long irradiation period adsorbed particles are photodesorbed from the glass walls or, more probably, from the fluorescent screen, and are collected on the tip causing the small work-function increase of the clean tip and the appearance of spots in the patterns. Qualitatively, the former conclusion of desorption of an adsorbed

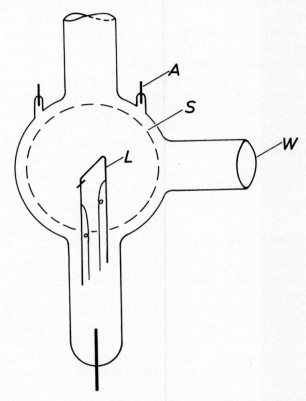

Fig. 5. Sketch of the field-emission microscope for photodesorption studies. L: tungsten loop with field-emission tip and potential leads; S: fluorescent screen; A: anode contacts; W: quartz window (Suprasil I).

species with a positive dipole is not affected by this contamination, as the effect on the clean tip is only a quarter of that on the CO-covered tip and is appreciable only after a very long time, while the main change on the covered tip occurs in the first 100 minutes.

A quantitative evaluation is not very meaningful because of this contamination. However, a rough estimate of the desorption cross section can be made in the following way. If it is assumed that the work-function changes caused by

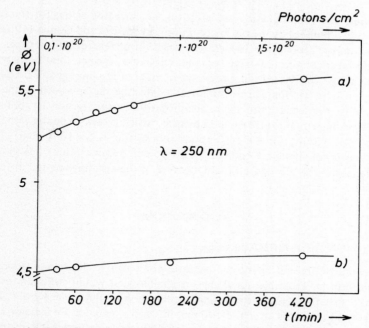

Fig. 6. (a) Variation of the work function of a W field-emission tip covered with CO, with time of irradiation or total number of photons impinged (wavelength 250 nm). (b) Same for a clean field-emission tip.

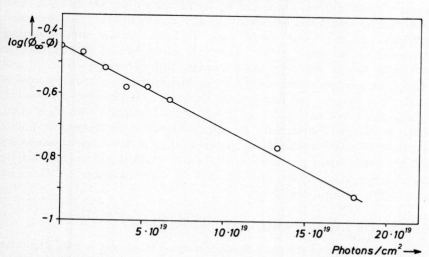

Fig. 7. Semilogarithmic plot of the data of curve a of Figure 6, corrected by those of curve b (see text).

desorption and contamination are additive, the corrected values for the former can be extracted. Following the analysis given by Menzel and Gomer (1964a) for electron impact desorption (which assumes a linear relationship between work function change and coverage of the respective species, and a first-order dependence of the desorption rate on the number of the impinging particles) the desorption cross section can be obtained by plotting the approach of the work function to the asymptotic value, semi-logarithmically, versus the number of photons impinged. This plot should be linear, which is seen to be the case (figure 7). The value found for the desorption cross section is 6×10^{-21} cm^2. (If the correction is not applied, a value of 8×10^{-21} cm^2 results.) This value should be considered only as a rough estimate, because of the assumptions and difficulties involved.

III. DISCUSSION

EXCLUSION OF THERMAL EFFECTS

The use of the automatic temperature controller excludes any appreciable contribution by bulk heating of the metal ribbon. It would still seem possible, however, that the comparatively small penetration of the metal by the photons (probably in the order of 20 to 100 nm) leads to a temperature increase only of the surface layer of the metal, which would not become noticeable in the bulk temperature. The inverse variation of desorption signal and power input with wavelength, and the very strong wavelength dependence of the former provides an experimental argument against such a local heating effect. On the other hand, these findings would still be compatible with such an effect, if the absorption coefficient of the metal would show a strong increase with decreasing wavelength in the region below 350 nm. It would then be possible that the dissipation zone of the metal would decrease with decreasing wavelength and lead to an increase of the local heating. The available data on the optical constants of tungsten and molybdenum (Roberts, 1959; Waldron and Juenker, 1964) indicate that no pronounced change of the absorption takes place in the region in question. Furthermore, a rough calculation of the temperature profile perpendicular to the surface shows that no heating of the surface region occurs with the power inputs used here, even if the total dissipation takes place directly at the surface. We conclude, therefore, that the observed desorption effects are not caused by thermal desorption.

NATURE OF THE ADSORBED SPECIES CONCERNED

The elimination of the photodesorption signal by heating to 600°K and the increase of the work function by photodesorption show that the species active in photodesorption is the weakly bound α-CO. Comparison with electron desorption experiments (Menzel, 1968) suggests that, under the conditions used in this work, only that part of the α-layer is present which leads to O^{+}-desorption under electron impact and which has been termed α_2-CO. If a coverage of 1×10^{14} per cm^2 is assumed for this species, a desorption cross section of about 4×10^{-21} cm^2 results for 250 nm. The value of 6×10^{-21} cm^2 estimated from the FEM experiments compares quite well with this result, if the difficulties mentioned above, the somewhat different temperatures (about $311°K$ in the measurements with the ribbon, about $296°K$ in the FEM measurements), which could lead to different coverages, and the different angles of incidence are taken into account.

CONCLUSIONS FROM THE RESULTS AND COMPARISON WITH FORMER WORK

These experimental results and the discussion so far show that photo-desorption of α-CO on W is induced by light of wavelengths below 450 nm (2.75 eV), and that the desorption probability rises with decreasing wavelength to about 4×10^{-7} at 250 nm (about 5 eV), at an angle of incidence of $60°$. The desorption cross section at 250 nm is in the order of 5×10^{-21} cm^2. The desorbing species is found to be CO only. Effects other than desorption do not appear to be prominent.

The finding by Lange (1965) of strong thermal effects is compatible with our results, as is his result for the strongly bound CO. That Adams and Donaldson (1965) did not see any desorption from CO/W may be connected with their lower detection sensitivity and possibly also with the vertical incidence of the light beam used by them. Definite disagreement with the findings of Moesta *et al.* (1969) and Moesta and Trappen (1970) concerning the nature of desorbing particles and the quantum efficiency of the processes must be stated. With regard to the first point, our findings (see section II.C.) suggest that wall effects were prominent in their measurements; as to the second, an exact comparison is difficult, as they do not give exact numbers in their paper, but their effects seem to be much higher than even those reported here for wall effects. No explanation can be offered.

While the published work on CO on Ni cannot be compared with this work, it is still interesting to note that in the present system non-thermal photo-desorption has been found (in the case of CO/Ni Schram (1970) attributed all effects reported to thermal desorption), and that no dramatic work function increases as reported by Moesta and Breuer (1968, 1969) for CO/Ni have been found for CO/W.

That no photodesorption of oxygen from tungsten has been found, might mean that there is simply no molecular oxygen present on the surface under the conditions of the experiment (desorption of atomic oxygen would probably not be detectable). On the other hand it is interesting that up to now CO is the only adsorbate which has been found to be photodesorbed from any metal.

THE MECHANISM OF PHOTODESORPTION OF CO FROM W

A thermal nature for the observed effects has been excluded. Another mechanism which can be excluded is the emission of photoelectrons, which would lead to electron impact desorption: even at the higher energies used here, where emission of photoelectrons occurs at all, the energies are too small ($h\nu - \phi$) to lead to desorption.

It seems obvious, therefore, that the observed photodesorption must proceed via an electronic excitation. Two basic mechanisms appear to be possible, which differ in the kind of primary excitation by absorption of a photon.

(1) An impinging photon can be absorbed by the metal, resulting in the excitation of an electron from the conduction band to an unoccupied state above the Fermi energy. In two ways this excitation energy could be transferred to the adsorbate:

(a) If the final energy and momentum of the electron have an appropriate value, and the excitation took place close enough to the surface, the electron could tunnel elastically through the surface barrier into an empty level of the adsorbate complex above the Fermi energy. Similar processes have been invoked for photoemission of electrons from metals into molecular crystals (Baessler and Vaubel, 1968; Many et al., 1968). The threshold for such a process would be the difference between the energy of the empty adsorbate level to be filled and the Fermi energy of the metal: this is the minimum energy required to lift an electron from the top of the conduction band to the empty level.

(b) If an adsorbate level exists within the energy range of the conduction band of the metal (which would be filled, of course), an electron could tunnel from it into the hole created in the conduction band by photon absorption. In this case the characteristics of the hole would have to correspond to this level. Similar processes have been postulated for hole emission from metals into anthracene, by Williams and Dresner (1967). The threshold of such a process would be the distance between the energy of the filled adsorbate level and the Fermi energy: this is the minimum energy required to create a hole in the conduction band at the energy of the filled adsorbate level.

Of these two possibilities the first seems to be more likely, as empty adsorbate levels above the Fermi energy certainly exist, while this is uncertain for filled levels in the range of the conduction band. In both cases, the created charged adsorbate state could either be repulsive itself (which seems unlikely), or could cross into a repulsive state during its life time. This crossing would again involve the tunneling of an electron to or from the metal.

If one of these mechanisms would operate, there should be a large primary excitation, of which only a small part would result in energy transfer because of the special requirements on the tunneling electron or hole, resp. The final desorption yield would be further cut down by decay of the charged surface complex before desorption—this process, which again involves the tunneling of an electron back to or from the metal, should be rather effective. The small overall desorption yield would thus be understandable. The found energy threshold seems to be compatible with both processes.

(2) The primary absorption could also take place in the adsorbate complex itself, leading either directly to a repulsive or to a metastable state which can cross into a repulsive state. The state which would be produced by the primary absorption can not be derived from the (slightly perturbed) molecular CO, as assumed by Moesta et al. (1969), as the lowest excited state of CO, the $a^3\pi$, lies 6 eV above the ground state. The low threshold energy shows that the bond formation to the metal must lead to new, low-lying unoccupied MO's. Although it is not clear yet, to what extent a parallel may exist between CO-adsorption on transition metals and the respective carbonyls, it is interesting to note two features of $W(CO)_6$: (1) It possesses two very strong absorption bands around 287 and 224 nm, which are attributed to charge transfer from the metal to the ligand, and a weaker (d–d) band at 307 nm (Gray and Beach, 1963); and (2) it can be photolysed (i.e. one CO molecule can be split off) by light of 370 and even of 430 nm, in solution (see for instance Strohmeier, 1964; Koerner von Gustorf and Grevels, 1969). If the bonding of α-CO is similar to that in $W(CO)_6$, which is quite likely, then one should expect similar behavior. The observed photodesorption would then correspond to photodissociation of the carbonyl. It is seen that the energy ranges for these two processes coincide roughly. However, both the mechanism of photolysis of the carbonyl and the bonding of the adsorbed CO will have to be better understood, before a more detailed discussion of these parallels is possible. Also, measurements of the photo-desorption at lower wavelengths should be informative here; these are planned.

It is seen that unfortunately no decision between the two basic absorption mechanisms is possible on purely energetic grounds. The fact that adsorbed CO seems to be especially prone to photodesorption, contrary to all other adsorbates investigated so far, suggests that the second mechanism is more likely, as it contains more prominently the properties of the adsorbate. But it should be noted that the difference of the two absorption mechanisms is not as large as may appear at first sight: if a strong localization of the excitation in the metal is required, the first mechanism can go over into the second.

More experimental and conceptual work is required for a better understanding of these processes. Such investigations may help in the understanding of the surface bond.

ACKNOWLEDGEMENTS

We thank Dr. A. M. Bradshaw and Dr. E. Michel-Beyerle for valuable discussions. Financial support from the Deutsche Forschungsgemeinschaft and the Fonds der Chemischen Industrie is gratefully acknowledged.

REFERENCES

Adams, R. O. and Donaldson, E. E. (1965). *J. Chem. Phys.* **42**, 770.

Baessler, H. and Vaubel, G. (1968). *Solid State Commun.* **6**, 97.

Ford, R. R. (1970). *Advan. Catal. Relat. Sub.* **21**, 51.

Gomer, R. (1961). "Field Emission and Field Ionization". Harvard University Press, Cambridge, Mass.

Gray, H. B. and Beach, N. A. (1963). *J. Amer. Chem. Soc.* **85**, 2922.

Hansen, R. S. and Gardner, N. C. (1968). *In* "Experimental Methods in Catalytic Research" (R. B. Anderson ed.), p. 189. Academic Press, New York and London.

Koerner von Gustorf, E. and Grevels, F. W. (1969). *Fortschr. Chem. Forsch.* **13**, 366. Springer Verlag, Berlin.

Lange, W. J. (1965). *J. Vac. Sci. Technol.* **2**, 74.

Lange, W. J. and Riemersma, H. (1962). 1961 Transactions American Vacuum Society, p. 167. Pergamon Press, New York.

Many, A., Levinson, J. and Teucher, I. (1968). *Phys. Rev. Lett.* **21**, 1161.

Menzel, D. (1968). *Ber. Bunsenges. Phys. Chem.* **72**, 591.

Menzel, D. (1970). *Angew. Chem. Int. Ed. Engl.* **9**, 255.

Menzel, D. and Gomer, R. (1964a). *J. Chem. Phys.* **41**, 3311.

Menzel, D. and Gomer, R. (1964b). *J. Chem. Phys.* **41**, 3329.

Moesta, H. and Breuer, H. D. (1968). *Naturwissenschaften.* **55**, 650.

Moesta, H. and Breuer, H. D. (1969). *Surface Sci.* **17**, 439.

Moesta, H. and Trappen, N. (1970). *Naturwissenschaften* **57**, 38.

Moesta, H., Breuer, H. D. and Trappen, N. (1969). *Ber. Bunsenges. Phys. Chem.* **73**, 879.

Redhead, P. A. (1967). *Nuovo Cimento, Suppl.* **5**, 586.

Roberts, S. (1959). *Phys. Rev.* **114**, 104.

Schram, A. (1970). *Nederlands Tijds. Vacuümtech.* **8**, 85.

Strohmeier, W. (1964). *Angew. Chem.* **76**, 873.

Waldron, J. P. and Juenker, D. W. (1964). *J. Opt. Soc. Amer.* **54**, 204.

Williams, R. and Dresner, J. (1967). *J. Chem. Phys.* **46**, 2133.

KINETICS OF ADSORPTION AND DISPLACEMENT IN THE INTERACTION OF H_2 AND CO WITH CLEAN NICKEL SURFACES

A. M. HORGAN* and D. A. KING

School of Chemical Sciences, University of East Anglia, Norwich, England

I. INTRODUCTION

Nickel is widely used as a hydrogenation catalyst in industry, and its particular application to the formation of methane from hydrogen and carbon monoxide at elevated temperatures has been extensively investigated (Greyson, 1956). In moving towards a fundamental understanding of the mechanism of catalytic reactions it is necessary to examine the surface interactions between the reactants under well-controlled conditions. The study of interactions between different chemisorbed species also serves as a basis for a better understanding of the physical and chemical nature of each of the chemisorbed components. Thus a study of the interaction between CO and O_2 on Ni films (Horgan and King, 1971) revealed that different adsorbed states formed from one gas may be distinguished by their chemical reactivity, and Yates and Madey (1971a; 1971b) have recently used N_2 and CO as "chemical probes" to study the adsorption of H_2 on the (100) face of tungsten. The adsorption of both H_2 and CO on clean Ni surfaces has been extensively investigated (see Wedler, 1970), but a detailed kinetic study of the sequential adsorption of these gases has not been previously reported.

II. EXPERIMENTAL

The reflexion detector technique for determining adsorption efficiencies for the collision of gases with ultra-high-vacuum-deposited metal films, which has been described in detail (Horgan and King, 1968; 1970), is readily adapted to studies of the sequential adsorption of different gases on metal surfaces (Horgan and King, 1971). The sticking probability profile, as a function of coverage of a

* Present address: Space Sciences Laboratory, Marshall Space Flight Center, Huntsville, Alabama, USA

gas interacting with a surface precovered to a measured degree with an alien gas, can be measured and overall surface coverages in both gases and displacement or catalytic processes can be continuously monitored. Nickel films weighing between 30 and 50 mg were formed by vapour-deposition at pressures below 10^{-9} Torr (1 Torr $\equiv$ 133.3 Nm^{-2}) on the uncooled walls of a spherical 1 litre pyrex glass vessel (ultimate pressure $\sim$5 x 10^{-11} Torr). During adsorption, the gas is allowed to flow into the reaction vessel through a centrally positioned spherical diffuser, situated so as to achieve a uniform impingement rate across the metal surface, at a measured and constant rate, equivalent to an adsorption rate of about 10^{12} molecules cm^{-2} sec^{-1}. (This figure, and all coverages reported herein, refers to the geometric area of the film.) The partial pressures of the impinging gas and displaced species are determined with a 180° magnetic sector miniature mass spectrometer, calibrated with a Redhead-modulated ionization gauge, which is positioned so as to sample only gaseous molecules reflected from the metal surface and not molecules coming directly from the diffuser.

Using a bare tungsten filament as the mass spectrometer cathode small amounts of CH_4 were formed from H_2 and CO gas mixtures in the absence of a Ni film. The filament was thus coated with LaB_6 to reduce its operating temperature, and at 5 μA emission current no further CH_4 production was observed: mass spectra obtained from H_2 (B.O.C. high purity) and CO (prepared *in situ* by thermal decomposition of $Mo(CO)_6$) separately in the cell were, when superimposed, almost identical to those obtained from gas mixtures.

III. RESULTS AND DISCUSSION

ADSORPTION OF HYDROGEN

Results obtained for H_2 are in close agreement with those reported by Taylor and Creasey (1971) at this Symposium. Sticking probability profiles for H_2 adsorption on Ni films at adsorbent temperatures between 77 and 373°K are presented in figure 1. The sticking probability extrapolated to zero coverage, s_o, decreases with increasing temperature, from 0.65 at 77°K to 0.35 at 300°K, and falls monotonically with increasing coverage. On terminating the gas supply during adsorption the background pressure was only re-attained at low coverages and at temperatures below 290°K. Otherwise the H_2 pressure fell to a steady equilibrium value, indicating appreciable desorption from the primary adlayer. Sticking probabilities in figure 1 refer to the net rate of adsorption; when desorption from the adlayer is appreciable this is not the same as the absolute adsorption rate. The effect of desorption on the sticking probability has been examined in detail by Hayward *et al.* (1967). The total uptake for Ni films at

$300°K$ is about 30×10^{14} molecules cm^{-2} (geometric area); Eley and Norton (1970) report an uptake on Ni filaments of 5.7×10^{14} molecules cm^{-2}, yielding a roughness factor of about 5 for the films; the same figure was obtained from a consideration of the amount of oxygen adsorbed on Ni films (Horgan and King, 1970).

The hydrogen uptake at $77°K$ is smaller than that obtained at $195°K$; a similar result is reported by Savchenko and Boreskov (1968). This is clarified by

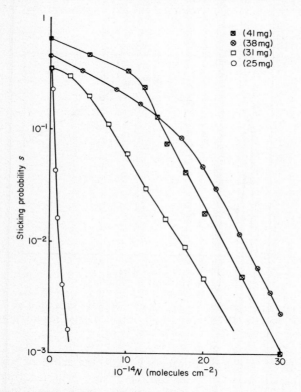

Fig. 1. Sticking probability profiles for H_2 on Ni films: $-\odot- 373°K$; $-\square-$ $300°K$; $-\otimes- 195°K$; $-\boxtimes- 77°K$.

the desorption spectrum (I) shown in figure 2, obtained after H_2 saturation of a Ni film at $77°K$ by removing the cold bath and recording the H_2 pressure changes in the cell, with the cell open to the pumps. Results are similar to those obtained by Hayward *et al.* (1966) for H_2 on Mo films: two peaks are observed, the major peak showing desorption over the range 180 to $280°K$ and corresponding to the desorption of 2.2×10^{14} molecules cm^{-2}. Following Hayward *et al.* (1966), the first peak is not necessarily attributable to the existence of a distinct low heat binding state on the surface: the fall in pressure in the range 110-180°K is more readily ascribed to an activated rearrangement of the original ("virgin") H_2 adlayer with the creation of new H_2 adsorption sites. This model is compatible with the larger uptake observed at 195°K over that obtained at 77°K. After the desorption spectrum I had been obtained, the film was re-cooled to 77°K. An additional 5.0×10^{14} molecules cm^{-2} H_2 were adsorbed to give an equilibrium pressure of 10^{-6} Torr; the desorption spectrum (II) subsequently obtained by warming the film again to room temperature is

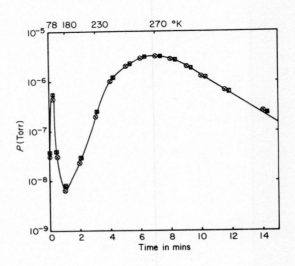

Fig. 2. Desorption spectra for H_2 from a Ni film saturated at 77°K. I ($\otimes$): first desorption spectra. II ($\boxtimes$): spectrum obtained after recooling to 77°K and adding a further 5×10^{14} H_2 molecules cm^{-2}.

shown in figure 2, and is identical to I—again the activated conversion was observed, and again 2.2×10^{14} molecules cm^{-2} were desorbed. On cooling once more to 77°K, a further 6×10^{14} molecules cm^{-2} were added to restore the equilibrium pressure to 10^{-6} Torr. By means of this re-cycling process, the H_2 coverage can thus be quite substantially increased.

An equilibrium isotherm was determined for the H_2/Ni system, and this is displayed as $\log_{10} P_{H_2}$ against θ_{H_2} in figure 3. For comparison purposes, isotherms constructed from the data of Rideal and Sweett (1960) and Savchenko and Boreskov (1968) are included, the data being normalized according to the (somewhat arbitrary) definition of Rideal and Sweett, that the fractional coverage θ is unity at 77°K and a H_2 pressure of 1.5×10^{-3} Torr after cooling in H_2 from room temperature at a uniform rate over a period of 3 hours.

Agreement between results obtained in different laboratories is excellent. The isotherms are analytically described by the Temkin relationship

$$\log aP_{H_2} = \kappa\theta_{H_2} \tag{1}$$

which may be derived from the assumption that the adsorption heat q is a linear function of θ (Hayward and Trapnell, 1964)

$$q = q_0(1 - b\theta) \tag{2}$$

where q_0 is the adsorption heat at zero coverage and b is a constant. The four isotherms shown are consistent with $q_0 = 27,350$ (±700) cal/mole and $b = 0.80$ (±0.15), and are described by the relationship

$$\theta_{H_2} = 1\cdot25 + (T/4760)(\log P_{H_2} - 7.3). \tag{3}$$

ADSORPTION OF CARBON MONOXIDE

Detailed results for the interaction of CO with Ni films have been given (Horgan and King, 1969), and we reproduce here (figure 4) the sticking probability profiles for comparison with the profiles obtained on H_2-saturated Ni surfaces. At temperatures between 77 and 373°K, s_0 is close to unity (>0.99) and initially independent of coverage. Desorption spectra are similar to those observed for H_2 on Ni (figure 2), showing readsorption due to an activated rearrangement of the adsorbate in the temperature range 130-190°K.

INTERACTION OF HYDROGEN WITH PREADSORBED CARBON MONOXIDE

A Ni film presaturated with CO to an equilibrium pressure of 10^{-7} Torr at 300°K does not take up any measurable quantity of H_2 ($<10^{13}$ molecules cm^{-2}) at either 300 or 77°K. In contrast, Ni films saturated with CO in the same way were found to take up, at 77°K, a further 2×10^{14} molecules cm^{-2} of CO; or 3.3×10^{14} molecules cm^{-2} of O_2; or 3×10^{14} molecules cm^{-2} of N_2. Thus sites are available on the surface for the low heat γ states of CO, O_2 and N_2, but not for H_2. It may reasonably be assumed that the γ states of the former gases are non-dissociatively adsorbed and that these are adsorbed on isolated sites on the surface, in which case non-adsorption of H_2 would be attributable to the requirement for two vacant neighbouring sites on the surface. However, Eley and Norton (1966) concluded that γ-H_2 is also a non-dissociatively bound species, since the para-ortho H_2 and para-ortho D_2 conversions at low temperatures were 10^3 times faster than the H_2–D_2 exchange reaction at 77°K, and it is unlikely that the requirement for two vacant neighbouring sites would exist for a non-dissociatively adsorbed species. A more attractive explanation may be offered in terms of the exclusion of suitable sites for γ–H_2 adsorption, through collective adsorbate-adsorbate interactions in the CO adlayer.

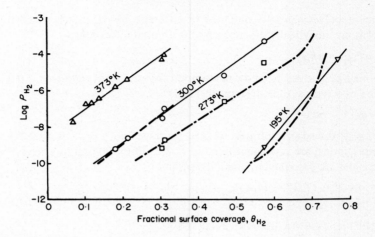

Fig. 3. Adsorption isotherms for H_2 on Ni films. — — — —, present work; — — — —, from Savchenko and Boreskov (1968); experimental points, constructed from Rideal and Sweett (1960).

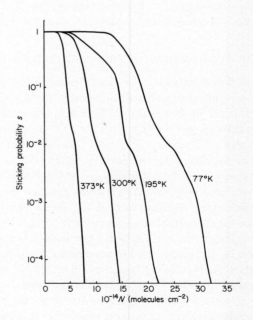

Fig. 4. Sticking probability profiles for CO on Ni films.

INTERACTION OF CARBON MONOXIDE WITH PREADSORBED HYDROGEN

After saturation with H_2, Ni films remained receptive towards CO, the nature of the rather complex interaction being markedly temperature dependent. Under all conditions examined, only H_2 and CO were detected in the gas phase: methanation was not observed, as expected at the low temperatures employed.

Cooperative Adsorption at 77°K

A Ni film was saturated with H_2 at 77°K until a steady equilibrium pressure of 7×10^{-8} Torr was attained; the coverage was 25×10^{14} H_2 molecules cm^{-2}. Results obtained when CO was allowed to flow into the cell at a steady rate are shown in figure 5(i). The CO sticking probability profile is very similar to that

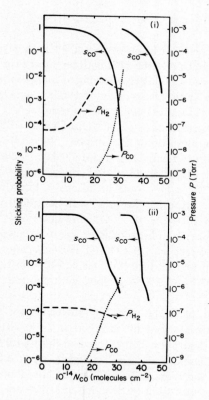

Fig. 5. Interaction of CO with preadsorbed H_2 on Ni films at 77°K. (i) H_2 preadsorbed at 77°K to coverage of 25×10^{14} H_2 molecules cm^{-2}. (ii) H_2 preadsorbed at 300°K to coverage of 23×10^{14} H_2 molecules cm^{-2}.

obtained for a clean Ni film at $77°K$ (figure 4): thus, the initial value of s is close to unity and is independent of coverage up to $\sim 10 \times 10^{14}$ CO molecules cm^{-2}; and the uptake required to reduce s to 10^{-4} is 30.5×10^{14} CO molecules cm^{-2}, compared with 31×10^{14} for a clean film. As shown, a small amount of H_2 is displaced from the surface during the latter course of the CO uptake; from the integrated form of equation (1) the total amount displaced was determined as 7×10^{13} H_2 molecules cm^{-2}. Eley and Norton (1970) concluded that the molecular γ state of H_2 on Ni, which desorbs at $\sim 120°K$, was about 10% of the total adsorbed H_2; the amount displaced at $77°K$ by CO is about 3% of the total, which suggests an identity with the γ state. This was verified in a further experiment, in which a Ni film was saturated with H_2 at $300°K$ and cooled to $77°K$ before introducing CO. Results are shown in figure 5(ii), and indicate very similar behaviour to that for H_2 preadsorption at $77°K$, except that no H_2 was displaced by CO.

Desorption Spectra

After a mixed adlayer had been formed at $77°K$ as described above, the cell was opened to the pumps and the cold bath removed so as to obtain desorption spectra for both H_2 and CO during warming to room temperature. A very small CO desorption peak was observed at $\sim 100°K$, equivalent to $<10^{10}$ CO molecules cm^{-2}; this contrasts with the substantial CO desorption observed from a CO adlayer formed on a clean Ni surface at $77°K$ (Horgan and King, 1969). Substantial H_2 desorption was observed, however, and the spectra were similar to those shown in figure 2 for a clean Ni film, including the smaller peak at low temperatures; a total of 5.5×10^{14} H_2 molecules cm^{-2} were desorbed. On re-cooling to $77°K$, no H_2 would be adsorbed, but a further 16×10^{14} CO molecules cm^{-2} were taken up (figure 5) with an initial sticking probability of unity, giving a total of 47×10^{14} CO molecules cm^{-2} and 19×10^{14} H_2 molecules cm^{-2}. Since the film roughness factor is about 5 (see above), this implies a surface coverage of 1.3 CO and H_2 *molecules* per surface Ni atom. On warming to room temperature, CO was desorbed in two well-defined states, with maxima at $130°K$ and $270°K$, involving a total desorption of 5×10^{14} CO molecules cm^{-2}.

In a further experiment, H_2 (32×10^{14} molecules cm^{-2}) and CO (35×10^{14} molecules cm^{-2}) were sequentially adsorbed at $77°K$. After warming to $300°K$, only negligible CO desorption being observed during this thermal treatment, a further 18×10^{14} CO molecules cm^{-2} could be adsorbed at $300°K$ before the sticking probability was reduced to 10^{-4}, giving a total uptake of 53×10^{14} CO molecules cm^{-2} and again illustrating that the presence of the H_2 adlayer is capable of inducing a substantial degree of CO-supersaturation of the surface.

Co-adsorption at 195-373°K

The results of experiments on three Ni films, respectively saturated with H_2 at 195, 300 and 373°K prior to CO adsorption at these temperatures, are presented in figure 6, and summarized in table I together with results obtained at 77°K. CO adsorption at these temperatures is accompanied by an initial increase in the H_2 equilibrium pressure in the cell as H_2 is displaced from the surface, but at high coverages the H_2 equilibrium pressure falls. This may either be due to depletion of the H_2 adlayer as pumping through the diffuser becomes appreciable, or, more likely, it is due to the formation of a CO-H surface complex with a higher adsorption heat and hence lower equilibrium pressure

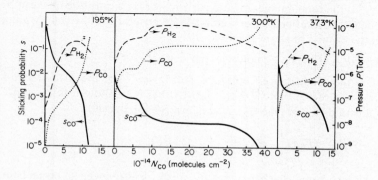

Fig. 6. Interaction of CO with preadsorbed H_2 on Ni films at 195, 300 and 373°K.

than that for the simple H adatom. At 195°K, the CO uptake is less than that for a clean film, but at 300 and 373°K there is an appreciable enhancement in the CO uptake due to the H_2-adlayer. At all temperatures between 195 and 373°K the CO sticking probability profiles are significantly different from those obtained on clean films (figure 4).

Model for Hydrogen Displacement

The data for the displacement of H_2 have been analysed by the following procedure. Two basic assumptions are made.

(i) During CO adsorption, at a rate of $\sim 10^{12}$ molecules cm^{-2} sec^{-1}, the H_2 gas phase pressure remains in equilibrium with the H_2 adlayer. From the Knudsen collision factor and the H_2 sticking probability this can be shown to be a reasonable assumption provided that $P_{H_2} > 10^{-7}$ Torr.

(ii) The Temkin isotherm, equation (1), relating the H_2 pressure to the fractional H_2 coverage θ_{H_2} at temperatures between 195 and 373°K and derived from data for clean Ni surfaces, is assumed to be applicable to surfaces partially covered with CO. θ_{H_2} is then the fractional H_2 coverage for the remainder of the surface, and the effect of CO adsorption is simply to reduce the effective area for H_2 adsorption.

If we define α as the number of CO molecules which occupy one H_2 *molecule* site (or two H atom sites), then at a total CO coverage N_{CO} (molecules cm^{-2}) the number of sites per cm^2 remaining for H_2 adsorption is

$$N_s = N_s{}^0 - \alpha N_{CO} \tag{4}$$

where $N_s{}^0$ is the total number of sites per cm^2 for H_2 adsorption on clean Ni.

During CO adsorption, H_2 is continually removed from the cell by pumping through the diffuser at a rate given by

$$-dN_{H_2}/dt = (F/A)P_{H_2} \quad \text{molecules cm}^{-2} \text{ sec}^{-1} \tag{5}$$

where F is the conductance (molecules Torr^{-1} sec^{-1}) of the diffuser and A the geometric area of the film (500 cm^2). If CO has been allowed to flow into the cell for a period of t sec, the H_2 surface coverage is reduced to

$$N_{H_2} = N_{H_2}^0 - (F/A) \int_0^t P_{H_2} \, dt - P_{H_2}(V/kT) \tag{6}$$

where the last term, which is generally small, is introduced to correct for the number of H_2 molecules in the gas phase in the cell. At constant temperature the H_2 equilibrium is described by the Temkin isotherm

$$\log aP_{H_2} = \kappa \theta_{H_2}$$
$$= \kappa(N_{H_2}/N_s).$$

Inserting equations (4) and (6) into this equation, we have on rearranging

$$1 - \frac{\kappa\left[\theta_{H_2} - (F/AN_s{}^0) \int_0^t P_{H_2} \, dt - P_{H_2}(V/kT)\right]}{\log aP_{H_2}} = \alpha\theta_{CO} \tag{7}$$

where $\theta_{H_2}^0$ is the fractional coverage in H_2 at $t = 0$ and $\theta_{CO} \ (= N_{CO}/N_s{}^0)$ is the fractional coverage in CO at time t after starting CO adsorption. The left-hand side of equation (7) should thus be a linear function of θ_{CO}, with slope α. Results from the data in figure 6 are plotted according to equation (7) in figure (7); for each run the value of $N_s{}^0$ was determined from the H_2 equilibrium pressure and coverage prior to CO adsorption by reading off the fractional coverage given by the relevant isotherm in figure 3. Temkin constants κ and a were also derived from these isotherms.

At $195°K$ equation (7) yields a good straight line of unit slope up to a coverage of 5×10^{14} CO molecules cm^{-2}; at higher coverages, however, the slope tends to zero and equation (7) is no longer a useful description of the results. At $300°K$ equation (7) is a good description of the displacement process up to a coverage of 22×10^{14} CO molecules cm^{-2}, with a slope of unity, but again at higher coverages the slope tends to zero. The results obtained at $373°K$ do not yield a satisfactory plot, but in the fractional coverage range 0.07 to 0.34 the curve again has unit slope.

We conclude that at relatively low CO coverages and pressures the model is in accord with the data, with $\alpha = 1$: one CO molecule utilizes one H_2 molecule

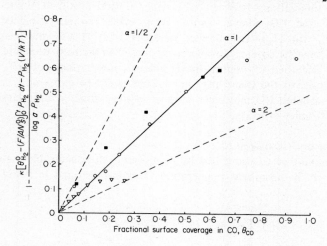

Fig. 7. Analysis of H_2 displacement from Ni surfaces by CO according to equation (7). $\triangledown$, $195°K$; $\circ$, $300°K$; $\blacksquare$, $373°K$.

adsorption site. Since the H_2 adlayer is probably dissociatively adsorbed it could be argued that at low coverages the CO molecule is bridge-bonded to two surface

Ni atoms, i.e. $\begin{matrix} Ni \\ \\ Ni \end{matrix} \diagdown \diagup C=O$, thereby excluding two H adatoms from the surface. At

higher CO coverages the H_2-induced CO-supersaturation process becomes important, and the simple displacement model breaks down. Lapujoulade (1971) has recently obtained evidence for complex formation at high overall coverages by observing desorption spectra from mixed H_2 and CO adlayers on Ni (111). However, the applicability of the model even over a restricted coverage range is surprising, since it implies, from assumptions (i) and (ii) above, that the adsorption heat q against fractional coverage θ relationship for H_2 from which the Temkin isotherm was derived (equation (2)) is completely unaltered by the presence of a substantial amount of CO. Thus, for example, since both the CO and the H_2 adlayers are mobile at $300^\circ K$, if the observed q against θ relationships for H_2 and CO were attributable to *a priori* heterogeneity in the metal surface as a result of crystallographic specificity, the q against θ relationship for H_2 should be progressively altered by an encroaching CO adlayer which, having the higher adsorption heat, would preferentially seek out the higher binding-energy sites. It is concluded, therefore, that the fall in q with increasing θ is an induced effect for either (or both) CO or H_2–H, as the smaller adatom, is more likely to be crystallographically non-specific. This means that interactions between H adatoms are strong and repulsive (q falls as θ increases), and that interactions between adsorbed CO and adsorbed H are negligible (otherwise the CO adlayer would again effect an alteration to the q–θ relationship for H_2). The absence of appreciable H–CO interaction on the surface at low CO coverages could be attributed to the formation and growth of islands of adsorbed CO within the H adlayer, interactions between CO and H being reduced to the island periphery. This would be expected if interactions between adsorbed CO molecules were attractive, which is quite feasible in the model of Grimley (1967).

CO Uptake on H_2-covered Ni

The amount of CO adsorbed by H_2-covered Ni is a sensitive and complex function of the adsorbent temperature. This is illustrated in figure 8 for the

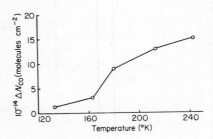

Fig. 8. CO uptake by a H_2-covered Ni film as a function of temperature; each point represents a running total of the successive amounts of CO required to reduce the CO sticking probability to 0.08 as the film was warmed up in steps.

range 130 to $240^\circ K$, where the total amount of CO adsorbed before the sticking probability falls to 0.08 is seen to undergo a significant increase at about $180^\circ K$. There is a minimum in the uptake at about $100^\circ K$, since it was shown (figure 5) that at $77^\circ K$ the equivalent CO uptake is 27×10^{14} molecules cm^{-2}. It is remarkable, therefore, that no CO desorption occurs on warming to $300^\circ K$. Apparently CO adsorption at $77^\circ K$ occurs into a binding state with an

Table 1. *Interaction of carbon monoxide with preadsorbed hydrogen*

Temperature (°K)	H_2 preadsorbed (molecules cm^{-2})†	CO adsorbed on H_2-covered Ni (molecules cm^{-2})	CO adsorbed on clean Ni (molecules cm^{-2})	H_2 displaced by CO (molecules cm^{-2})	Total H_2 + CO adsorbed (molecules cm^{-2})
77	25×10^{14}	30.5×10^{14}	31×10^{14}	0.7×10^{14}	55×10^{14}
195	27×10^{14}	12×10^{14}	22×10^{14}	0.8×10^{14}	38×10^{14}
300	18×10^{14}	40×10^{14}	15×10^{14}	11.0×10^{14}	47×10^{14}
373	3.7×10^{14}	13.5×10^{14}	7.5×10^{14}	2.1×10^{14}	15.1×10^{14}

† All coverages refer to the geometric area of the adsorbent: the roughness factor of the Ni films is about 5.

adsorption heat of about 10 kcal/mole, this figure being too low for stable occupation at 130°K. On warming, the mixed adlayer undergoes a rearrangement before desorption can occur, a stable surface complex is formed with an adsorption heat >20 kcal/mole, and a fraction of the H_2 adlayer is desorbed. The rearrangement of the mixed adlayer may coincide with the rearrangement of the "virgin" H_2 adlayer, which takes place in the absence of CO at ~110°K (figure 2). At 300°K this film, with a total uptake of 62×10^{14} CO and H_2 molecules cm^{-2}, was capable of adsorbing a further 18×10^{14} CO molecules cm^{-2}; allowing for a roughness factor of 5, this total coverage is well above that required for a monolayer. Following Holscher and Sachtler (1966), this could be attributed to corrosion of the Ni surface, with the formation of a "surface compound" (Grimley, 1969); adsorbed hydrogen would appear to promote this corrosive process. Bulk nickel hydrocarbonyls such as $H_2Ni(CO)_6$ have been prepared (Behrens and Lohofer, 1961). Fixed stoichiometric ratios may be indicated by the results in table 1: thus the ratio of (CO enhancement) to (H_2 retained) at both 300 and 373°K is 3.6. Alternatively, the excess uptake may be attributed to solution of the H adlayer into the metal surface. The presence of positively charged protons between chemisorbed CO species, as in the model of Siddiqi and Tompkins (1962), would serve to reduce dipole-dipole interactions between neighbouring CO molecules and hence yield a more crowded surface layer. Further speculation at this stage, without a detailed characterization of the mixed adlayer, is not justified.

ACKNOWLEDGEMENT

Thanks are expressed to the Science Research Council for a Research Studentship to A.M.H.

REFERENCES

Behrens, H. and Lohofer, F. (1961). *Chem. Ber.* **94**, 1391.

Eley, D. D. and Norton, P. R. (1966), *Discuss. Faraday Soc.* **41**, 135.

Eley, D. D. and Norton, P. R. (1970). *Proc. Royal Soc. (London) Ser. A* **314**, 319.

Greyson, M. (1956), *In* "Catalysis" (P. H. Emmett, ed.) Vol. IV, chap. 6. Reinhold Publishing Company, New York.

Grimley, T. B. (1967). *Proc. Phys. Soc. London* **90**, 751.

Grimley, T. B. (1969). *In* "Molecular Processes on Solid Surfaces" (E. Drauglis, R. D. Gretz and R. I. Jaffee, eds.), p. 299. McGraw Hill, New York.

Hayward, D. O. and Trapnell, B. M. W. (1964). *In* "Chemisorption", p. 176. Butterworth, London.

Hayward, D. O., Taylor N. and Tompkins, F. C. (1966). *Discuss. Faraday Soc.* **41**, 75.

Hayward, D. O., King, D. A., Taylor N. and Tompkins, F. C. (1967). *Nuovo Cimento, Suppl.,* **5**, 374.

Holscher, A. A. and Sachtler, W. M. H. (1966). *Discuss. Faraday Soc.* **41**, 29.

Horgan, A. M. and King, D. A. (1968). *Nature,* **217**, 60.

Horgan, A. M. and King, D. A. (1969). *In* "The Structure and Chemistry of Solid Surfaces" (ed. G. A. Somorjai), Chap. 57, p. 1.

Horgan, A. M. and King, D. A. (1970). *Surface Sci.* **23**, 259.

Horgan, A. M. and King, D. A. (1971). *Trans. Faraday Soc.* **67**, 2145.

Lapujoulade, J. (1971). *J. Chim. Phys. Physicochim. Biol.* **68**, 73.

Rideal, E. K. and Sweett, F. (1960). *Proc. Royal Soc. London. Ser.* A**252**, 291.

Savchenko, V. I. and Boreskov, G. K. (1968). *Kinet. Katal.,* **9**, 142.

Siddiqi, M. M. and Tompkins, F. C. (1962), *Proc. Royal Soc. London. Ser.* A**268**, 452.

Taylor, N. and Creasey, R. (1971). This volume III, 7, p. 293.

Wedler, G. (1970). *In* "Adsorption", Chap. 5. Verlag Chemie., Weinheim.

Yates, J. T. Jr. and Madey, T. E. (1971, a). *J. Vac. Sci. Technol.* **8**, 63.

Yates, J. T. Jr. and Madey, T. E. (1971, b). *J. Chem. Phys.* In press.

ADSORPTION STUDIES WITH A Pd(111) SURFACE

G. ERTL and J. KOCH

Institut für Physikalische Chemie und Elektrochemie
Technische Universität, Hannover, Germany

I. INTRODUCTION

The "clean single crystal approach" has over the past few years been a rather successful and rapidly growing field in the study of phenomena at gas/solid interfaces. The most powerful tool for such investigations seems to be the low-energy electron diffraction (LEED) technique. However, LEED studies alone fall far short of being able to supply the complete picture of the complex surface processes. It is therefore necessary to combine different methods to supplement the LEED-derived information.

In the following a study of the interaction of oxygen and carbon monoxide with a palladium(111) surface by means of LEED, Auger electron spectroscopy, mass spectrometry and work-function measurements is reported. It will be shown that for these systems the situation is not too complex and therefore a relatively complete picture could be obtained. The experiments were performed with a four grid LEED system (Varian 120) with a glancing angle electron gun for Auger electron spectroscopy and a quadrupole mass spectrometer (Finnigan). Some measurements of the change in work function were made with the diode method but, in the main, an electronically self-compensating Kelvin method was used. The results of both methods agreed to within a few millivolts.

After crystallographic orientation the sample was cut as a cylindrical slice (diameter 6 mm), mechanically polished, and etched in aqua regia. The final cleaning process consisted of extensive cycles of Argon ion bombardment and heating until the Auger spectrum showed no detectable impurities, and the LEED pattern consisted of sharp spots from the Pd lattice. The main contaminants removed by this procedure were carbon and sulphur.

II. ADSORPTION OF CARBON MONOXIDE

CO is rapidly adsorbed on a clean surface at room temperature. The sticking coefficient s_0 at $\theta = 0$ is near to unity. The increase $\Delta\varphi$ of the work function as a function of exposure is shown in figure 1 and reaches a maximum value of 985 mV. Flash desorption experiments show a single peak at about 200°C. The area below a desorption peak is a measure of the adsorbed amount, i.e. the coverage. From such measurements it follows that $\Delta\varphi$ is proportional to θ. The relative sticking coefficient s/s_o as a function of coverage has been derived from figure 1

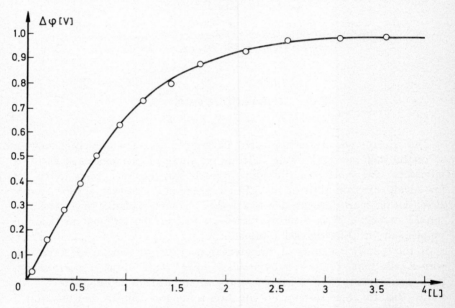

Fig. 1. Increase of the work function $\Delta\varphi$ as function of CO exposure.

and is shown in figure 2. The sticking coefficient s is constant up to about $\theta = 0.4\,\theta_{max}$ and then decreases linearly with coverage. We conclude from these observations that there may be a weakly bound second state, with a residence time long enough to allow the CO molecules to reach a free chemisorption site by surface diffusion. At higher coverages the mean diffusion path increases and competition with the desorption from the physisorbed layer takes place. Similar observations have been made for several other systems; a detailed discussion of such phenomena is given by Tracy and Blakely (1969).

At higher temperatures CO desorbs. For each temperature the equilibrium coverage for a certain CO partial pressure can be evaluated from work-function

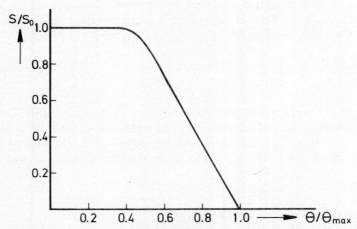

Fig. 2. Relative sticking coefficient s/s_o at room temperature of CO as function of coverage.

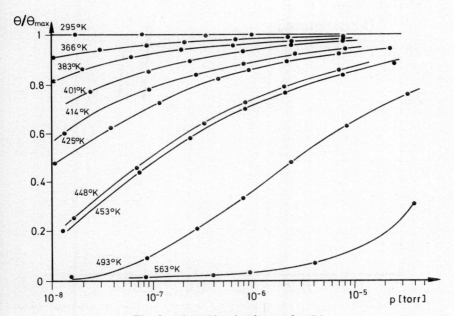

Fig. 3. Adsorption isotherms for CO.

measurements. Similar experiments have been performed by Tracy and Palmberg (1969) with the system CO/Pd(100). From these adsorption isotherms the differential isosteric heat of adsorption can be obtained by applying the Clausius-Clapeyron equation

$$\frac{d\ln p}{d(1/T)}\bigg|_{\theta\,=\,\text{const}} = -\frac{E_{ad}}{R}$$

A series of our experimental adsorption isotherms is shown in figure 3 and the evaluated isosteric heat of adsorption E_{ad} as a function of coverage in figure 4. It can be seen that E_{ad} is 34 kcal/mole independent of the coverage up to $\theta = \frac{2}{3}\theta_{\max}$ at which point there is a sudden decrease of 2 kcal/mole and thereafter a

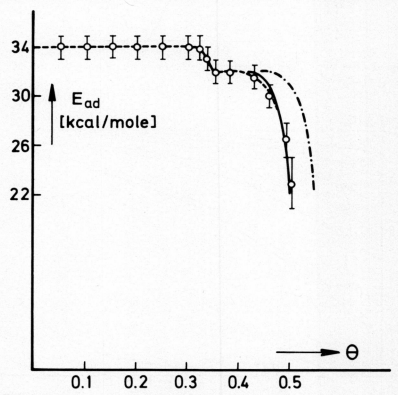

Fig. 4. Isosteric heat of adsorption of CO as a function of coverage. (The maximum coverage $\theta = \frac{1}{2}$ has been derived from the structural model as discussed below.) (a) $\circ$ experimental values; (b) $-\cdot-\cdot-$ Theoretical curve with the Lennard-Jones parameters for gaseous CO; (c) ————— Theoretical curve with σ 3.95 Å (see text).

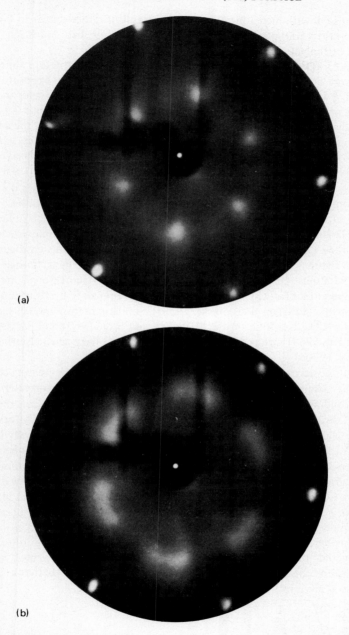

(a)

(b)

Fig. 5. LEED patterns after CO adsorption. $U = 34$ V. (a) $\sqrt{3} \times \sqrt{3}/R30°$ structure corresponding to $\theta = \frac{1}{3}$; (b) Pattern for maximum coverage $\theta = \frac{1}{2}$.

continuous fall with increasing coverage. This result is rather similar to the values found for the (100)-plane (Tracy and Palmberg, 1969; Ertl and Koch, 1970a) and the (110) plane (Ertl and Rau, 1969).

In the LEED pattern extra spots of a $\sqrt{3} \times \sqrt{3}/R\ 30°$ structure appear after an exposure of about 0.5 L (figure 5a). These spots split at an exposure leading to about $\frac{2}{3}$ of the maximum coverage and move continuously with increasing coverage until the final stage is reached after an exposure of about 3 L (figure 5B). Analysis of the diffraction pattern in figure 5(b) reveals that the adsorbed layer consists of three domain orientations of ordered structures which are rotated by 60° with respect to each other. The size of the CO molecule is such that only *one* adsorbed particle may be contained in the unit cell of the adsorption structure. Unfortunately, current theoretical treatments do not yet allow complete structural analysis from the diffraction intensities, i.e. the position of the surface mesh relative to the Pd atoms cannot be determined from the LEED pattern. However, only one possible surface structure exists in which the adsorbed CO molecules are all in equivalent positions. This configuration is shown in figure 6(c): at maximum coverage each CO molecule is "bridge" bound to two Pd atoms. (It is assumed that the molecule is fixed through the carbon atom to the surface for which strong evidence is given mainly by IR spectroscopic measurements (Eischens and Pliskin, 1958)). Similar structure models have been derived for CO adsorption on Pd(110) (Ertl and Rau, 1969) and Pd(100) (Park and Madden, 1968) surfaces. A detailed discussion has been given elsewhere (Ertl and Koch, 1970b).

As no drastic changes occur in the adsorption state with increasing coverage, it can be assumed that at lower coverages the adsorbed molecules exhibit similar binding configurations. The resulting structure model for the $\sqrt{3} \times \sqrt{3}/R30°$ structure corresponding to $\theta = \frac{1}{3}$ is given in figure 6(a). Increasing the coverage leads to a compression of the unit cell as shown schematically in figure 6(b) and causes the observed continuous splitting of the diffraction spots until saturation is reached at $\theta = \frac{1}{2}$. With these structure models the variation of the heat of adsorption E_{ad} with coverage can be interpreted easily: Below $\theta = \frac{1}{3}$ the adsorbed CO molecules are far apart without any significant mutual influence. Therefore E_{ad} remains constant. At $\theta > \frac{1}{3}$ the molecules become displaced from their symmetric positions, hence the sudden decrease of E_{ad} by 2 kcal/mole at $\theta = \frac{1}{3}$. A further increase of the coverage leads to van der Waals repulsion. (It can be estimated that the dipole-dipole interaction can be neglected at these intermolecular distances.) For $\frac{1}{3} \leqslant \theta \leqslant \frac{1}{2}$ the distance y between two CO molecules is given by

$$y = \frac{a_0}{2} \cdot \sqrt{3 + \frac{1}{\theta^2}}$$

where $a_0 = 2.73$ Å is the Pd–Pd distance.

Assuming a pairwise Lennard-Jones potential for the interaction between the adsorbed particles the repulsion energy as a function of coverage can be calculated. Using a value for the parameter σ—describing the "size" of the molecule—as found for gaseous CO (Corner, 1946), $\sigma = 3.76$ Å, we obtained the

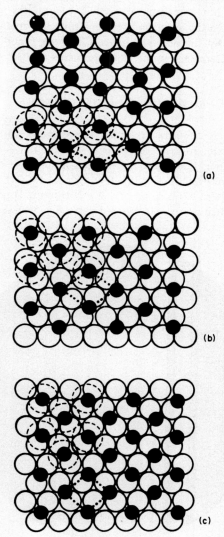

Fig. 6. Structure models for CO adsorption. (a) $\theta = \frac{1}{3}$ (three domain orientations). (b) $\frac{1}{3} < \theta < \frac{1}{2}$. (c) $\theta = \frac{1}{2}$ (Only one domain orientation is shown in (b) and (c)).

curve *b* in figure 4. On this basis the heat of adsorption should begin to decrease at somewhat closer distances than observed, i.e. higher coverage. The best fit to the experimental data is obtained by using $\sigma = 3.95$ Å (curve *c* in figure 4). This would mean that the adsorbed CO is slightly "bigger" than the gaseous molecule. This assumption is not unreasonable when one considers that the adsorption process involves a partial electron transfer from the metal to the adsorbed molecule as shown by the observed increase in work function and lowering of the C=O stretching frequency in the IR experiments.

III. ADSORPTION OF OXYGEN

Like CO, oxygen adsorbs at room temperature with a high sticking coefficient. The maximum increase of the work function is 610 mV. This value is considerably higher than that for oxygen adsorption on Pd(100), where $\Delta\varphi_{max} = 180$ mV has been observed (Ertl and Koch, 1970a).

In the LEED pattern extra spots of a 2 × 2 structure appear after adsorption of O_2 (figure 7). When an oxygen-covered surface is heated the adsorbate spots

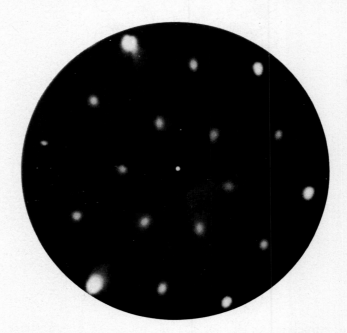

Fig. 7. LEED pattern of the 2 × 2-structure after oxygen adsorption. $U = 34$ V.

disappear at about 250°C as well as $\Delta\varphi$ decreases, but desorption does not occur at this temperature. It must therefore be concluded that oxygen diffuses into the bulk at moderate temperatures. The activation energy for this diffusion process could be estimated by observing the LEED pattern and by following the rate of change of the work function at different temperatures, and was found to be about 20 kcal/mole. Desorption of adsorbed oxygen starts at about 400°C. The evaluation of adsorption isotherms through the measurement of work-function

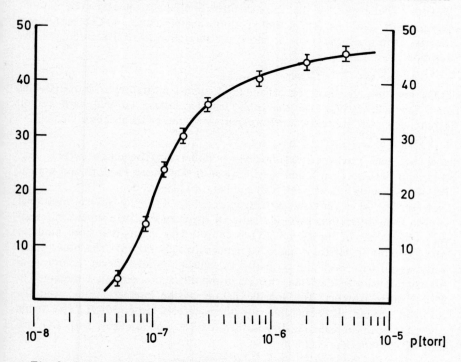

Fig. 8. Auger spectra for adsorbed oxygen. Peak height (arbitrary units) as a function of oxygen pressure at 400°C.

changes was complicated by the competition of the diffusion processes. However from observations at higher temperatures the adsorption energy could be estimated to be in the range of 50-60 kcal/mole. This value is in accord with the findings for the (110)- (Ertl and Rau, 1969) and (100)-plane (Ertl and Koch, 1970a). However bulk diffusion was only observed with the (111) surface.

In the Auger spectrum oxygen causes a strong transition at 520 V. The height of these signals is a measure for the concentration. As shown in figure 8 the oxygen coverage reaches saturation at a pressure of some 10^{-6} Torr at 400°C.

IV. INTERACTION BETWEEN OXYGEN AND CARBON MONOXIDE

The interaction between oxygen and carbon monoxide can be observed directly in the LEED pattern since both gases form different ordered adsorption phases. In addition the rate of catalytic reaction could be measured by means of a quadrupole mass spectrometer. When an oxygen-covered surface is exposed to gaseous CO the spots of the 2 x 2-O-structure disappear quickly even at room temperature and the diffraction pattern characteristic of CO adsorption appears. The mass spectrometer shows the formation of CO_2 according to the reaction

$$O_{ad} + CO_{gas} \rightarrow CO_{2,gas}$$

(CO_2 does not chemisorb on palladium at room temperature or above (Collins and Trapnell, 1957)). On the other hand a surface covered with carbon monoxide is very inactive towards oxygen, which means the reaction

$$CO_{ad} + \tfrac{1}{2}O_{2,gas} \rightarrow CO_2$$

does not take place or needs perhaps a rather long induction period. This behaviour has also been found previously with evaporated Pd films and normal Pd catalysts, as well as with Pd(110) and Pd(100) planes.

A surface which is only partially covered by CO may adsorb additional oxygen. The diffraction pattern (figure 9) then consists of a superposition of diffraction spots of the O- and CO-structures. This means that both kinds of particles exist in different patches (competitive adsorption). The other possibility would have been the formation of a mixed adsorbed layer (co-operative adsorption) with fundamental changes in the diffraction pattern as observed in several other systems. The specific nature of the interactions between the adsorbed species is responsible for these effects (Ertl, 1969). The stable co-existence of domains with adsorbed oxygen and domains with CO also demonstrates that the reaction

$$O_{ad} + CO_{ad} \rightarrow CO_2$$

does not play the dominant role in the catalytic oxidation of carbon monoxide.

The main steps in the steady-state catalytic reaction are therefore the following

$$O_2 + * \xrightarrow{k_1} 2O_{ad} \tag{1}$$

$$O_{ad} + CO_{gas} \xrightarrow{k_2} CO_{2,gas} \tag{2}$$

* represents the configuration of surface atoms which is responsible for the oxygen chemisorption. The proposed reaction mechanism can be checked by measuring the reaction rate r as a function of temperature and pressure. Figure 10 shows the variation of r as a function of the crystal temperature T using a 2:1

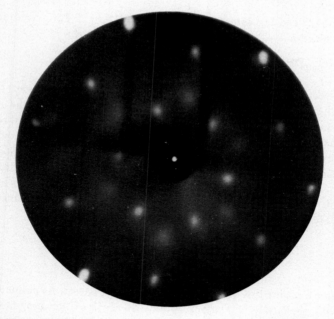

Fig. 9. LEED pattern of a Pd(111) surface with adsorption of O_2 and CO.

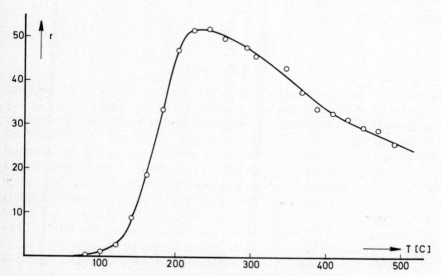

Fig. 10. Rate of catalytic formation of CO_2 as a function of temperature.

mixture of CO and O_2 with a total pressure of $2 \cdot 10^{-6}$ Torr. According to the discussed reaction mechanism the reaction rate is given by

$$r = k_2 [O_{ad}] \cdot p_{CO} \tag{3}$$

At lower temperatures the surface is completely covered by CO which inhibits the adsorption of oxygen. With increasing temperature the concentration of CO_{ad} decreases. The concentration of O_{ad} and therefore r then increase until a maximum is reached between 200 and 250°C where the adsorbed CO is completely removed. At even higher temperatures the bulk diffusion and the desorption of oxygen start and lower the reaction rate. This observation—as well as other experiments on the partial pressure dependency which are not shown here—is in full agreement with earlier measurements on other planes where a detailed discussion has been given (Ertl and Rau, 1969).

V. CONCLUSIONS

The above results can be compared with other systems. As pointed out there exist strong similarities in the adsorption behaviour of the three most densely packed planes (111), (100), and (110) of palladium: There are no significant differences between them in the adsorption energies of carbon monoxide and oxygen as well as in the mechanism and activity of catalytic reaction. Structure models for the adsorption of CO have been derived in which in all three cases the same configuration

$$
\begin{array}{c}
O \\
\| \\
C \\
Pd^{\diagup} \quad ^{\diagdown} Pd
\end{array}
$$

appears with an equal distance between the Pd atoms of $a_o = 2.73$ Å. As a first approximation these adsorption complexes can be considered as surface compounds, which are coupled to the rest of the crystal. The concept of surface compounds has been introduced by Grimley (1969) and is justified if the interaction from d orbitals of the metal atoms to orbitals of the adsorbed molecule is stronger than to the d orbitals of neighbouring metal atoms. Striking similarities also appear in a comparison with other platinum metals: Adsorption of oxygen causes a 2×2 structure as on Pd(111) on Rh(111) and Ru(0001) (Grant and Haas, 1970), Pt(111) (Tucker, 1964), and Ir(111) (Grant, 1971). The $\sqrt{3}$ structure of CO adsorption has also been reported for Rh(111) and Ru(0001) (Grant and Haas, 1970), and Ir(111) (Grant, 1971).

The indication is that the close relationship between the platinum metals is expressed in their surface chemistry too.

REFERENCES

Collins, A. C. and Trapnell, B. M. W. (1957). *Trans. Faraday Soc.* **53**, 1476.
Corner, J. (1946). *Proc. Roy. Soc. Ser. A***58**, 737.
Eischens, R. P. and Pliskin, W. A. (1958). *Advan. Catal. Relat. Subj.* **10**, 1.
Ertl, G. (1969). *In* "Molecular Processes on Solid Surfaces" (Drauglis, Gretz, Jaffee, eds.), p. 143, McGraw Hill, New York.
Ertl, G. and Koch, J. (1970a). *Z. Phys. Chem. Frankfurt am Main.* **69**, 323.
Ertl, G. and Koch, J. (1970b). *Z. Naturforsch. A.* **25**, 1906.
Ertl, G. and Rau, P. (1969). *Surface Sci.* **15**, 443.
Grant, J. T. (1971). *Surface Sci.* **25**, 451.
Grant, J. T. and Haas, T. W. (1970). *Surface Sci.* **21**, 76.
Grimley, T. B. (1969). *In* "Molecular Processes on Solid Surfaces" (Drauglis, Gretz, Jaffee, eds.), p. 299, McGraw Hill, New York.
Park, R. L. and Madden, H. H. (1968). *Surface Sci.* **11**, 158.
Tracy, J. C. and Blakely, J. M. (1969). *Surface Sci.* **15**, 257.
Tracy, J. C. and Palmberg, P. W. (1969). *J. Chem. Phys.* **51**, 4852.
Tucker, C. W. (1964). *J. Appl. Phys.* **35**, 1897.

ON THE IDENTIFICATION OF THE CHEMICAL STATE OF ADSORBED SPECIES WITH AUGER ELECTRON SPECTROSCOPY

T. W. HAAS, J. T. GRANT and G. J. DOOLEY

Aerospace Research Laboratories (LJ)
Wright-Patterson AFB, Ohio, USA

I. INTRODUCTION

Recently, our understanding of surface phenomena has advanced quite rapidly due to the widespread use of sophisticated experimental techniques, as well as to a renewed interest in the theory of surface processes. One goal of such work is an understanding of the chemical bonds formed between adsorbate-adsorbent atoms during the adsorption process. In the study of other states of matter, this information has been gained by applying such techniques as soft X-ray spectroscopy (Fabian, 1968), photoelectron emission spectroscopy (or electron spectroscopy for chemical analysis, ESCA) (Siegbahn *et al.*, 1967; Siegbahn *et al.*, 1969), as well as other spectroscopic techniques. It seems reasonable to expect that similar studies could be carried out on surfaces, if spectroscopy measurements could be developed which emphasize primarily surface atoms so that the observed effects are not masked by contributions from the bulk. Two techniques which have recently proved suitable are ion neutralization spectroscopy (Hagstrum and Becker, 1969) and soft X-ray appearance potential spectroscopy (Houston and Park, 1971). Since Auger electron spectroscopy has also been shown to be dominated by surface effects (Palmberg and Rhodin, 1968; Palmberg, 1968) it would seem an ideal tool for such studies. Preliminary studies have shown that such is indeed the case (Haas and Grant, 1969; Haas and Grant, 1970; Haas *et al.*, 1970; Grant and Haas, 1971). It will be the purpose of this work to report some recent results which will illustrate how Auger electron spectroscopy can be useful in defining the chemical state of adsorbed species.

There are two main effects one looks for in Auger spectra which are useful in defining the chemical state of an adsorbed atom. The first of these is a shift in the energy of an Auger electron. This effect is observed when charge transfer occurs. If electrons are transferred from a less to a more electronegative element

359

then the binding energies of the remaining electrons in the positively charged atom are increased and the energies at which the Auger electrons will appear is decreased. The greater the amount of charge transfer (change of valence state) the greater will be the shift. Unfortunately, crystal field effects enter in and one cannot simply equate energy shift with charge transfer.

The second important chemical effect in Auger spectroscopy obtains when the electrons involved in the Auger process include the valence electrons. Here a change in the type of chemical bonding requires a change in the wavefunction used to describe the valence electrons and this in turn means a different transition probability for a given Auger process. This case will lead to a change in the shape of a complex Auger spectrum giving a fingerprint for the species involved which can be useful in identifying the chemical state of certain surface atoms.

II. EXPERIMENTAL

Most of the experimental results were obtained using a 4-grid LEED system operated as a retarding potential energy analyzer (Palmberg, 1968). A side electron gun operated at 1.5 KeV with approximately 50 μA current into a 1 mm diameter spot size was used to excite the Auger transitions. Modulating voltages were usually 2 VRMS at 1 KHz although lower voltages were sometimes used to improve resolution. Resolution was between 0.5 and 1% depending on the modulating voltage used.

The vacuum system was a typical all metal ion pumped system. Base pressure (uncorrected gauge readings) were $2\text{-}5 \times 10^{-11}$ Torr and pressures with the electron guns operating were generally an order of magnitude higher. The samples were single crystal metals that had been employed in previous LEED studies and were oriented and prepared using standard techniques. The gases used in adsorption studies were reagent grade purity.

Some of the data reported here was produced using a coaxial cylindrical analyzer. It was similar to that described previously (Palmberg et $al.$, 1969) except that a double focussing model was constructed to improve resolution (Zashkvara et $al.$, 1966). A schematic diagram of the device is shown in figure 1. The resolution of this electron spectrometer was measured at 0.25%. The signal to noise ratio and transmission remains high enough so that the Auger spectrum may be displayed continuously on an oscilloscope. Improved versions are expected to show a resolution of 0.1% which will allow more adequate investigations of fine structures in Auger spectra. The increased operating speed of these spectrometers is a distinct advantage in avoiding electron beam degradation of samples.

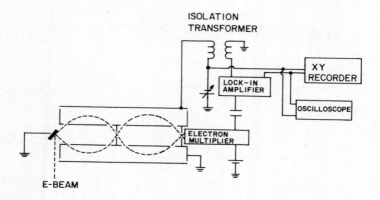

Fig. 1. Diagram of coaxial cylindrical electron spectrometer with double focusing to improve resolution.

III. ENERGY SHIFTS OF AUGER ELECTRONS

We have reported previously (Haas and Grant, 1969) on the shifts in energy of Auger electrons which occur when a clean refractory metal surface (viz. Nb, Mo, Ta, W) is oxidized. As expected, progressive oxidation gives a more or less continuous shift in the energy at which the metal Auger electrons appear. This result clearly indicates that the adsorption of oxygen onto these surfaces is ionic. As anticipated, the shift of the metal Auger electrons is to lower voltages indicating a more positively charged metal atom in the surface region. There is no observable equal and opposite shift in energy of the oxygen Auger spectrum with coverage, indicating little change in valence state of the oxygen with coverage. One should compare the energies of Auger electrons from gaseous molecular or atomic oxygen to get some idea of the ionic state of chemisorbed oxygen. A comparison of the Auger mechanism with photoelectric emission spectroscopy (ESCA) shows that the shifts in energy for the two measurements should be of similar size. Some of the maximum shifts due to the oxidation of refractory metals observed by us are given in table 1. It can be seen that the shifts, which are a few eV in energy, are comparable to those reported from ESCA measurements (Siegbahn *et al.*, 1967).

During the course of these experiments there was a shift to lower energies of the metal Auger electrons when CO was chemisorbed at room temperature. This somewhat unexpected result indicates that electrons are transferred from the metal to the adsorbed CO. The shifts observed for approximately 1 monolayer

Table 1. *Energy shifts (eV) of metal Auger electrons due to oxidation and to adsorption of CO. The column (a) lists the maximum shifts observed from extensive oxidation. Column (b) lists the shifts observed for approximately one monolayer coverage of CO. The shifts are such that the measured Auger electron energies are lowered.*

Material	(a) Oxidation	(b) CO Adsorption
W(112)	−4	−0.5
Mo(110)	−3.5	−
Nb(110)	−4	−1.0
Ta(112)	−6	−

adsorption of CO (assuming saturation coverage to be 1 monolayer) are summarized also in table 1. The shifts of 0.5 to 1 eV would seem to indicate a fairly extensive shift of the electron density around the surface metal atoms toward the CO. As we shall show later, it appears that the CO is decomposed into C and O on the Ta and Nb samples even at room temperature. Previous work had indicated that this decomposition took place at elevated temperatures

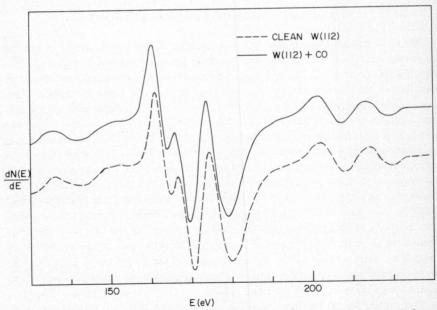

Fig. 2. Derivative, $dN(E)/dE$, of the energy distribution, $N(E)$, versus E from a W(112) crystal before and after adsorption of approximately 1 monolayer CO. A shift of about 0.5 eV in energy is observed in the W Auger transitions shown in the figure.

(Haas *et al.*, 1967). For the W and Mo samples, this decomposition does not seem to obtain, and here the observed energy shifts are smaller. An Auger spectrum from clean and CO covered W(112) is shown in figure 2. It can be seen that the shape of the spectrum is not altered but that a small but measurable shift in energy has occurred.

IV. CHANGES IN SHAPES OF COMPLEX AUGER SPECTRA

A second important observation is the change in shape of a complex spectrum when the chemical bonding of an atom changes. This is generally observed when the valence electrons are involved in the Auger process and one of the best candidates for such studies is the carbon spectrum. The Auger spectrum from carbon is the KLL spectrum. The L electrons in carbon are the $2s$ and $2p$ electrons. Changes in the hybridization of these orbitals will change the shape of the complex KLL spectrum due to changes in the transition probabilities for the various configurations of the final state of the doubly ionized carbon atoms after the Auger transition is finished. Some examples of the types of spectra observed are shown in figure 3. Each form of carbon (metal carbide, diamond, graphite, adsorbed CO) gives an identifiable spectrum which can be used as a fingerprint for the state of the carbon, assuming that a given form of carbon gives the same spectrum whether in a thin layer at a surface or as the surface of the same bulk material. This assumption seems valid because a bulk sample of molybdenum carbide gives the same spectrum as what is thought to be a thin carbide layer on molybdenum metal and similarly for a bulk and a surface layer of graphite (Grant and Haas, 1971).

This observation of the fine structure in the carbon spectrum can be used to follow effects of the electron beam on an adsorbed CO layer. It is often found that the beam decomposes the CO quite readily. In the case of CO on a W(112) surface, for example, the shape of the carbon peak changes from that of adsorbed CO to metal carbide, although the peak-to-peak height of the carbon line is roughly constant. If the LEED pattern is examined where the high-energy beam was incident on the surface, a well-formed c(4X6) pattern was found which could not be removed by heating but could be removed by oxidation and heating. This carbide pattern on W(112) has been reported previously (Chen and Papageorgopoulos, 1970). On the other hand, if CO is adsorbed on the W(112) crystal with all electron beams off, then flashing the sample removes all the CO as shown by the Auger spectra. This electron beam cracking is a serious problem in studies of this sort and must be carefully watched. The increased speed of data taking, with the coaxial cylindrical spectrometer, is a great advantage here.

The case of the adsorption of CO on Ta and Nb surfaces is an interesting one. Adsorption of CO at room temperature gives a diffuse LEED pattern indicative

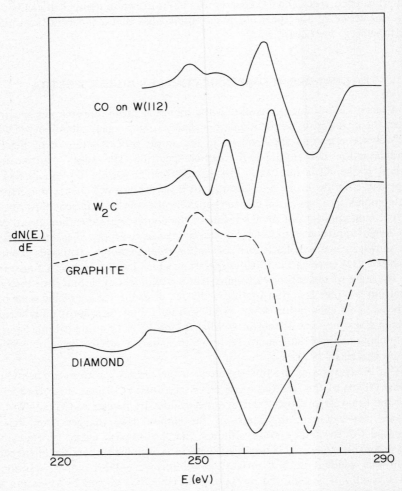

Fig. 3. Derivative of the energy distribution, *dN(E)dE,* versus *E* showing the Auger transitions from carbon in various chemical forms. A shift in energy for the diamond spectrum may have occurred due to the insulating nature of the crystal studied.

of an amorphous layer, while heating to 600°-1500°C gave sharp LEED patterns that were identical to those obtained by oxygen exposure. This evidence, coupled with the fact that only a very small pressure burst is observed on flashing a preadsorbed CO layer on Ta or Nb, led to the conclusion that the CO had decomposed on the surface at elevated temperature (Haas *et al.*, 1967, and Haas, 1969). These conclusions were checked using Auger spectroscopy with the following results. The spectrum of roughly 1 monolayer of CO on Ta(100)

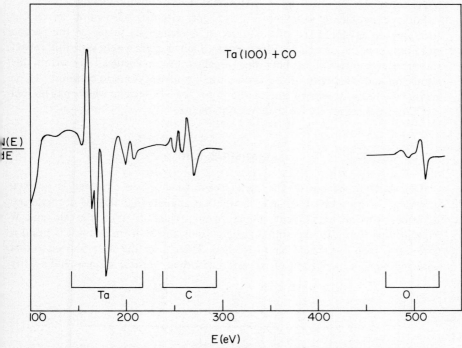

Fig. 4. Derivative of the energy distribution, $dN(E)/dE$, versus E from a Ta(100) crystal following exposure to CO. The omitted part of the spectrum form 300 to 450 eV was swept very fast in order to minimize electron beam effects.

adsorbed at room temperature is as shown in figure 4. Note that the carbon line shape is very similar to that of tungsten carbide and also to that of molybdenum carbide (Haas and Grant, 1970). We do not have a spectrum from tantalum carbide, but if we assume it to be similar to the other carbides, then we can conclude that the CO must be decomposed on the Ta surface at room temperature. This conclusion is further borne out by the heating experiment on preadsorbed CO. Heating for a minute or two at ~400°C causes the carbon peak to diminish slightly but with no change in shape while the oxygen peak stays constant. Continued heating causes the carbon peak to get smaller and eventually disappear but no detectable change in shape is observed. At the same time, both the shape and relative size of the oxygen peak are constant. Apparently, the diffuse pattern observed with room temperature adsorption of CO is a mixed oxide-carbide film. Heating to moderate temperatures then causes the carbon to diffuse into the bulk leaving an ordered oxide layer. Very similar results to these were obtained for Nb samples as well, and the same has been found for the (100), (110), and (112) crystal faces of both materials.

Some experiments were also carried out with CO adsorption at various coverages on a Mo(112) surface to see if a change in shape of the carbon spectrum could be detected with coverage. Such changes would probably reflect strong lateral interactions between adsorbed CO molecules. To within the resolution and sensitivity of the system used, no changes could be found. These measurements were carried out in the 4-grid LEED system which has limited resolution and sensitivity for these measurements.

V. DISCUSSION

The shifts in energy of the metal peaks when CO is adsorbed is not too surprising for the case of Ta and Nb where it appears that the CO decomposes. In effect it is an experiment similar to an oxidation. With the Mo and W experiments, however, the shift in energy indicates that the metal CO bond is more ionic than one would have expected. The size of the shift is close to that observed when a monolayer of oxygen is adsorbed on these surfaces and a fairly

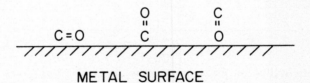

METAL SURFACE

Fig. 5. Illustration of several ways that CO molecule may be oriented on metal surfaces.

ionic bond is expected. For W and Mo, however, the available evidence does not suggest extensive decomposition of the CO on the metal surface in the absence of an electron beam interaction. It would be most interesting to repeat these experiments with other metals such as Pt, Ni, etc.

One question which also comes up is whether the CO lies flat on the surface or whether the O or C is bonded to the metal in some way. Some of the ways that a CO molecule may orient itself on a metal surface are shown in figure 5. If the electrons are scattered strongly by even one layer of atoms, then one might get different ratios for the relative intensities of C to O Auger lines for the various orientations shown. We tried to find some evidence for one or the other configuration by looking at the ratios of the C to the O peak on various metals for various coverages. The ratios were always close to 2:1 with the C peak being the larger. This result is inconclusive since it may be that having the O or the C on the metal side would not result in measurable differences in the C to O peak ratio. One expects the Auger yield for C to be slightly higher than for O but using a semi-empirical relationship for the Auger yield (Hagedoorn and Wapstra, 1960) indicates that the difference would not be measurable. Again it would be most interesting to make this same measurement on other surfaces such as Pt and Ni.

VI. CONCLUSION

These results suggest that Auger electron spectroscopy should become an important tool for the study of gas-solid interactions. To obtain the most from this technique, however, one should look at the fine details of the spectra using the highest resolution and sensitivity available. It is absolutely essential that one watch for electron beam interactions with adsorbed species and to this end the coaxial cylindrical analyzers are recommended for their increased speed and resolution.

Within reasonable limits it appears that one can determine the presence of electron transfer in adsorption by means of energy shifts in the Auger electrons, while fingerprints which seem to identify the chemical state of certain atoms can be obtained. The fine details of the Auger spectra are strongly affected by the disposition of the electrons around the atom or, put another way, by the wave functions which describe these electrons. It is perhaps not too optimistic to hope that eventually we may be able to obtain details of the chemical bonding of surface atoms from these spectra.

ACKNOWLEDGEMENT

The authors wish to acknowledge the helpful comments and assistance of Dr Simon Thomas during the course of this work.

REFERENCES

Chen, J. and Papageorgopoulos, C. A. (1970). *Surface Sci.* **20**, 195.
Fabian, D. J., ed. (1968). "Soft X-Ray Band Spectra and the Electronic Structure of Metals and Materials". Academic Press, New York and London.
Grant, J. T. and Haas, T. W. (1971). *Surface Sci.* **24**, 332.

368 T. W. HAAS, J. T. GRANT AND G. J. DOOLEY

Haas, T. W. (1969). "The Structure and Chemistry of Solid Surfaces". (G. Somorjai, ed.). John Wiley, New York.
Haas, T. W. and Grant, J. T. (1969). *Phys. Let.* **30A**, 272.
Haas, T. W. and Grant, J. T. (1970). *Appl. Phys. Lett.* **16**, 172.
Haas, T. W., Grant, J. T. and Dooley, G. J. (1970). *J. Vac. Sci. Technol.* **7**, 43.
Haas, T. W., Jackson, A. G. and Hooker, M. P. (1967). *J. Chem. Phys.* **46**, 3025.
Hagedoorn, H. L. and Wapstra, A. H. (1960). *Nucl. Phys.* **15**, 146.
Hagstrum, H. D. and Becker, G. E. (1969). *Phys. Rev. Lett.* **20**, 1064.
Houston, J. E. and Park, R. L. (1971). *J. Vac. Sci. Technol.* **8**, 91.
Palmberg, P. W. (1968). *Appl. Phys. Lett.* **13**, 183.
Palmberg, P. W. and Rhodin, T. N. (1968). *J. Appl. Phys.* **39**, 2425.
Palmberg, P. W., Bohn, G. K. and Tracy, J. C. (1969). *Appl. Phys. Lett.* **15**, 254.
Siegbahn, K., Nordling, C., Fahlman, A., Nordberg, R., Hamrin, K., Hedman, J., Johansson, G., Bergmark, T., Karlson, S., Lindgren, I. and Lindberg, B. (1967) "ESCA Atomic, Molecular, and Solid State Structure Studied by Means of Electron Spectroscopy". Almquist and Wiksells Boktryckeri AB, Uppsala.
Siegbahn, K., Nordling, C., Johanasson, G., Hedman, J., Heden, P. F., Hamrin, K., Gelius, U., Bergmark, T., Werme, L. O., Manne, R. and Baer, Y. (1969). "ESCA: Applied to Free Molecules". North-Holland Publishing Co., Amsterdam-London.
Zashkvara, V. V., Korsunskii, M. I. and Kosmachev, O. S. (1966). *Soviet Phys. – Tech. Phys.* **11**, 96.

THE BEHAVIOUR OF THE Mo(100) SURFACE PLASMON WITH CO ADSORPTION

J. LECANTE

*Centre d'Études Nucléaires de Saclay, Section
d'Études des Interactions Gaz-Solides, Gif-sur-Yvette, France*

I. INTRODUCTION

In experimental studies of chemisorption on metallic surfaces information is sought on substrate or adsorbate species and on their modifications during the reaction. Such information may be obtained either by destructive means: thermal or electron induced desorption, sputtering; or by non-destructive means: Auger spectroscopy, ion neutralization spectroscopy (Hagstrum, 1969), infra-red adsorption or inelastic electron scattering (Propst and Piper, 1967) for vibronic species.

Information on metal and its modifications by the adsorption phenomenon may be obtained by: work-function measurements, photoemission spectroscopy (Callcott and Mac Rae, 1969), ion neutralization spectroscopy and, more recently, by surface plasmon studies (Raether, 1967; Bennett, 1970). The latter method, providing a useful probe of the electronic density in the vicinity of the surface (Bennett, 1970), will be here applied to the case of chemisorption of carbon monoxide on a Mo(100) face.

II. PLASMA EFFECTS

Since the work of Bohm and Pines we know that the free electrons of a metal submitted to an external perturbation may perform collective oscillations similar to the longitudinal oscillations of a gaseous plasma, with an angular frequency

$$\omega_p = \left(\frac{ne^2}{\epsilon_0 m} \right)^{1/2}$$

where e and m are the charge and mass of the electron, n the bulk density of the free electrons and ϵ_0 the vacuum dielectric constant. These are the oscillations of the volume electronic density. As in a gaseous plasma we find, due to the boundary conditions, a longitudinal oscillation of the surface free-electron density with a reduced frequency ω_s: for a flat metal surface bounded by vacuum, Ritchie (1957) found the relation

$$\omega_s = \frac{\omega_p}{\sqrt{2}}$$

These results are only valid for "free-electron" metals but for the transition metals which present interesting features in chemisorption, the d-electrons can no longer be considered as entirely free. In this case interband transitions become important and act so as to shift the free-electron value of the plasma frequencies ω_p and ω_s: following Pines (1954) we can write for the dispersion relation of the volume plasmon

$$1 - \frac{e^2}{\epsilon_0 m} \sum_n \frac{f_{n0}}{\omega^2 - \omega_{n0}^2} = 0$$

where f_{n0} is an oscillator strength for the electronic transitions of frequency ω_{n0} from the ground state to the excited state n. In this formula if we separate the free electron contribution to the bound electron contribution, as Ehrenreich *et al.* (1963) (and also Nozières and Pines, 1958) have done, there are two possibilities.

(a) If $\omega_{n0} > \omega_{p0}$ (i.e. if the single particle transition has a frequency ω_{n0} higher than the free-electron plasma frequency ω_{p0}): the actual plasma frequency will be lowered.

(b) If $\omega_{n0} < \omega_{p0}$: the plasma frequency will be increased by the low-lying transition.

Pines (1954) using a similar qualitative discussion had predicted a position in the periodic classification where the shifts of the plasma frequency due to interband transitions lying above or below ω_{p0} compensate and restore the theoretical position. He found that this position would be obtained for six valence electron elements for chromium: molybdenum and tungsten, for example. This mechanism also explains the width of volume plasmon for these metals. Similar effects will be observed on the surface plasmon peak but due to the lower surface plasmon frequency the shift will be different. Thus the simple relation $\omega_s = \omega_p/\sqrt{2}$ between the two frequencies will be no longer valid for d-metals.

It is necessary to find other criteria for identifying collective oscillations among the single-particle transitions. The comparison with optical data when available provides us with a powerful tool for such identification. These oscillations may be excited by a beam of electrons either traversing a thin metallic film or scattered by a thick crystal. In analysing the energy of transmitted or back-scattered electrons, we can observe discrete peaks, due to electrons that have suffered energy losses by exciting interband transitions or collective oscillations of the metal's electrons. Reflection method is more sensitive to the surface state than transmission experiments because the electrons are nearer the surface longer.

III. APPARATUS

Figure 1 shows the experimental tube. It consists mainly of a pyrex vessel divided into two independently pumped sections: the oil diffusion pumped electron gun assembly and the ionically pumped target chamber. They are linked

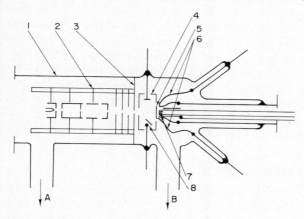

Fig. 1. Schematic diagram of the experimental tube: (A) and (B) to pumping lines, (1) glass cell; (2) electron gun; (3) aperture lens separating electron gun from target cell; (4) collector; (5) target; (6) thermocouples; (7) filament for flash-desorption; (8) ion gun.

by one of the 1 mm aperture lenses of the gun. The ultimate pressure obtained was in the 10^{-8} N/m^2 range after a 450°C bake-out for eight hours. Partial pressures were measured by a quadrupole mass filter, the main residual gases being H$_2$ 80%, CO 18% and traces of CO$_2$, Ar. . . . With a hairpin tungsten filament the electron gun can deliver a current in the range 10^{-8} to 10^{-7} A at an energy between 50 to 800 eV. The primary electrons strike the sample surface at normal incidence and are then collected by a cylinder surrounding the target. A small ion gun, placed inside the collector is used to clean the front face of the target by ion bombardment. A tungsten filament close to the back face of the target heats the crystal by electron bombardment to temperatures up to 2000°C. We can thus clean the surfaces by heating or measure roughly the population of the adsorbed species and their binding energy by the flash filament technique.

The crystal target, a 6 mm diameter disk of thickness 0.2 mm, is supported by three tungsten wires, a W/W-Rh thermocouple is used for temperature measurements.

Three Helmholtz coils surrounding the experimental tube reduce the earth's magnetic field to less than 0.01 gauss in the region of the collector as the retarding potential method needs a minimal value of stray magnetic fields.

IV. EXPERIMENTAL PROCEDURES

SAMPLE TREATMENT

The crystal target of orientation (100), to within ±1° as checked by X-ray diffraction, was electropolished before mounting. The cleaning procedure consisted of repeated ion bombardment, 10 to 50 $\mu A/cm^2$ during 10 to 15 minutes with 500 eV Ar ions, followed by annealing at 500 to 700°C. After several such cycles measurements of work-function changes and flash-desorption show a good reproducibility. This cleaning procedure is known to produce a clean Mo(100) surface (Hayek et al., 1968). We have also observed that, after some initial ion bombardments heating alone to above 1300°C restores the clean surface, i.e. in our case the most reactive one.

WORK-FUNCTION MEASUREMENTS

Variations in the Mo(100) surface potential were determined during the experiments and following the cleaning procedures with the help of the electron beam. A retarding potential was applied to the target and the current electronically differentiated. Changes in the work-function were measured by the shift of the peak thus obtained. In this method, the reference electrode was the gun filament whose high temperature (2000°C) prevents any change in work function due to modification of the gas pressure. In addition, the differential pumping in our system allows us to maintain a low pressure on the electron gun side during the adsorption processes on the target side. Thus we have no variations in the work function of the filament nor in the electron intensity delivered by this gun which guarantees a good appreciation of the surface potential shifts as checked by the reproducibility of similar measurements. The energy distribution of the primary electrons was made as narrow as possible, with the help of a lens gun to increase the accuracy of the measurement.

The gun current was maintained below 10^{-8} A and the applied retarding potential swept over a small voltage range to avoid appreciable desorption of any adlayer. Surface potential variations of less than 20 mV were determined by this method, and the degree of surface cleanliness estimated from these measurements.

ENERGY LOSS MEASUREMENT

A retarding potential was applied to the collector, and the energy distribution of the backscattered electrons was obtained by electronic differentiation of the collected current. The resolution of the analysing apparatus was improved by a suitable electron lens system (Simpson, 1961) placed in front of the collector. For primary energies from 250 eV to 500 eV we obtained a fairly constant resolution represented by a half height width of 0.8 eV for the elastically reflected peak.

V. EXPERIMENTAL RESULTS

SURFACE POTENTIAL VARIATIONS

The work-function variations of the target due to carbon monoxide adsorption are shown in figure 2, the same curve being obtained for the two

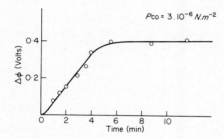

Fig. 2. Work-function changes due to CO adsorption on (100) Mo versus time at constant CO pressure.

initial cleaning techniques mentioned above. The total change $\Delta\phi$ of 0.40 eV was reproducible to within ±3% for similar experimental conditions. The degree of surface cleanliness was determined after each cleaning procedure, from the value of the surface potential relative to that of the cathode gun.

CHARACTERISTIC ENERGY LOSS SPECTRA

Curve 1 in figure 3 represents the energy distribution of back-scattered electrons from a clean Mo(100) surface for a primary energy of 250 eV. There are two main features: a very broad peak (~8 eV half height width) approximately centered at an energy loss of 23 eV; and a relatively sharp 11.6 eV peak (~1 eV half height width) showing a shoulder at 13.3 eV.

Curves 2, 3 and 4 correspond to primary energies of 330, 430 and 530 eV respectively. With increasing primary energy the 23 eV peak remains essentially unchanged but the 11.6 eV decreases in magnitude and the 13.3 eV peak becomes better resolved. The effect of CO adsorption on the 11.6 eV peak for a

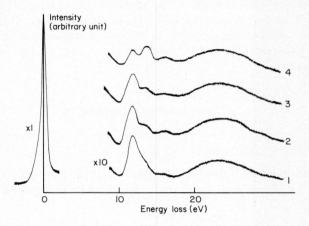

Fig. 3. Energy-loss spectrum for different primary energies: curve 1: 250 eV; curve 2: 330 eV; curve 3: 430 eV; curve 4: 530 eV.

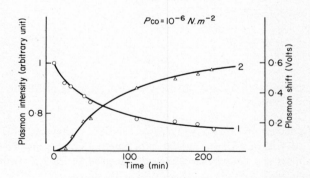

Fig. 4. Curve 1: relative change of surface plasmon intensity; curve 2: surface plasmon shift versus time at constant CO pressure.

primary energy of 250 eV is illustrated in figure 4. We have plotted the decrease in intensity (curve 1) and the shift towards greater value of energy loss (curve 2) of the 11.6 eV peak, versus time of adsorption at constant CO pressure. The position of the 13.3 eV peak remains unchanged but its resolution improves as the 11.6 eV peak decreases.

VI. DISCUSSION

VOLUME PLASMON

The 23 eV energy loss may be attributed to volume plasmon (Lynch and Swan, 1968). It corresponds roughly to the collective oscillations of the six valence electrons of molybdenum ($4d^5$, $5s^1$). The theoretical free-electron value of the plasma frequency for six electrons per molybdenum atom. is

$$\omega_p = \left(\frac{ne^2}{\epsilon_0 m}\right)^{1/2} = 3.5 \times 10^{16}\ s^{-1}$$

corresponding to an energy $\hbar\omega_p$ = 23 eV. This free-electron behaviour for a d-metal has been explained above by the position of molybdenum in the periodic table.

SURFACE PLASMON

For a free-electron metal the energy loss due to the excitation of surface oscillations would be given by $\hbar\omega_s = \hbar\omega_p/\sqrt{2}$ = 16.2 eV. However, no peak is observed in our loss spectra for this value of $\hbar\omega_s$ but the main feature appears at 11.6 eV. Due to the lower value of the surface plasmon this is not surprising as it is more probable for its value to be lowered by high-lying interband transitions than in the case of the higher value volume plasmon. We will now try to attribute the observed 11.6 eV energy loss to the excitation of surface oscillations.

COMPARISON WITH OPTICAL DATA

For a metal bounded by vacuum the excitation probability for surface plasmons is proportional to the function

$$- Im\ \frac{(1 - \epsilon)^2}{\epsilon(1 + \epsilon)}\quad \text{(Ritchie, 1957; Raether, 1967)}$$

where $\epsilon(\omega)$ is the complex dielectric constant, depending on ω, of the free-electron gas. The maximum of this function of ω for $\epsilon(\omega) = -1$ gives the surface plasmon energy value for a clean surface. The value of the complex dielectric constant $\epsilon(\omega) = \epsilon_1(\omega) + i\ \epsilon_2(\omega)$ obtained from light reflectance measurements on metal surface may be used. In doing this we identify the transverse dielectric constant $\epsilon_\perp$ describing the response of the free-electron gas to the transverse electric field of the incoming photons, with the longitudinal dielectric constant ϵ set in action by the longitudinal electric field of the incoming electrons. This assumption, valid in the R.P.A. approximation (Pines, 1964), is well justified by experiment; however some discrepancies between the two sets of data (photons or electrons) may be explained by this different behaviour.

For molybdenum, tantalum and tungsten optical data exists: see Juenker *et al.* (1968). From their results we can plot the curve of the surface loss function versus photon energy as in figure 5. A maximum of the surface loss function is observed near 11 eV where we find $\epsilon_1 \sim -1$, representing the dispersion relation of the surface plasmon and where ϵ_2 is sufficiently small to ensure no large damping of the collective oscillations. There is close agreement with the observed experimental peak at 11.6 eV. The decrease of the 11.6 eV peak with

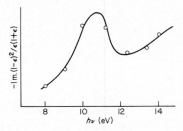

Fig. 5. Surface energy-loss function versus photon energy.

increasing primary energy represented in figure 3 may also strengthen the idea of a surface peak: the cross section of these excitations seems to present a maximum for a given value of energy which is consistent with a surface phenomena.

DEPENDENCE ON CO ADSORPTION

Curves 1 and 2 of figure 4 show the dependence of intensity and position of the plasmon surface peak on CO adsorption. The small decrease in intensity may be due in part to the increase of the potential barrier measured by the work-function change during CO adsorption. Several reasons may be advanced for the explanation of the surface plasmon shift, they are: modification of the dielectric constant of the boundary surface, appearance of the true adsorbate plasma frequency, modification of the dielectric constant of the metal, variation of the superficial electronic density. We may briefly review these reasons:

(a) Modification of the boundary dielectric constant.

In the case of a thick dielectric layer Stern and Ferrell (1960) have shown that the surface plasma dispersion relation became

$$\epsilon = -\eta$$

where η is the dielectric constant of the layer. They found the new surface plasma frequency for the metal substrate

$$\omega_s = \frac{\omega_p}{\sqrt{1+\eta}}.$$

This effect has been experimentally proved on aluminum at various degrees of oxidation by Powell and Swan (1960). But in the case of Al-Al$_2$O$_3$ system the oxide layer is relatively thick and we cannot compare it with the case of chemisorption where we have only a more or less continuous adsorbed monolayer to which we cannot attribute any dielectric constant value.

(b) Appearance of the adsorbate plasma frequency.

In the case of Cs adsorbed on W, Mac Rae *et al.* (1969) have found an energy loss which they attribute to the excitation of the true surface plasmon in the Cs layer. But this case does not apply to chemisorption because it needs several layers of adsorbate.

(c) Modification of the metal dielectric constant.

The dipolar layer adsorbed on the metal may produce a sufficient electric field to modify the dielectric constant $\epsilon(\omega)$: a modification of surface potential of 0.5 eV over distances in the range of 1 Å may produce a surface electric field of 5×10^7 V/cm. What will be the influence of this field on $\epsilon(\omega)$? Buckman and Bashara (1968), using a modulated ellipsometry method, have measured for Au and Ag the changes in ϵ_1 and ϵ_2 induced by an electric field (10^5-10^7 V/cm) normal to the metal surface. They found changes in ϵ_1 and ϵ_2 of the order of 10^{-2} to 10^{-3} which are too small to explain our plasmon shift.

(d) Variations of the surface electron density.

One may explain the change of the surface plasmon frequency by the variation of the free-electron density in the vicinity of the surface (Bennett 1970). Adsorption on a clean metal surface is accompanied by a charge transfer between the foreign atom and the metal. It results in a modification of the surface electronic states. Hagstrum and Becker (1968) have shown, by ion neutralization spectroscopy, changes in the density of states and the amplitude of wave function in the vicinity of the adsorbed atom. Gadzuk (1970) has interpreted the results of Mac Rae *et al.* (1969) in terms of variations of the Cs layer electronic density due to charge transfer from the cesium to the metal.

In this experiment the adsorption of CO on transition metals is accompanied by a charge transfer in both directions: CO → metal and metal → CO. Following Blyholder (1964) in his interpretation of CO chemisorption, we can say that the lone pair of electrons from the carbon hybrid orbital sp_z will be accommodated by an empty d-orbital of the metal thus creating a σ-bond. This bond is stabilized by back transfer from a full d-orbital of the metal to the empty antibonding orbital $2\pi^*$ of the CO molecule. Thus the net transfer of charge is still controversial and the experimental results are not easily interpreted. We can only say that the free-electron density near the surface of the metal has increased during the process.

In roughly applying the free-electron formula we find that the total shift of 0.6 eV for $\hbar\omega_s$ corresponds to an equivalent increase of 0.3 free-electrons per molybdenum atom. However this increase is probably not due to a simple transfer from the CO molecule to the metal: the bonding electrons do not participate in the collective oscillations of the metal's free-electrons. A more probable explanation may be in the free or bound character of the d electrons. These electrons may have their bond to the ion cores modified by the charge transfer occurring near the surface and so some of them, originally too tightly bound to be considered as free, may participate in the collective oscillations when the chemisorbed layer is growing. If we compare the curves of figure 2 with those of figure 4 we see that change in the work function is more rapid than change in the surface plasma loss. This apparent discrepancy is probably due to two different electronic rearrangements occurring: (a) in the dipole layer, (b) in the uppermost atomic layers of the metal.

The above ideas will be corroborated by considering information obtained from flash desorption and work-function measurements. Work is already in progress along these lines.

VII. CONCLUSION

The energy spectrum of electrons, inelastically scattered by metallic surfaces, may provide interesting information on these surfaces when the surface plasmon loss is sufficiently well defined and unambiguously identified. This study may lead to a more complete knowledge if compared with data obtained in the same apparatus such as: work-function changes, population of adsorbed species and a rough evaluation of their binding energies.

REFERENCES

Bennett, A. J. (1970). *Phys. Rev. B.* **1**, 203.
Blyholder, G. (1964). *J. Phys. Chem.* **68**, 2772.
Buckman, A. B. and Bashara, N. M. (1968). *Phys. Rev.* **174**, 719.
Callcott, T. A. and MacRae, A. U. (1969). *Phys. Rev.* **178**, 966.
Ehrenreich, H., Phillipp, H. R. and Segall, B. (1963). *Phys. Rev.* **132**, 1918.
Gadzuk, J. W. (1970). *Phys. Rev. B.* **1**, 1267.
Hagstrum, H. D. (1969). *J. Appl. Phys.* **40**, 1398.

Hagstrum, H. D. and Becker, G. E. (1968). *In* "The Structure and Chemistry of Solid Surfaces" (Somorjai G. A. ed.), John Wiley, New York.

Hayek, K., Farnsworth, H. E. and Park, R. L. (1968). *Surface Sci.* **10**, 429.

Juenker, D. W., Le Blanc, L. J. and Martin, C. R. (1968). *J. Opt. Soc. Am.* **58**, 164.

Lynch, M. J. and Swan, J. B. (1968). *Aust. J. Phys.* **21**, 811.

Mac Rae, A. U., Muller, K., Lander, J. J., Morrison, J. and Phillips, J. C. (1969). *Phys. Rev. Lett.* **22**, 1048.

Nozières, P. and Pines, D. (1958). *Phys. Rev.* **109**, 741.

Pines, D. (1954). *Rev. Mod. Phys.* **28**, 184.

Pines, D. (1964). "Elementary Excitations in Solids". W. A. Benjamin, New York.

Powell, C. J. and Swan, J. B. (1960). *Phys. Rev.* **118**, 640.

Propst, F. M. and Piper, T. C. (1967). *J. Vac. Sci. Technol.* **4**, 53.

Raether, H. (1967). *Surface Sci.* **8**, 233.

Ritchie, R. H. (1957). *Phys. Rev.* **106**, 874.

Simpson, J. A. (1961). *Rev. Sci. Instrum.* **32**, sr,1283.

Stern, E. D. and Ferrell, R. A. (1960). *Phys. Rev.* **120**, 130.

INTERACTION OF OXYGEN WITH Mo(111)

R. M. LAMBERT, J. W. LINNETT and J. A. SCHWARZ

Department of Physical Chemistry, University of Cambridge, England

I. INTRODUCTION

A considerable amount of recent work has been devoted to studies of the (110) and (100) faces of Mo by LEED and also Auger electron spectroscopy (AES). By comparison, the (111) face has received relatively little attention, the only substantial work being the purely LEED investigation of Barton and Ferrante (1968). We have studied the room temperature interaction of oxygen with Mo(111) using LEED, AES, and flash desorption mass spectrometry, with the object of determining the initial sticking coefficient (S_0), the saturation coverage (θ_{max}) and the nature of the desorption product. Barton and Ferrante did not obtain a value for S_0 and were therefore unable to decide between two alternative interpretations of a low-coverage LEED pattern.

II. EXPERIMENTAL

A standard 3-grid post acceleration and display LEED apparatus was used, the LEED optics also being employed for energy analysis in the Auger experiments in the usual manner. Initially, the LEED gun was used for Auger excitation to ensure that both the LEED and Auger spectra arose from the same region of sample surface. When further experiments had established that the distribution of contaminant species was essentially uniform across the sample surface, a glancing incidence gun which gave increased sensitivity was used for quantitative measurements. Molybdenum samples 0.2 mm thick were cut oriented to within $\frac{1}{2}°$ of the (111) plane. They were polished as described by Haas and Jackson (1966), and mounted in a stainless steel holder between molybdenum shims. The sample was heated resistively by passing a large DC current, and temperatures monitored by a W–W/26% Re thermocouple. Power limitations restricted the maximum temperature to $\approx 1700°K$. A quadrupole mass spectrometer was used for residual gas analysis, detection of desorption products, and as an absolute

381

pressure gauge during oxygen exposures. For the latter purpose the output at $m/e = 32$ was calibrated against a Bayard-Alpert gauge at high enough pressures such that oxygen atom desorption effects in the gauge were unimportant. The low-pressure exposure sequences were then measured accurately by displaying the $m/e = 32$ signal on a $y-t$ recorder. During oxygen exposures the $m/e = 28$ signal increased slightly but was never more than 5% of the oxygen signal. All gases were spectroscopically pure grade and introduced through a variable leak valve; the system base pressure was $<10^{-10}$ Torr.

III. RESULTS

CLEANING

AES showed that initial thermal treatment led to the segregation of carbon and a great deal of sulphur at the surface, a fairly common observation for transition metals which have been studied by this method (Haas et al., 1970). Heating in vacuo produced no significant cleaning, while bombardment with 450 eV argon ions reduced the amount of sulphur but had little effect on the carbon, a behaviour which parallels that observed on Ni(110) (Sickafus, 1970). Heating to 1700°K in oxygen at a pressure of 5×10^{-8} Torr resulted in complete removal of sulphur and a drastic reduction in carbon, and a small oxygen signal appeared. Repeated attempts to remove the last traces of these remaining impurities were unsuccessful, and in one case led to irreversible facetting to (112) planes. Because of this, the partially contaminated surface corresponding to the Auger spectrum (b) of figure 1 was taken as the starting point for further experiments. The composition of this surface was reproducible throughout the adsorption-desorption experiments.

LEED AND AES MEASUREMENTS

The LEED patterns produced by the surface of figure 1(b) showed the characteristic symmetry of the bulk (111) planes, with sharply defined spots and low background intensity. Exposure to oxygen at room temperature resulted in a uniform decrease in intensity of all spots, and all coherent features were practically extinguished after ≈ 5 Langmuirs. At no stage did any new diffraction features occur. Heating to $\approx 1700°K$ restored the surface to the initial "clean" condition as evidenced by both the LEED pattern and Auger spectrum. The intensity of the (0,0) beam at the 180 V maximum was monitored as a function of oxygen exposure, a typical result is shown in figure 2.

In a second series of experiments, the increase in the 516 V oxygen Auger signal was followed as a function of exposure. Previous work by Musket and Ferrante (1970) on the W(110)$-O_2$ system has shown that the shape and width of the 516 V line is independent of coverage, and other work has demonstrated a linear relationship between Auger signal amplitude and concentration of surface species (Weber and Peria, 1967; Florio and Robertson, 1969). We have therefore used the peak-to-peak amplitude of the 516 V transition in the $dN(E)/dE$ spectrum as a direct measure of the concentration of surface oxygen. There were no detectable chemical shifts in this signal during exposure, nor did the amount

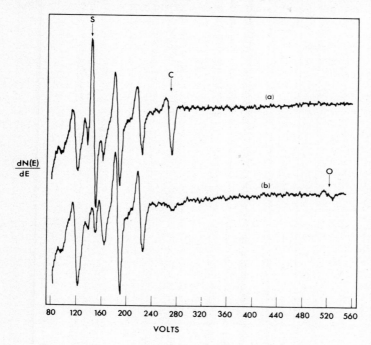

Fig. 1. Auger spectrum of surface: (a) before, (b) after cleaning procedure. (Reproduced with permission from Lambert *et al.*, 1971.)

adsorbed depend on electron-beam irradiation of the surface. Figure 3 shows the collected data from three experiments for the relative oxygen Auger signal as a function of exposure.

When the sample was flashed to 1700°K prior to each exposure run, the background oxygen Auger signal was observed to increase from a very small value to a limiting value as the crystal cooled over a period of several minutes. The limiting value was constant over many days of experiments, and over different regions of the surface studied with the electron beam. This observation is not fully understood. It is unlikely to be the result of a temperature-dependent Auger cross-section, since none of the other peaks behaved in this manner. Simple background adsorption of CO must be ruled out because the pressure was too low and there was no increase in the carbon signal. Coad *et al.* (1970) reported a time-dependent increase in the oxygen Auger signal from a Si(111) in ultra high vacuum, which they associate with electron beam induced dissociation of physisorbed CO. It is unlikely that such an effect can account for our observations in which the accumulation of surface oxygen appears to be temperature dependent rather than time dependent, in addition to which we did

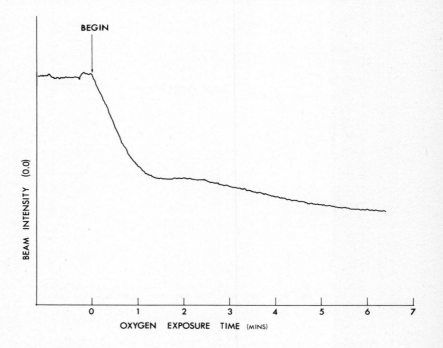

Fig. 2. (0, 0) Beam intensity versus time—raw data. (Reproduced with permission from Lambert *et al.*, 1971.)

not observe the concurrent increase in the carbon signal which was found in the Si(111) case. The only remaining explanation would appear to be that oxygen in the bulk migrates to the surface with cooling, to yield a stable room temperature configuration. This would also explain the failure to remove the last traces of oxygen from the surface by heating—on raising the temperature some of the oxygen always goes into solution rather than leaving as desorption product, only to reappear on cooling. Apparently similar temperature reversible migrations have been observed recently for hydrogen on W(100) (Yonehara and Schmidt, 1971) and for oxygen on W(110) and W(100) (Weinberg, 1971). In any event, it is clear from figure 3 that this surface is reproducible, since data taken many days apart normalize well to the same curve.

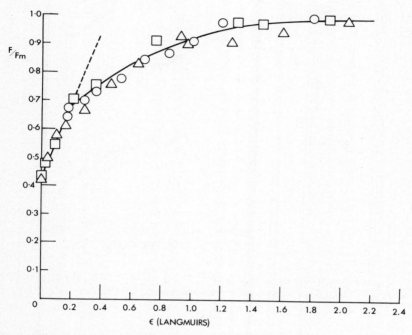

Fig. 3. Relative oxygen Auger signal versus exposure: △ ○ □ refer to different experiments. (Reproduced with permission from Lambert *et al.*, 1971.)

FLASH DESORPTION

In these experiments, the sample was rotated to face the mass spectrometer and rapidly heated after an oxygen exposure. The only significant desorbing species was at $m/e = 28$, presumably CO. Flash desorption spectra of this CO signal were then taken after varying oxygen exposures, and some representative results are shown in figures 4 (a), (b), (c). Although no precautions were taken to mask the crystal supports from line of sight with the mass spectrometer, we are confident that the desorption spectra of figure 4 refer to the crystal alone. The very large thermal capacity of the supports compared with the sample ensured a very slow temperature rise in the former while the sample heated up. In fact it was easily demonstrated that no detectable desorption from the crystal holder occurred until the sample had reached $\approx 1600°$K and desorption from it was complete.

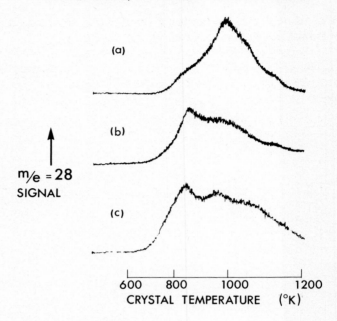

m/e = 28
SIGNAL

Fig. 4. Desorption spectra at m/e = 28: (a) after 0.18 L, (b) after 0.40 L, (c) after 1.30 (L). (Reproduced with permission from Lambert *et al.*, 1971.)

IV. DISCUSSION

INITIAL STICKING COEFFICIENT

The LEED intensity data for the 180 V (0, 0) beam was normalized to the intensity at zero exposure. Figure 5 shows the fractional change in intensity as a function of exposure (ϵ) in Langmuirs (1 L = 1 Langmuir = 10^{-6} Torr sec). Intensity data from exposure sequences at different total pressures normalized to this same curve with initial slope σ_1 = -0.95 L^{-1}. The relative oxygen Auger signal is expressed as F/F_m, where F_m is the saturation value obtained after about 2.0 L and is taken as a measure of the maximum coverage at room temperature. At exposures <0.25 L the slope of the F/F_m versus ϵ curve is constant at σ_2 = 1.42 L^{-1}.

Either the AES or LEED data can be used to calculate S_0 if a value can be assigned to the coverage at saturation or at some intermediate stage. This has been done by assuming a particular coverage determined by the appearance of new diffraction features, or by arbitrarily assigning a coverage to correspond to the maximum Auger signal. However, as will be shown here, when both kinds of data are available it is possible to formulate expressions for S_0 in terms of both the experimental quantities. These may be solved simultaneously to obtain S_0 and the saturation coverage θ_{max}. The number density of Mo atoms on the ideal (111) surface ($N_t = 5.84 \times 10^{14}$ atoms/cm^2) is used as the basis for monolayer

Fig. 5. Normalized (0, 0) beam intensity versus exposure. (Reproduced with permission from Lambert *et al.*, 1971).

coverage. In what follows, θ_0 refers to the fractional coverage at $\epsilon = 0$ and θ_1, $n1$ to the coverage and the number of oxygen species respectively, which accumulate during exposure. At any time during exposure

$$F/F_m = (\theta_0 + \theta_1)/\theta_m \tag{1}$$

and for the initial stage after time t at pressure p

$$n_1 = \frac{2S_0 pt}{\sqrt{(2\pi MkT)}} = \frac{2S_0 \epsilon}{\sqrt{(2\pi MkT)}} \tag{2}$$

where M and T refer to the mass and temperature of the gas molecules.

Therefore,

$$\theta_1 = CS_0\,\epsilon \tag{3}$$

where

$$C = \frac{2}{N_t\sqrt{(2\pi MkT)}} \tag{4}$$

and from (1)

$$F/F_m = (\theta_0 + CS_0\,\epsilon)/\theta_m \tag{5}$$

$$S_0 = \theta_m\,\sigma_1/C \tag{6}$$

where

$$\sigma_1 = \frac{d(F/F_m)}{d\epsilon}\bigg|_{\epsilon\,\to\,0}$$

is the experimentally measured quantity.

The LEED data are treated by taking the intensity of the (0, 0) beam to be the sum of contributions from covered and uncovered portions of the Mo mesh. This leads to an expression for the normalized intensity as follows

$$I/I_0 = 1 - (2\beta/\alpha)\,\theta_1 + (\beta/\alpha)^2\,\theta_1^2 \tag{7}$$

where

$$\beta = k_2 - k_1$$
$$\alpha = k_2 - (k_2 - k_1)\,\theta_0$$

and k_1/k_2 is the ratio of amplitude reflection coefficients for the covered and uncovered portions of the surface. Assuming for the moment $k_1/k_2 \ll 1$ then

$$\beta/\alpha = 1/(1 - \theta_0) \tag{8}$$

and using (3), (7), and (8) yields

$$I/I_0 = 1 - \frac{2CS_0\,\epsilon}{1 - \theta_0} + \frac{CS_0^2\,\epsilon^2}{1 - \theta_0} \tag{9}$$

and

$$S_0 = -\sigma_2(1 - \theta_0)/2C \tag{10}$$

where

$$\sigma_2 = \frac{d(I/I_0)}{d\epsilon}\bigg|_{\epsilon\,\to\,0}$$

is also known from experiment.

The value of θ_0 in (10) is given by

$$\theta_0/\theta_m = F_0/F_m$$

so that (10) and (6) may now be solved for θ_m and S_0. This gives the values

$$\theta_m \simeq 0.3; \quad S_0 \simeq \tfrac{1}{3}.$$

CONCLUSION

The observed desorption of CO can only result from the reaction of surface carbon with adsorbed oxygen. It can be seen that there is a good correlation between the AES data and the complexity of the flash desorption spectrum. At exposures $\leqslant 0.25$ L characterized by a high sticking coefficient a single peak occurs in the flash desorption spectrum. For $0.25 < \epsilon < 0.8$ there is a distinct change of slope in the curve of figure 3 corresponding to an approximately fourfold decrease in the sticking coefficient and a second desorption peak occurs at a *lower* temperature. This behaviour is consistent with the presence of two kinds of binding site for oxygen on this surface—one associated with a high sticking coefficient and a high characteristic reaction temperature, and the other, more loosely bound, corresponding to a lower sticking coefficient and a reduced reaction temperature. It is interesting that bulk thermodynamic data (Spencer and Justice, 1934) give $\Delta G = 0$ for the reaction

$$MoO_2 + 2C \rightarrow 2CO + Mo$$

at $1000°K$; this is very close to the temperature of the low exposure desorption peak.

The correlation between LEED and AES data indicates that $\approx \tfrac{2}{3}$ of the surface is inactive in oxygen chemisorption at room temperature. This particular value has arisen from the assumption $k_1/k_2 \ll 1$; however, θ_m and S_0 are both insensitive functions of k_1/k_2 (e.g. for $k_1/k_2 = 0.5$, $\theta_m = 0.62$, $S_0 = 0.62$) and in particular the value of S_0/θ_m, the sticking coefficient per unit active surface area, is unchanged at unity. Thus, if the foregoing treatment is correct in principle, it is clear that a large fraction of the surface has been rendered inactive. Both LEED and AES data indicate that the level of carbon contamination is low, and certainly not as high as a large fraction of a monolayer, so that masking of the Mo surface by a large number of foreign atoms must be ruled out as a possible explanation for the reduced activity. If the interaction between a surface carbon atom and its nearest neighbour Mo atoms is sufficient to modify substantially the chemisorption properties of the latter, then impurity concentrations as low as $\tfrac{1}{25}$ of a monolayer could affect the activity of as much as 40% of the surface. At even lower levels of contamination it is possible that large regions of the surface may still be affected by long-range interactions taking place through the metal lattice. This concept has been advanced by Schram and others (1970) to explain the anomalously low adsorption of H_2 on Ni, caused by very small traces of CO. By using the saturation oxygen Auger signal as a guide to instrumental sensitivity in conjuction with the relative O and C Auger yields deduced by Coad *et al.*, we calculate that the residual C contamination in this work was $\leqslant \frac{1}{30}$ monolayer.

REFERENCES

Barton, G. C. and Ferrante, J. (1968). *NASA Tech. Note* TN D-4735.
Coad, J. P., Bishop, H. E. and Rivière, J. C. (1970). *Surface Sci.* **21**, 253.
Florio, J. V. and Robertson, W. D. (1969). *Surface Sci.* **18**, 398.
Haas, T. W., Grant, J. T. and Dooley, G. J. (1970). *Phys. Rev., B.* **1**, 1449.
Haas, T. W. and Jackson, A. G. (1966). *J. Chem. Phys.* **44**, 2921.
Lambert, R. M., Linnett, J. W. and Schwarz, J. A. (1971). *Surface Sci.* **26**, 1285.
Musket, R. G. and Ferrante, J. (1970). *J. Vac. Sci. Technol.* **7**, 14.
Schram, A. (1970). Proceedings of the International NEVAC—Symposium, p. 85.
Sickafus, E. N. (1970). *Surface Sci.* **19**, 181.
Spencer, H. M. and Justice, J. L. (1934). *J. Amer. Chem. Soc.* **56**, 2301.
Weber, R. E. and Peria, W. T. (1967). *J. Appl. Phys.* **38**, 4355.
Weinberg, W. H. (1971). Private communication.
Yonehara, K. and Schmidt, L. D. (1971). *Surface Sci.* **25**, 238.

CRYSTALLOGRAPHIC ANISOTROPIES IN CHEMISORPTION STRUCTURES AND KINETICS ON BCC METALS

L. D. SCHMIDT

*Department of Chemical Engineering and Materials Science,
University of Minnesota, Minneapolis, USA*

I. INTRODUCTION

A knowledge of the role of crystallographic anisotropies in adsorption is obviously a prerequisite to any detailed interpretations of processes which occur at solid surfaces. We have recently been studying the interactions of simple gases (H_2, N_2, CO, O_2, CO_2, NH_3) on individual crystal planes of bcc transition metals (W, Mo, Nb, Ta) using flash desorption. The main object of this work is the interpretation of the structures and kinetics involved on an atomic scale. The gases were chosen because of their simplicity (s and p orbitals), and the substrates were chosen because they possess identical crystal structures, so that differences between substrates must be attributed to the electronic properties of the substrates.

We shall summarize some of the results which most clearly illustrate the role of structural and chemical influences on chemisorption structures and kinetics. We shall first consider the binding states and desorption kinetics of H_2 and CO on individual crystal planes of W and Mo and then the kinetics of condensation of N_2 and H_2 on several planes of W.

II. EXPERIMENTAL

Measurements were made using primarily flash desorption mass spectrometry. The apparatus and procedure has been described in detail previously (Tamm and Schmidt, 1969, 1970; Clavenna and Schmidt, 1970). For quantitative studies of the amounts adsorbed and the desorption-rate parameters, a crystal exposing only one crystal plane, a uniform crystal temperature and a high pump-out rate are necessary. Crystals were oriented, cut and polished to expose only the plane desired and were heated by electron bombardment to avoid temperature non-uniformities. Tantalum evaporated film getters connected to the crystal chambers by short tubes provided pumpout times as short as 10 msec.

Amounts adsorbed, and desorption rate parameters, were determined by flash desorption. For a binding state with a desorption activation energy E_d, pre-exponential factor $v_0^{(m)}$, and order of desorption m, the desorption rate is given by the expression

$$\frac{dn}{dt} = -v_0^{(m)} e^{-E_d/RT} n^m \tag{1}$$

where n is the surface atom density. Now if a surface at temperature T is allowed to adsorb an amount of gas n_0 and is then heated rapidly such that $T = T_0 + \beta\tau$, the rate of desorption versus temperature will exhibit a peak whose shape and position are determined by the parameters. If the pumping speed is high, the pressure will be proportional to dn/dt and one will observe a peak for each binding state. Data was obtained by displaying the partial pressure of a particular species versus time on an oscilloscope for heating rates β between 50 and 1000°/sec. Amounts in each binding state were determined from the area under each peak, and desorption-rate parameters were determined by fitting the experimental desorption traces with computer-generated curves.

Sticking coefficients were determined both by measuring the rate of adsorption for a known pressure and by measuring the pumping by the crystal during adsorption (Tamm and Schmidt, 1970). Both methods gave general agreement although the latter is more accurate because absolute values of s do not depend on a mass spectrometer sensitivity calibration. Constant crystal temperatures below 400°K were provided by constant temperature baths and those above 400°K by a focused light beam or electron bombardment. It was established that different methods gave identical results, thus eliminating artifacts of particular heating methods. Care was also used to show that electron impact effects had a negligible influence on the adsorbate; this is especially important for CO for which cross-section data (Menzel and Gomer, 1964) indicate that some conversion from the α state might be expected.

LEED data were obtained in a conventional four-grid display system. Diffraction spot intensities and shapes were determined by scanning the photographs with a photometer. Simultaneous flash desorption measurements were used to determine the coverages corresponding to particular LEED structures.

III. BINDING STATES OF H_2 ON W AND Mo

The (110) and (100) planes are the most densely packed and highly symmetric planes of bcc crystals and therefore should exhibit the fewest possible adsorption sites. Figure 1 shows the flash desorption spectra of H_2 on the (100) and (110) planes of W and Mo following saturation with H_2 at $\sim 1 \times 10^{-8}$ Torr

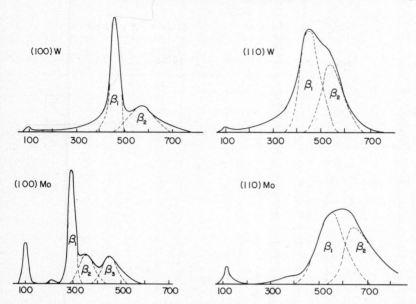

Fig. 1. Flash desorption traces of H_2 after saturation at $78°K$ on the (110) and (100) crystal planes of W and Mo. Peaks corresponding to the major states are indicated by dashed lines.

and $78°K$ (Tamm and Schmidt, 1971; Han and Schmidt, 1971; Mahnig and Schmidt, 1971).

It is evident that there are several binding states on each substrate. Computer fits to these spectra were used to determine the rate parameters and saturation amounts. Also "second-order plots" (Tamm and Schmidt, 1969) were used to determine the pre-exponential factors for those states which obey second-order desorption kinetics. These parameters are summarized in table 1. The most tightly bound states on all planes obey second-order kinetics as expected for a dissociated species; however on the (100) planes of W and Mo there is a fairly tightly bound state which obeys "first-order" desorption kinetics. Hydrogen on the (111) plane of W has also been examined. Here observe four atomic states, the most weakly bound of which has an activation energy of 14 kcal mole^{-1}.

The flash desorption spectra can be fitted almost quantitatively, assuming constant values of ν_0 and E_d for all states. In fact for those states which obey first-order kinetics the spectra can be fitted to within a few percent taking $\nu_0 = 10^{13}$ sec^{-1} and E_d a constant. However, while the second-order states can be fitted satisfactorily for $\theta < 0.2$ (θ is the fraction of the saturation coverage in that state), near saturation all second-order states deviate significantly from the curves predicted with constant parameters.

Table 1. *Binding states and desorption-rate parameters*

Adsorbate	Substrate	State	m	E_{d0} (kcal mole^{-1})	α (kcal monolayer^{-1})
H_2	(100)W	β_1	1	26	<1
		β_2	2	32	$\leqslant 3.5$
H_2	(110)W	β_1	2	27	$\leqslant 2$
		β_2	2	33	$\leqslant 3.3$
H_2	(100)Mo	β_1	1	16	<1
		β_2	2	20	—
		β_3	2	27	—
H_2	(110)Mo	β_1	2	~ 29	—
		β_2	2	~ 34	—
CO	(100)W	α	1	~ 22	—
		β_1	1	57	—
		β_2	1	62	—
		β_3	2	74	—
CO	(100)Mo	α	1	~ 17	—
		β_1	1	~ 54	—
		β_2	1	~ 76	—
		β_3	2(?)	~ 88	—

One can obtain quantitative agreement assuming that E_d or $\nu_0^{(2)}$ are functions of coverage. Some variations in E_d with θ should arise from repulsive interactions between adsorbate atoms. The simplest functional form for this variation is

$$E_d = E_{d0} - \alpha\theta, \tag{2}$$

with α a constant. Values of α for most states are shown in table 1. An alternative explanation in that $\nu_0^{(2)}$ may vary with coverage. It has been shown (Clavenna and Schmidt, 1970) that, if desorption involves recombination of atoms via a random-walk surface-diffusion process, $\nu_0^{(2)}$ should be given by an expression of the form

$$\nu_0^{(2)} = \frac{a^2 \nu_0}{\left(1 - \dfrac{a_c}{a}\theta^{1/2}\right)^2} \tag{3}$$

where a is the jump length, ν_0 the vibrational frequency and a_c the critical distance for recombination. The desorption spectra can also be fitted quantitatively using equation 3. It is not possible to decide between these mechanisms, and the values of α in table 1 should be regarded only as upper bounds on the variation E_d with coverage. It is also found that $\nu_0^{(2)}$ at low coverage is ~ 0.01 for all second order states of N_2 and H_2 examined so far. This is approximately $a^2\nu_0$ and thus supports the hypothesis that desorption of atoms is limited by surface diffusion recombination of atoms.

It is evident from figure 1 and table 1 that there is considerable variation in binding states and their desorption activation energies between different planes. However the most tightly bound states on all planes have approximately the same binding energies on the (100), (110), and (111) planes. This shows that the strength of the adsorption bonds bears no simple relationship to the number of nearest neighbors involved on a particular plane. In fact for Mo the binding energy is significantly higher on the close-packed (110) plane than on the more open (100) plane.

It is also interesting to compare adsorption of W and Mo. These metals have identical lattice constants (3.15 Å versus 3.14 Å) and are isoelectronic, each having six outer electrons. While there is a general similarity between binding states, there are definite differences. First, bonding on the (100) plane is stronger on W than on Mo, while one the (110) plane it is slightly stronger on Mo. Second, on the (110) planes there are two identical atomic states with equal densities while on the (100) planes there are corresponding β_1 and β_2 states but there is in addition a β_3 state on (100)Mo which is not observed on W.

Hydrogen on the (100) planes of W and Mo also exhibits extra diffraction beams in LEED, showing that there are ordered structures with periodicities greater than those of the substrates. On (100)W H_2 adsorption leads initially to a c(2 x 2) structure with alternate sites occupied (Estrup and Anderson, 1966; Yonehara and Schmidt, 1971). The coverage corresponding to this structure is ~0.2, indicating that it coincides with population of the atomic β_2 state. Detailed examination (Yonehara and Schmidt, 1971) of the intensity, shape, coverage and temperature dependences of the extra diffraction spots reveals greater complexity than was assumed. No completely acceptable explanations for the structures have yet been given. On (100)Mo Dooley and Haas (1970) showed that at low coverage a c(4 x 2) LEED pattern is obtained while at higher coverage the pattern reverts to (1 x 1). From desorption measurements the c(4 x 2) structure should correspond to occupation of the β_3 state, while population of the β_2 state leads to the (1 x 1) structure. Estrup (1971) has shown that the c(4 x 2) pattern may result from a situation in which one-fourth of the sites on the surface are occupied.

Another conclusion to be drawn from these measurements is that the surfaces are in fact quite different. There are no direct experimental measurements to determine whether all of the surface substrate atoms are actually in the bulk lattice configuration or whether an appreciable fraction of surface disorder exists. However, the data of figure 1 show that the sites on the different planes are quite different. For example there is no peak from (110)Mo at a temperature corresponding to the β_1 state on (100)Mo; this shows that (100) type sites on the (110) plane comprise at most 1% of the total hydrogen sites. Close examination of the flash desorption spectra shows the presence of several small (less than 1% of the total adsorbed hydrogen) states on all planes, but the combined amounts in these states is in all cases less than 5% of the amounts in the major states. This observation, and the fact that the saturation atom densities are in a ratio of small integers, indicates that the adsorption sites for these gases on these planes are those of "ideal crystal planes".

It is instructive to consider bonding on a one-dimensional potential energy curve, figure 2. The potential curves describing these systems are of course not one dimensional, since sites for different states will occur on different regions of the surface and will probably involve surface diffusion processes. These potential

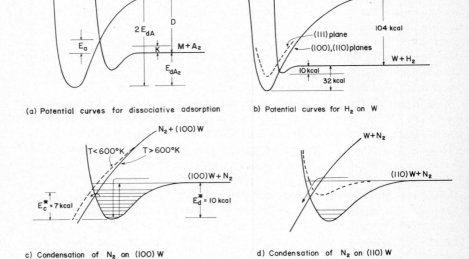

(a) Potential curves for dissociative adsorption   b) Potential curves for H_2 on W

c) Condensation of N_2 on (100) W   d) Condensation of N_2 on (110) W

Fig. 2. Schematic potential curves for dissociative adsorption. Curves for H_2 on W are drawn to scale in (b) and curves pertaining to N_2 on the (100) and (110) planes are indicated in (c) and (d) respectively.

curves should therefore be regarded as the "reaction coordinate" for the system. In the absence of an activation energy of adsorption E_a (which we shall later show to be negligible for the present case) the activation energy of desorption E_{dA_2} of molecule A_2 on substrate M is related to E_{dA} the binding energy of atom A by the expression

$$E_{dA} = \tfrac{1}{2}(D + E_{dA2}).$$
(4)

The potential curves for H_2 on W are shown in figure 2b. The energy scale is drawn correctly but, the distance coordinate can only be estimated. The experimental activation energies from the (100), (110) and (111) planes are indicated, and the depth of the H_2 curve is drawn as 10 kcal mole^{-1}, the activation energy of the molecular γ state; this could also be a physically adsorbed state which would have a somewhat lower binding energy. One point, obvious from equation 4 and these curves, is that the binding energies of the atoms do not vary much: from 69 kcal mole^{-1} for the most tightly bound states to 59 kcal mole^{-1} for the most weakly bound atomic state on the (111) plane.

To determine the structures and bonds involved one attempts to develop models which correlate the states (stoichiometries and kinetics) observed in flash desorption, with information on sticking coefficients: LEED patterns, electron-impact desorption, mixed adsorption, etc. We and others have speculated on possible structures for these states but we shall not consider them here. Clearly, however, the bonds formed are highly specific to each crystal plane and to each substrate. For example the results for H_2 on the (100) plane of W can be rationalized assuming substrate bonding with only s and d character, while H_2 adsorption on (100)Mo can be rationalized assuming substrate bonds with p character (Han and Schmidt, 1971).

IV. CARBON MONOXIDE ON (100)W AND (100)Mo

In conjunction with studies of CO_2 decomposition we have examined the flash desorption spectra of CO from (100)W and (100)Mo. CO does not dissociate appreciably though its adsorption properties are quite complex (Ehrlich, 1966). This is not a complete study of this system, for adsorption was only examined at 300 and 78°K and no readsorption measurements were made (Bell and Gomer, 1966; Kohrt and Gomer, 1970). However the flash desorption spectra do reveal some interesting features of this system.

The saturation flash desorption spectra of CO from (100)W and (100)Mo are shown in figure 3. There are several binding states evident, with two peaks at ~400°K, usually termed the α states, and two or more peaks between 1000 and 1500°K, usually termed β states. From the peak shape and the absence of a peak temperature shift with increasing coverage it is found that most of the states obey first-order desorption kinetics as expected for a non-dissociated species. Assuming $\nu_0^{(1)} = 10^{13}$ sec^{-1}, we obtain activation energies shown in table 1.

However the most tightly bound state on the (100) plane of W, labelled β_3, exhibits a peak width and a peak temperature shift with increasing coverage characteristic of second-order kinetics. A second-order plot for this state gave a straight line for $0.1 < \theta_{\beta_3} < 0.9$ from which the desorption activation energy was determined to be 75 kcal mole^{-1}. For $\theta_{\beta_3} < 0.1$ the peak no longer shifted but remained at 1660°K down to a coverage of 0.004, typical of first-order desorption.

L. D. SCHMIDT

Care was used to assure that this state was not another species such as N_2 or a state of CO induced by the presence of another species. The partial pressures of other species (primarily H_2 and CO_2) were always less than a few percent of the total. Also flash desorption spectra at masses 12 and 16 (C and O) were identical

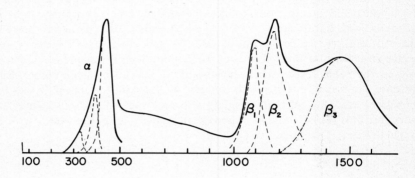

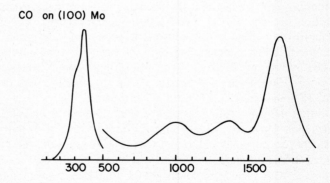

Fig. 3. Flash desorption traces of CO on the (100) planes of W and Mo after saturation at 78°K. There are several states evident on both surfaces.

in shape to that at mass 28 while that at mass 14 (N) was at least three orders of magnitude smaller. In this and all other adsorption studies there is the possibility of surface contamination by other metals, carbon, or sulfur. Data reported here was completely reproducible, chemical analysis revealed no detectable bulk impurities, and other investigators report clean surfaces under identical or less stringent cleaning procedures.

Second-order desorption would be expected if desorption required collision of two adsorbed CO molecules according to the sequence

$$2CO(s) \underset{k_2}{\overset{k_1}{\rightleftarrows}} (CO)_2(s) \xrightarrow{k_{d2}} CO(g) + CO(s) \text{ or } 2CO(g). \tag{5}$$

This is presumed to be in competition with the direct process

$$CO(s) \xrightarrow{k_{d1}} CO(g). \tag{6}$$

If the concentration of $(CO)_2(s)$ is small enough that the steady-state approximation is valid, one obtains

$$\frac{dCO(g)}{dt} = k_{d2}(CO)_2(s) + k_{d1}CO(s)$$

$$= \frac{k_{d2}k_{d1}}{k_2 + k_{d2}}[CO(s)]^2 + k_{d1}CO(s). \tag{7}$$

At high coverage the first term should predominate giving second-order kinetics, at low coverage the first-order term should predominate to give first-order kinetics.

It has been observed that, while CO apparently does not dissociate, it does exhibit partial isotope exchange (Madey, Yates and Stern, 1965). This has been interpreted as implying a dimer complex on the surface with the structure $\begin{smallmatrix} O-C \\ C-O \end{smallmatrix}$. It is interesting that this complex could also be the intermediate state for desorption with second order kinetics according to equation 5.

One notable difference between the flash desorption spectra of CO from those of H_2 and N_2 is the considerably greater complexity for CO. For H_2 and N_2 we argued that the surface sites were those of the ideal planes because the spectra were relatively simple in number of states and their kinetics and saturation amounts. For CO this is not the case: there are many binding states with appreciable amounts but these amounts bear no simple relation to each other. Also all peaks appear to be smeared out much more than those of H_2 and N_2. This could result either from many small states on the surface or some sort of conversion between states competing with desorption. It may be that CO adsorption, because of its multiple bonding possibilities, is capable of existing with many types of bonds rather than the relatively simple configurations observed for the monatomic species. The greater complexity for CO may also arise from conversion from the original state which exists on the cold surface. It has been found (Bell and Gomer, 1966) that different work-function changes and flash desorption spectra are observed when CO is deposited on a surface which had been saturated and then heated, compared to one saturated at low temperatures.

V. CONDENSATION

For condensation to occur upon collision of a gas molecule with a surface, the molecule must transmit the normal component of its kinetic energy $K \simeq kT$ to the substrate. Also there must be binding states available from which the molecule will not evaporate at the surface temperature T_s.

The simplest picture of condensation envisions a single-step process with the sticking coefficient s determined by the probability of energy transfer s_0 (equal to s at $\theta = 0$) multiplied by the probability $g(\theta)$ of the incoming molecule striking a vacant site: $g(\theta) = 1 - \theta$ for single-site adsorption, $(1 - \theta)^2$ for random two-site adsorption, etc. It is obvious that for dissociative adsorption one should expect more complexity because dissociation must occur upon impact. The one-dimensional potential-energy diagram, figure 2a shows that the molecule, even if accommodated, must cross to the atomic curve for dissociative adsorption. It has been known for some time (Ehrlich, 1956) that N_2 and H_2 condensation on polycrystalline metals can only be explained by assuming the existence of a precursor state.

We have measured sticking coefficients of H_2 and N_2 on several crystal planes of W as functions of T_s and coverage. Our objective is to examine in detail the processes involved in condensation. As we shall show, these processes are quite complex and vary greatly between different crystal planes.

Figure 4 shows measured sticking coefficients for H_2 and N_2 at 300°K on the (100), (110), and (111) planes of W for $T_s = 300°K$ (Tamm and Schmidt, 1970; Clavenna and Schmidt, 1970; Tamm and Schmidt, 1971). The coverage θ in this case is the total fractional coverage at 300°K; for H_2 this represents all of the β states while for N_2 on the (100) and (110) planes there is a single atomic state with desorption activation energies of 79 and 80 kcal mole^{-1} respectively (Clavenna and Schmidt, 1970; Tamm and Schmidt, 1971).

For H_2, a system where there are several states, sticking coefficients provide information regarding the existence and properties of these states at the condensation temperature. This complements flash desorption data which only applies to the temperatures and coverages at which desorption occurs. We see from figure 4 that on (100)W there is a break in the curve at $\sim\frac{1}{3}$ of saturation, precisely the coverage corresponding to the β_2 state; this suggests that distinct states populate sequentially even at 300°K. On the (111) plane, where there are four atomic states at 78°K, there is a definite break in the curve upon completion of the two most tightly bound states. This suggests that at least two species populate sequentially on this plane. On the (110) plane s for H_2 varies precisely at $1 - \theta$, indicating that, if two states exist at low temperatures, they exhibit identical condensation kinetics.

For N_2 there is only one atomic state on the (100) and (110) planes, and one can therefore attempt to quantitatively compare $s(\theta)$ curves with theory. The only explanation yet offered for a sticking coefficient which is independent of coverage involves a precursor state whose properties (binding energy E_d^* and

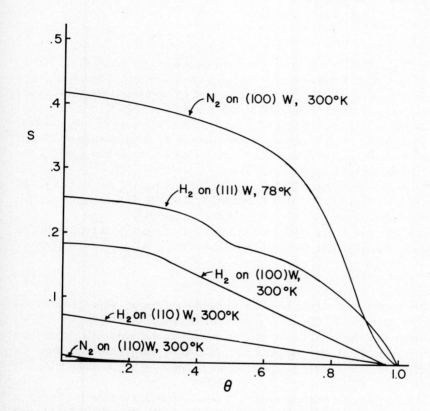

Fig. 4. Sticking coefficient versus coverage curves for N_2 and H_2 on the (100), (110), and (111) planes of W for a gas temperature of $300°K$ and substrate temperature of 300 or $78°K$.

sticking coefficient) are independent of the coverage θ in the tightly bound state. If condensation of a gas molecule $A_2(g)$ into the chemisorbed state $A(s)$ proceeds according to the sequence

$$A_2(g) \underset{k_d^*}{\overset{s^*f}{\rightleftarrows}} A_2^*(s) \xrightarrow{k_c g(\theta)} A(s), \tag{8}$$

with f the flux and k_d^* and k_c the rate constants for desorption and conversion from the precursor, then it can be shown (Clavenna and Schmidt, 1970) that s should be given by the expression

$$s = \frac{s^*}{1 + \dfrac{k}{g(\theta)}} = \frac{s_0(1 + K)}{1 + \dfrac{k}{g(\theta)}} .\qquad (9)$$

where $K = k_d^*/k_c$.

For N_2 on (100)W $s(\theta)$ curves have been determined for $195 < T_s < 1100°K$, and all data can be fitted quantitatively using equation 9. Figure 5 shows a plot of $\log K$ and $\log s_0$ versus $1/T_s$. It is seen that s_0 (approximately equal to s^*) decreases monotonically from ~ 0.4 at $195°K$ to 0.08 at $1100°K$.

For H_2 on (100)W data taken for the β_2 state in the range $78 < T_s < 400°K$ can be fit by equation 9, although the accuracy in the determination of K is not as high because of overlap with the β_1 state. Figure 5 shows that s_0 and K for this state are almost independent of T_s.

It is evident from figure 5 that a plot of $\log K$ versus $1/T_s$ for N_2 on (100)W gives straight lines for $T_s < 600°K$ and for $T_s > 600°K$. The activation energies for these segments are 0.8 and 3.0 kcal mole^{-1} respectively.

Some features of the states involved in N_2 condensation on (100)W can be inferred for these results. From the definition of K

$$K = \frac{k_d^*}{k_c} \sim e^{-(E_d^* - E_c)/RT} ,\qquad (10)$$

it is evident that K should obey an Arrhenius expression, with the activation energy being $E_d^* - E_c$. Thus, referring to the potential diagram in figure 2c, we can determine the point of crossing of the atomic and molecular curves from these results. We assume the precursor state to be the γ state of nitrogen for which $E_d^* \simeq 10$ kcal mole^{-1}. Therefore we obtain $E_c = 9$ kcal mole^{-1} for $T > 600°K$ and $E_c = 7$ kcal mole^{-1} for $T > 600°K$.

The change in the activation energy at $600°K$ evidently represents a change in the mechanism of condensation, and this is expected since the tightly bound state of N_2 becomes mobile on the surface at $600°K$. There should therefore be two parallel conversion paths from the precursor state to the final state, one where the final state is immobile with rate constant k_{c1}, and one where this state is mobile with rate constant k_{c2}. In this case equation 9 becomes

$$s = \frac{s_0(1 + K)}{1 + \dfrac{k_d^*}{(k_{c1} + k_{c2})g(\theta)}} .\qquad (11)$$

If the activation energies in k_{c1} and k_{c2} are 9 and 7 kcal mole^{-1} respectively, one obtains the observed behaviour.

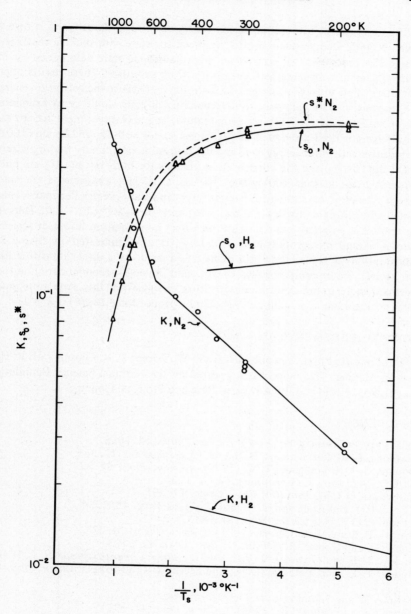

Fig. 5. Plot of s_0, s^*, and K versus $1/T_s$ for N_2 and H_2 on (100)W. The quantity K, defined by equation 10, gives straight line segments for N_2 from which activation energies of 3.0 and 0.8 kcal mole^{-1} are obtained.

Therefore we picture condensation on N_2 on (100)W occurring as shown in figure 2c with accommodation into the precursor state with sticking coefficient s^* and the probability of conversion into the atomic state determined by the relative rates of conversion and desorption from precursor. These results imply that there is no activation energy of adsorption E_a because the activation energy in K is positive. The only way to determine E_a directly would be by examining the dependence of s on the *gas* temperature, and this experiment has to our knowledge not been attempted as yet. However for both H_2 and N_2 on (100)W the positive activation energy in K and the high value of s_0 imply that E_a is zero.

On the (110) plane the condensation kinetics for both N_2 and H_2 are quite different than on the (100) plane. The initial sticking coefficients are much lower (by factors of 100 and 2 respectively), and the coverage dependences are completely different: for H_2 $s = s_0(1 - \theta)$ and for N_2 $s = s_0(1 - \theta)^2$. Thus for H_2 there is a precursor state because for direct condensation into a dissociated state s should be proportional to $(1 - \theta)^2$. As indicated in figure 2d, condensation of N_2 on the (110) plane could occur by a direct transition into the atomic state upon impact, or it could involve accommodation into a precursor state if the desorption activation energy from this state were small because, according to equation 9, s should be proportional to $g(\theta)$ if $K \gg 1$.

ACKNOWLEDGEMENTS

Data reported here was obtained by P. W. Tamm, L. R. Clavenna, H. R. Han and M. Mahnig. This work was supported by the National Science Foundation (Grant No. 16241) and the Advanced Research Projects Agency.

REFERENCES

Bell, A. E. and Gomer, R. (1966). *J. Chem. Phys.* **44**, 1065.
Clavenna, L. R. and Schmidt, L. D. (1970). *Surface Sci.* **22**, 365.
Dooley, G. J. III and Haas, T. W. (1970). *J. Chem. Phys.* **52**, 993.
Ehrlich, G. (1956). *J. Phys. Chem. Solids.* **1**, 3.
Ehrlich, G. (1966). *Ann. Rev. Phys. Chem.* **17**, 295.
Estrup, P. J. and Anderson, J. (1966). *J. Chem. Phys.* **45**, 2254.
Estrup, P. (1971). *J. Chem. Phys.* **54**, 1845.
Han, H. R. and Schmidt, L. D. (1971). *J. Phys. Chem.* **75**, 227.
Kohrt, C. and Gomer, R. (1970). *J. Chem. Phys.* **52**, 3283.
Madey, T. E. and Yates, J. T. Jr. (1967). *Nuovo Cimento, Suppl.* N. 2, **5**, 483.
Madey, T. E., and Yates, J. T. Jr., and Stern, R. C. (1965). *J. Chem. Phys.* **42**, 1372.
Mahnig, M. and Schmidt, L. D. (1971). In press.
Menzel, D. and Gomer, R. (1964). *J. Chem. Phys.* **41**, 3311.
Tamm, P. W. and Schmidt, L. D. (1969). *J. Chem. Phys.* **51**, 5352.
Tamm, P. W. and Schmidt, L. D. (1970). *J. Chem. Phys.* **52**, 1150.
Tamm, P. W. and Schmidt, L. D. (1971). *J. Chem. Phys.*, In press.
Yonehara, K. and Schmidt, L. D. (1971). *Surface Sci.* **25**, 238.

MEASUREMENT OF WORK-FUNCTION CHANGES (CO ON Mo)

D. MOUROT, Y. BALLU and D. A. DEGRAS

*Centre d'Études Nucléaires de Saclay, Section
d'Études des Interactions Gaz-Solides, Gif-sur-Yvette, France*

I. INTRODUCTION

In the case of metal surfaces, which will be considered exclusively, the adsorption of atoms or molecules usually induces a change in the metal work function. This is due either to charge transfer in either direction or, more generally, to local self-consistent perturbation of the adsorbate and adsorbent electrons. It is found experimentally that the work-function change is coverage dependent. A linear relationship holds in some cases, but more complex behaviour may arise which suggest co-operative phenomena, a significant fact for gas-metal interaction. Most of the recent theoretical descriptions of chemisorption on metals have not been explicitly focused on work-function changes, but accurate experimental data will be helpful when meaningful calculations are possible. In recent years the main techniques used for surface potential changes (s.p.) are: the condenser method, the so-called "retarding potential difference", the field emission microscope, the space-charge limited diode and photoelectric methods.

The basis of the condenser method is illustrated in figure 1, from Eberhagen (1960), cited by Hayward and Trapnell (1964). The electric field which exists between the two plates may be cancelled by a compensating potential which is such that

$$V_a = -V_{12} = \phi_1 - \phi_2 \qquad (1)$$

If plate 2 is taken as the reference electrode, then any change $\Delta\phi_1$, due to adsorption effects will be reflected through the appropriate ΔV_a. Obviously ϕ_2 must not be altered by adsorption. That the electric field between the plates is zero may be checked either by changing their capacity (vibrating method), or by using the current which flows between the electrodes during the building of the adsorbed layer to monitor the compensating potential V_a (steady method). The condenser method has been widely used by a number of workers: typical illustrations may be found in many papers of Hopkins *et al.* (1966, 1967).

405

The "retarding potential difference" technique uses an electron beam which is produced by a thermionic emitter and accelerated towards the sample (collector). A possible procedure (Madey and Yates, 1967) is allowed by the saturation region of the emitter, where the current is given by the Richardson equation

$$J_s = AT^2 \exp(-e\varphi_1/kT) \tag{2}$$

If a retarding potential ΔV_a is applied between the emitter and the sample, the net result is an increase of φ_1. Similarly the s.p. difference, say $\varphi_2 - \varphi_1$, which exists between the two electrodes must be added. The collector current is thus

$$J = J_s \exp\left[-e\frac{\varphi_2 - \varphi_1 + \Delta V_a}{kT}\right] \tag{3}$$

For a constant emitter temperature a plot of Ln J_s versus $-\Delta V_a$ is roughly represented by figure 2. The knee of the plot, where saturation occurs yields $\Delta V_a = \varphi_1 - \varphi_2$. So that any change in φ_2, due for instance to adsorption is equal to the observed subsequent change in ΔV_a, provided φ_1 is constant.

In the field emission microscope, the electron current is given by the Fowler-Nordheim equation (Gomer, 1961)

$$i = AV^2 \exp(-b\phi^{3/2}/V) \tag{4}$$

where A and b are constants, V the applied voltage on the tip, and ϕ the work function of this tip. This relation enables one to determine a change in ϕ upon adsorption. However the absolute magnitude of ϕ and in some respects of a change $\Delta\phi$ may be questionable on account of some experimental and theoretical uncertainties as pointed out by Gomer (1967).

The use of the space-charge limited diode has been illustrated for instance by P. A. Redhead (1963). Figure 3 shows the author's circuit for surface potential measurements. The voltage across the series resistance is V_L and the current i_a on the anode is such that:

$$i_a = CT^2 \exp\left[-\frac{e(\phi_a - V_a)}{kT}\right] \tag{5}$$

where V_a is the anode (sample) voltage. Then

$$\frac{dV_a}{d\phi_a} = \left(1 + \frac{kT}{eV_L}\right)^{-1} \tag{6}$$

and, provided $kT \ll eV_L$ (i_a is thus constant), any change in the measured anode voltage is equal to the change in the anode work function. The cathode temperature must be maintained constant and in this space charge regime the current i_a does not depend on the cathode work function. Some care has to be taken for the heat radiation of the latter.

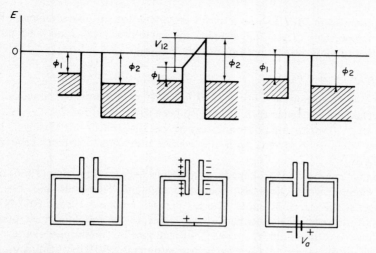

Fig. 1. Compensation of the surface potential by an applied voltage V_a.

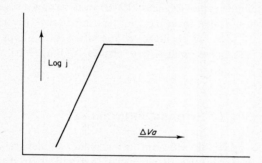

Fig. 2. Typical retarding potential plot.

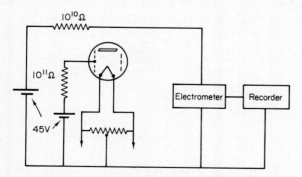

Fig. 3. Circuit for surface potential measurements.

Photoemission may also be a useful technique for s.p. measurements. For a given temperature T of the sample, the quantum yield I may be plotted according to the Fowler equation (Fowler 1931)

$$\text{Ln}\,(I/T^2) = B + F(x) \tag{7}$$

where B is a constant, $x = h(\nu - \nu_0)/kT$, and $F(x)$ is a universal function of x. The frequency ν_0 equals $e\phi/h$. It has also been shown that for $(h\nu - e\phi) < 1$ eV, a plot of $I^{1/2}$ versus the photon energy $h\nu$ yields a straight line (Eisinger, 1958) and $e\phi$ when it intercepts with the x-axis. Very small currents have to be measured.

Finally, mention must be made of the Holscher's modification (1966) of the RPD method which does not require a reference electrode and uses field emission as the electron source. The gas pressure must not exceed 10^{-4} N·m^{-2}. None of the methods described above seems to be suitable in all practical situations. But any of them, in well-defined specific conditions, can give very accurate results: a typical claim from most authors is a 5-10 mV accuracy.

Our previous work on electron monochromators (Ballu, 1968) and specially on the plane condenser type (Régnaut, 1970), was strongly suggestive of a new method which should be, in principle, less restrictive in the experimental conditions, and at least as accurate as the usual techniques. The design is described below and some preliminary results (CO on Mo) are given as an illustration.

II. BASIC PRINCIPLES

Figure 4 is a schematic diagram of a plane condenser electron mono-chromator. The width of the parallel slits is s and their distance is L. The distance of the electrodes is d. A circular portion B of electrode A (upper) is electrically insulated from the remainder of A. The normal projection of the centre of B on electrode C (lower) lies at $L/2$ from either slit.

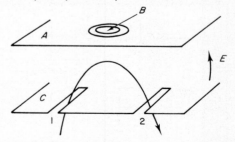

Fig. 4. Schematic diagram of the design: A—Upper plate; B—Sample; C—Lower plate; D—Electron beam.

Let V be the potential difference between A and C, βV, the potential difference between A and B, and eV_0 the electron energy of a narrow monoenergetic beam entering slit 1, with an angle of incidence $(b + \pi/4)$ from electrode C $(b < \pi/4)$. The plane of incidence is perpendicular to the slits. In this plane, the x-axis is in the plane C and the y-axis is the normal starting from the middle of slit 1. The assumptions are: (i) $\beta \ll 1$ and (ii) the space between the two plates A and C may be divided in three regions I, II, III (figure 5), such that

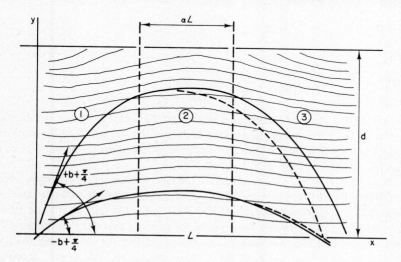

Fig. 5. Model of analyser showing different types of trajectories corresponding to: $\beta = 0$ for $\pm$ b (dotted line); $\beta \neq 0$ for $\pm$ b (full line).

the electric field is always parallel to the y-axis. The latter assumption is a useful approximation which allows analytical calculations as the actual potential map cannot be written in a simple way.

The components of an electron initial velocity are x_0' and y_0', such that

$$x_0'^2 + y_0'^2 = 2eV_0/m \tag{8}$$

$$y_0'/x_0' = tg(b + \pi/4) \tag{9}$$

With $E = V/d$ and $\beta = 0$, the trajectory of an electron is a parabola and the condition which is required for finding the exit slit 2 is

$$E = \frac{2V_0}{L}\cos 2b \tag{10}$$

The top of the parabola is:

$$y_{max} = \frac{L}{8} \frac{\cos 2b}{\cos^2 (b + \pi/4)} \tag{11}$$

Neglecting space charge effects—a narrow and low-density beam is used (Ballu, 1968)—it may be remarked that equation (10) yields the resolution versus the initial energy dispersion and the aperture of the beam. With E fixed

$$\cos 2b \cdot dV_0 - \frac{V_0}{L} \cos 2b \cdot dL - \frac{2V_0}{L} \sin 2b \cdot db = 0 \tag{12}$$

Thus

$$\frac{dL}{L} = \frac{dV_0}{V_0} - 2tg2b \cdot db \tag{13}$$

as long as the plane of incidence is normal to the slits. Equation (13) shows that if $b = 0$, then the energy resolution is defined by dL which is the sum of the widths of the two slits. In the first-order approximation a non-parallel beam is focused on slit 2 when b is quite small.

Now, if $\beta \neq 0$, we have to calculate the electron trajectory: (i) from entrance to the interface of regions I and II, (ii) in region II; and (iii) from the interface of regions II and III down to plate C. The electric field is again E in I and III and E $(1 - \beta)$ in region II. The motion continuity requires that the y ordinates on each part of each interface are respectively the same, and the same condition applies to the corresponding tangents $\dot{y}/\dot{x}$ to the trajectories, taking into account the potential jumps which occur on each interface.

Let ΔL be the excess of L due to β. Then a straightforward calculation yields:

$$\frac{\Delta L}{L} = \alpha\beta(1 + \tau^2)[1 - \alpha^2 \tau^2 - \beta\tau^2] \tag{14}$$

where $\tau = tg (b + \pi/4)$ and α is the ratio of the diameter of electrode B to the length L. The expansion of $\Delta L/L$ versus β and α has been limited here to first order in β and second order in α in the [ ]. As a matter of fact the usual dimensions of commercially available single crystals and reasonable values of L are such that α^2 is of the order of magnitude of β.

When $b = 0$, the relation (14) between β and ΔL becomes

$$\frac{\Delta L}{L} = 2\beta\alpha(1 - \alpha^2 - \beta) \tag{14bis}$$

If a sweep voltage is applied both on A and B, then the current peak is recovered for

$$\Delta V \simeq 2\alpha(1 - \alpha^2) \Delta\phi \tag{15}$$

where $\Delta\phi = \beta V$, the s.p. change of B. If the sweep voltage is applied only on electrode A, then

$$\Delta V \simeq \frac{2\alpha(1-\alpha^2)}{1 - 2\alpha(1-\alpha^2)}\, \Delta\phi \tag{16}$$

With $\beta \neq 0$ and $b = 0$, it is always possible to compensate $\Delta\phi$ by adding a suitable voltage on electrode B in order to keep constant the electron current at the exit slit.

However, in all cases, a slight shift of the s.p. of the reference electrodes A and C which should be the same for both electrodes would alter the measured $\Delta\phi$. This is because the effective energy V_0 of an electron would change and the same is true for the effective β: the two effects must be added. So that we are drawn back to the usual problem of reference electrodes in condenser methods. This may be avoided in the following way.

First, equation (10) shows that changing b into $-b$ still ensures the exit of the beam at slit 2 (figure 5). We can then use alternatively two beams, the angles of incidence of which are respectively symmetrical against the $\pi/4$ direction in the incidence plane. The ratio of the corresponding y_{max} of equation (11), say $(y^+/y^-)_{max}$ is

$$\left(\frac{y^+}{y^-}\right)_{max} = tg^2(b + \pi/4) \tag{17}$$

For $\alpha^2 \ll 1$, the effect of β on the two trajectories will be roughly proportional to this ratio, so that for a reasonable value of $tg(b + \pi/4)$ the effect of β on the lower trajectory may be experimentally negligible. However, in choosing b, care must be taken to avoid diffraction effects at the slits. Practically $tg^2(b + \pi/4)$ may be set equal to 10 which yields $b = 28°$. A more exact calculation should start from equation (14); neglecting in the [ ] the terms in α^2 and β and assuming $\Delta L^+ / \Delta L^- \simeq 10$ yields almost the same value of b.

The advantage of the two beams technique is that if a s.p. shift of the reference electrodes occurs the applied V_0 may be re-adjusted, all things being equal, on the lower beam to recover its normal intensity. By the same token, the s.p. change of the reference electrodes can be evaluated and the necessary correction of the $\Delta\phi$ which is subsequently measured can be made. According to equation (13) the resolution of the monochromator is even better for the lower trajectory than for the higher when the beams are not strictly parallel.

Turning now to the accuracy $\delta(\Delta\phi)$ which may be expected from this sort of design we have to define the shape of an ideal peak for the output current, the allowed amplification in regard of the signal/noise ratio, and the response of the observer looking at two close peaks which are recorded on a chart, in a step-by-step sweeping voltage procedure for instance. We assume that the peak shape is parabolic: the top curvature of the parabola is not strictly independent of the input energy distribution in the beam if this distribution is large. We assume also that the observer is able to separate two peaks when the distance of their respective tops equals at least the depth of their valley. The calculation yields

$$\delta(\varDelta\phi) = \lambda V_0 \left(\frac{\varDelta V_0}{V_0}\right)^2 \tag{18}$$

where λ depends on the associated electronics. In practice, with standard modern equipments (essentially the $X-Y$ recorder), λ should lie between 5.10^{-1} and 2. Equation (18) shows that $\delta(\varDelta\phi)$ is: (i) proportional to the energy of the electron probe, which is expected, as this energy reflects the magnitude of β, and (ii), proportional to the square of the normal resolution of the monochromator which roughly equals the ratio of the sum of the two slit widths to their distance L. As a numerical application, with $V_0 = 5$ volts and $\Delta V_0/V_0 = 2.10^{-2}$

$$1\,\mathrm{mv} < \delta(\varDelta\phi) < 4\,\mathrm{mv}$$

III. EXPERIMENTAL

Due to the incidence of magnetic fields on low energy electrons, the plates A and C, distant of 15 mm, are made from a copper-aluminium alloy gold-plated on all faces. The slits are 0.7 mm wide and 4 mm long. Alternatively holes of 0.7 mm diameter could be used. The distance of the slits is 50 mm. The sample is a disk 10 mm diameter, 0.2 mm thick, inserted in a circular aperture (12 mm diameter) in the upper plate. All electrical insulations are made with alumina and, where thermal expansions are expected, the junctions of the separate parts involve tungsten springs. In the present set-up the beam, with an angle of incidence of 45°, is fixed in space. Consequently the possibility of a translation of the monochromator has been found useful along its longitudinal axis to obtain the maximum input intensity. The assembly can be moved to and fro on small Cu-Al balls in alumina grooves. The sample may be heated with an electron gun mounted on backside. Its temperature can be measured by a thermocouple (initially chromel-alumel, and now Pt/Pt-Rh 5% which obviously allows higher temperatures). The cleaning procedure of the sample is thus restricted to high temperature heating, eventually in a residual pressure of O_2 or H_2. This procedure is generally efficient for refractory metals, specially for carbon contamination.

The electron source, or more exactly the beam, results from the combination of an electron gun and another electron monochromator (selector) the parameters of which are identical to those given above for the first monochromator. Nevertheless, the resolution of the selector would be slightly better—for a given V_0—for an improvement which has been described elsewhere (Régnaut, 1970). An electronic lens is mounted between the exit of the selector and the entrance of the main monochromator. It makes the decrease of the electron energy eV_0 easier for the latter, and for focusing of the beam on its input slit. The initial beam intensity is in the low 10^{-9} A range. The vacuum vessel into which the whole design is placed can be evacuated down to $\sim 10^{-6}$ N · m^{-2}. At this stage no special effort has been devoted to achieving lower pressures although a titanium ion pump (100 l · s^{-1}) is used. Within the same line, bake-out cycles have been made up to only 550°K for a few hours each. The main objective was then the evaluation of the instrument. A quadrupole mass spectrometer indicated the residual gas pressure, mainly CO and H$_2$ and traces of CO$_2$ and H$_2$O. Most of the hydrogen seemed to be due to the mass spectrometer itself. Temporary changes in the CO partial pressure could be achieved by increasing the electron current (higher cathode temperature) of the BA-gauge, which gives total pressure measurements.

All voltages were stabilized at 10^{-3} or more. For the output current of the monochromator, the partial pressures of CO or H$_2$ and the sample temperature, we used respectively a X—Y Moseley recorder and two Cimatic X—Y recorders.

IV. RESULTS

Until now two different samples have been used. A third one—the [100] face of a Mo single crystal—is presently mounted: correspondingly lower ultimate pressures ($\sim 10^{-8}$ N · m^{-2}) and improved cleaning procedures of the sample are the basis of this new step.

The first sample was made of the same Cu-Al alloy used for the plates of the monochromator. Our purpose was a simulation of a s.p. change by adding a small voltage to B relative to A. As adsorption effects should be identical on both electrodes, an absolute calibration of the instrument would be obtained. At this stage, however, and also for the second sample, only the 45° beam has been used. We intended also to check equations (15) and (16) of our calculations. Figure 6 sums up the data of this simulation process. The linear relationship between ΔV and $\Delta \phi$ holds at least up to $\beta = \pm 5.10^{-2}$ and the factor proportionality is within $\sim 10\%$ of the calculated value, in spite of the approximations which were used. With $V_0 = 12.5$ V, the sensitivity has been found better than 5 mV.

The second sample was a polycrystalline molybdenum disk which could be heated up to 550°K by the radiant heat from the cathode of the electron gun mounted on back side. Thus the α phase of CO could be completely evolved and the corresponding population is $\sim 2 \cdot 10^{13}$ cm^{-2}. The re-adsorption proceeded in the cooling period of the sample from 550°K down to 300°K. The thermal time constant is 280 seconds and the saturation of this α phase is obtained after 800-1000 seconds. The s.p. change $\Delta\phi$ is linear versus Ln t during cooling. When the time of exposure of the sample to a pressure of 10^{-5} N·m^{-2} does not exceed $5 \cdot 10^3$ seconds the total change of $\Delta\phi$ is 0.14 V. For exposure periods greater than some 10^4 seconds at ambient temperature and at the same pressure the total amount evolved on heating does not change appreciably but the total s.p. change is 0.22 V. These values are roughly in accordance with previous

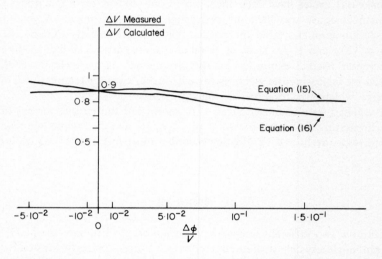

Fig. 6. Plots of ΔV versus $\Delta\phi$ in the simulation process as compared to equations (15) and (16).

measurements made in the laboratory (Degras and Lecante, 1967). The detailed kinetics of CO adsorption on the Mo surface have not been further investigated because greater interest had to be devoted to the (100)Mo crystal.

V. CONCLUSION

The instrument which has been described here has proved to be valuable for accurate s.p. measurements. Even with the one-beam technique, the efficiency of the gold-plating of the electrodes seems to yield good reproducibility of measurement. The results expected from the (100)Mo crystal, which has been recently investigated by J. Lecante (1971) in the laboratory, would provide further evaluation of the method.

REFERENCES

Ballu, Y. (1968). Thèse de Docteur-Ingénieur, Paris-Orsay.

Degras, D. A. and Lecante, J. (1967). *Nuovo Cimento, Suppl.* **V**, **2**, 598.

Eberhagen, A. (1960). *Fortschr. Phys.,* **8**, 245.

Eisinger, J. (1958). *J. Chem. Phys.* **29**, 1154.

Fowler, R. H. (1931). *Phys. Rev.* **38**, 45.

Gomer, R. (1961). "Field Emission and Field Ionization". Harvard Univ. Press, Cambridge, Mass.

Gomer, R. (1967). *In* "Fundamentals of Gas Surface Interactions" (H. Saltsburg *et al.,* eds.), Academic Press, New York and London.

Hayward, D. O. and Trapnell, B. M. W. (1964). "Chemisorption." Butterworths, London.

Holscher, A. A. (1966). *Surface Sci.,* **4**, 89.

Hopkins, B. J., Pender, K. R. (1966). *Surface Sci.* **5**, 155.

Hopkins, B. J. and Usami, S. (1967). *Nuovo Cimento, Suppl.* V, **2**, 535.

Hopkins, B. J., Pender, K. R. and Usami, S. (1967). *In* "Fundamentals of Gas Surface Interactions," (H. Salstburg *et al.,* eds.), 284. Academic Press, New York and London.

Lecante, J. (1971). This volume.

Madey, T. E. and Yates, J. T. Jr. (1967). *Nuovo Cimento, Suppl.* **V**, **2**, 483.

Redhead, P. A. (1963). *Proc. Symp. Electron Vac. Phy.* 89. Akademiai Kiado, Budapest.

Régnaut, C. (1970). Thèse de 3ème Cycle, Paris.

CHEMISORPTION OF OXYGEN
ON SILVER SINGLE CRYSTALS*

G. ROVIDA, E. FERRONI, M. MAGLIETTA
and F. PRATESI

Institute of Physical Chemistry, University of Florence, Italy

I. INTRODUCTION

Many studies have been carried out on the oxygen-silver system in order to explain the fairly complex chemisorption mechanism. This interest is due to the use of silver as catalyst in ethylene oxidation. Previous studies have been generally conducted on powders or evaporated films, using different techniques. The results are seldom in agreement, perhaps because of the different conditions of the surfaces under examination. We proposed to study chemisorption on single-crystal faces to display, if possible, different mechanisms of interaction on different faces. Our techniques allowed us to work on clean and more definite surfaces. Besides, the simultaneous use of several techniques on the same surfaces gives more complete and reliable results.

II. EXPERIMENTAL

For electron diffraction we used a standard Varian LEED system with three-grid optics, adapted for secondary electron analysis. We measured work-function variations by the electron beam method with an auxiliary electron gun (the LEED gun was found not to be adequate when working with oxygen). Single crystals were of variable purity (between 1 and 50 p.p.m. total impurities). For the (100) and (111) faces, three different samples were examined which also differed in orientation accuracy and had undergone different cleaning treatments. Only one crystal (50 p.p.m. impurities, ±2° orientation) was examined for the (110) face. The definite results relative to the (100) and (111) faces were obtained using single crystals prepared with a spark

* Work supported by the National Research Council of Italy (C.N.R.)

machine and having only 1 p.p.m. impurities (±1° orientation). We were able to clean the surface of the latter with controlled atmosphere heat treatments without having to use ion bombardment. We thus worked on crystals of equally clean surfaces and this increased the reliability of our desorption spectra.

Preliminary experiments were carried out in which oxygen was introduced at low pressures (less than 10^{-5} Torr). The observed variations with the various techniques with respect to the clean surface were small and of low reproducibility. In particular, ordered bi-dimensional phases were not observed with LEED.

To continue the study of the chemisorption of oxygen on silver we set up an auxiliary apparatus in which the sample could be exposed at a partial oxygen pressure much greater than that possible in the diffraction chamber. We connected a bakeable manifold to a crystal isolation valve. The manifold could be evacuated with a sorption pump and high-purity oxygen was introduced through a leak valve. Exposures at pressures varying from about 10^{-3} to 1 Torr were made, since it was not necessary to work at higher values. The residual static vacuum was in the 10^{-4} Torr range. The components were usually noble gases and oxygen, the partial pressure of other gases being much lower. Oxygen was introduced after the crystal had already reached thermal equilibrium at the desired temperature and was pumped off after the established time as the cooling of the crystal began. With this procedure only the effect of the oxygen that remained adsorbed after re-exposure to the ultra-high vacuum could be studied. It is possible that more labile adsorbed oxygen forms escaped our examination.

III. RESULTS

By rapidly increasing the temperature of the crystals after exposure to oxygen, similar desorption spectra were obtained for the various crystals examined. The common behavior of the desorption spectra after exposures at various temperatures is shown in figure 1. Exposure at low temperatures (up to 150°C) produces mainly a rapid desorption peak with a maximum at about 280°C. At higher exposure temperatures, a large peak appears corresponding to a high temperature slow desorption (maximum at about 500°C).

The low temperature desorption peak is observed after exposures at temperatures lower than 300°C while high temperature peak increases with the exposure temperature. Trials run in the absence of samples showed that the intermediate large peak (which is more evident in the bottom curve of figure 1) is probably due to crystal support. During heating of the sample, other than oxygen, only a certain quantity of carbon dioxide desorbs. We verified that the latter comes from the support and not from the crystal. Low temperature desorption peak can be attributed to adsorbed oxygen. In fact, as can be seen in figure 2, the quantity of oxygen that desorbs as a function of exposure time tends to a limit, in agreement with what has been found by Bagg and Bruce (1963) and Czanderna (1964). Under our experimental conditions the quantity of chemisorbed oxygen, as a function of exposure temperature, shows a maximum at around 180-200°C. This result, shown in figure 3, is in agreement with Czanderna's (1964) data.

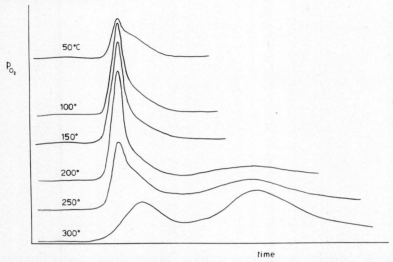

Fig. 1. Oxygen desorption spectra after exposures to 10^{-1} Torr oxygen, at the indicated temperatures, for 5 minutes.

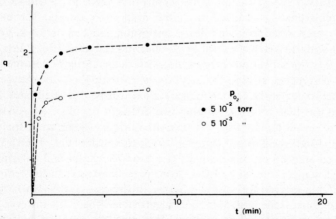

Fig. 2. Quantity (q, arbitrary units) of oxygen adsorbed as a function of time of exposure to oxygen at 150°C. Exposures to 5×10^{-1} Torr oxygen showed about the same limit as found at 5×10^{-2} Torr.

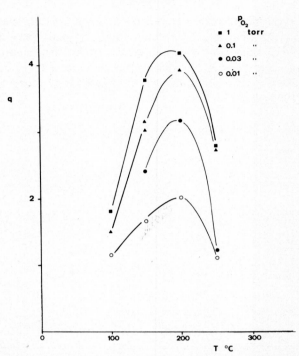

Fig. 3. Quantity (q, arbitrary units) of oxygen adsorbed as a function of exposure temperature at different pressures for 10 minutes.

An activation energy for the initial desorption between 35 and 45 kcal/mole can be approximately deduced from the desorption spectra. The best value, which has been obtained from slow desorption kinetics, is 36 kcal/mole, in agreement with the value obtained by Sandler and Durigon (1965). It is possible, however, that the activation energy varies with the quantity of adsorbed oxygen. The slow desorption at high temperature is attributable to the oxygen that is diffused below the surface. This interpretation will be discussed later. It should be pointed out that we are describing two different forms of oxygen which are referred to in different ways by various authors (Bagg and Bruce, 1963; Czanderna, 1964; Sandler and Durigon, 1965; Rudnitskii *et al.*, 1968). When the surface is covered by certain elements (for example, sulfur and tellurium) there is only high temperature desorption. After pretreatment at higher temperatures (450-500°C) in oxygen for several minutes, in the case of the (100) crystal, the low temperature peak is almost indistinguishable. There is, however, a much stronger high temperature desorption. This result is reported in figure 4. It appears that the low temperature peak remains, after the above treatment, for (111) crystals.

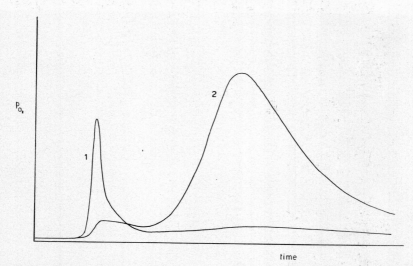

P_{O_2}

Fig. 4. Oxygen desorption spectra obtained with the (100) crystal, after exposure to 10^{-1} Torr oxygen: (1) at 150°C for 5 minutes; (2) at 500°C for 3 minutes and then at 150°C for 5 minutes.

On the (100) face LEED has in no case indicated the existence of a surface superstructure due to oxygen. The chemisorption instead produces a distinct faceting (see figure 5) in specific directions. An analysis of the patterns indicates that it is the (410) facets, inclined at 14°, that are involved. On the intensity-potential plots relative to the (00) spot (shown in figure 6) the adsorption of oxygen causes a general intensity decrease. It is mainly the maxima corresponding to voltages at which the faceting spots pass through the (00) which remain. This is a confirmation that the principal phenomenon is a decrease in the (100) domains.

After having desorbed the oxygen, the intensities return only slowly to their initial values. This slowness seems to be attributable to the kinetics with which the facets disappear after the elimination of oxygen. Treatments in oxygen at 500°C cause no faceting, but rather a small intensity decrease. Similar results were obtained with other LEED spots. Measurements of the work-function variations indicate an increase of about 0.5 eV. In this case we have a simultaneous variation of the reflection coefficient corresponding to the decrease of the LEED spot intensities (see figure 7). It should be noted that, after the desorption of the chemisorbed oxygen, the initial value of the work function is not re-attained, even if the LEED diagram indicates a clean surface. The initial value is reached only after long heating at temperatures above 500°C. This seems to be related to the slow desorption at such temperatures of the internal oxygen.

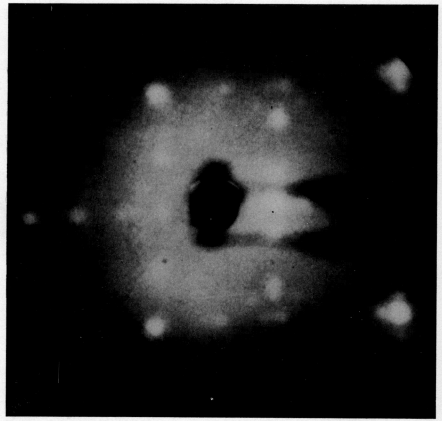

Fig. 5. LEED pattern of the Ag(100) surface after exposure to 10^{-1} Torr oxygen at 150°C for 5 minutes. (59 V).

We examined only one crystal for the (110) face. The initial surface was already faceted—mainly (100) and (111) facets—and at elevated temperatures there was a notable diffusion of tellurium towards the surface (Maglietta *et al.*, in press).

However, the behaviour with respect to oxygen seems to be similar to that of the (100) face, particularly as concerns the work-function variations and the faceting (see figure 8).

On the (111) face, the formation of a 4 x 4 superstructure (shown in figure 9) can be observed with LEED. This corresponds to the chemisorption of oxygen. In figure 10 we have reported the approximate stability field of this superstructure. The centre coincides with the maximum of oxygen adsorption.

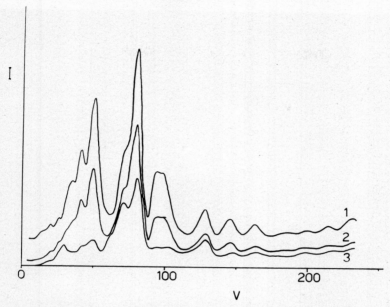

Fig. 6. Ag(100) surface. (00) spot intensity versus voltage at 5° incidence: (1) clean surface; (2) after exposure to 10^{-1} Torr oxygen at 500°C for 3 minutes; (3) after exposure to 10^{-1} Torr oxygen at 150°C for 5 minutes.

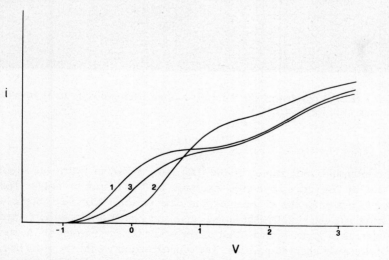

Fig. 7. Ag(100) surface. Retarding potential curves: (1) clean surface; (2) after exposure to 10^{-1} Torr oxygen at 150°C for 5 minutes; (3) after oxygen desorption by heating up to 450°C.

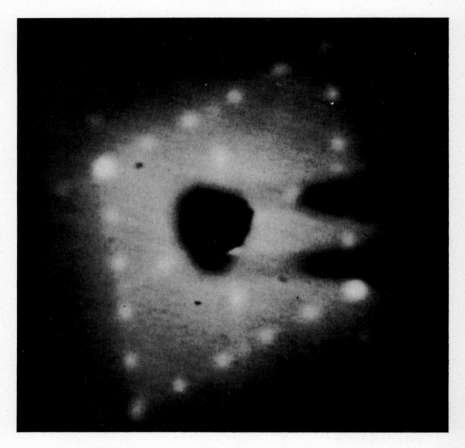

Fig. 8. LEED pattern of the Ag(110) surface after exposure to oxygen (same conditions as in figure 5). 48 V.

In the intensity–voltage plot of the (00) spot, shown in figure 11, a general decrease of the intensities after exposure at 150°C can be noted. This is, however, less than the decrease in the case of the (100) face. Some weak superstructure signs are visible in the curve and similar results are obtained for the other silver spots. It seems that on the whole the silver atoms on the surface have not greatly changed position. The superstructure would therefore be given by a pure layer of oxygen. After treatment in oxygen at 500°C, only a very slight decrease of all intensities is observed. As for the (100) face, this can be explained by the presence of oxygen below the surface. The return of the

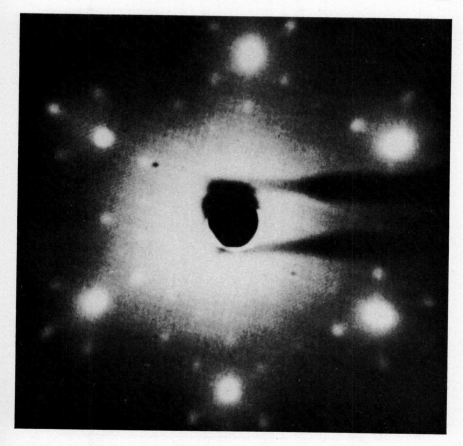

Fig. 9. LEED pattern of the Ag(111) surface after exposure to oxygen (same conditions as in figure 5). 48 V.

intensities to those of the clean surface is rapid and, as opposed to what happens on the (100) face, the final intensities are immediately reached. In particular, as shown in figure 12, the intensity variation occurs almost exactly in correspondence to the oxygen desorption peak.

The Auger peaks in the high energy zone are reported in figure 13, corresponding to both before and after oxygen exposure. Desorption of oxygen under electron bombardment cannot be excluded. After high temperature oxygen exposure a weak Auger peak is still visible, due to the internal oxygen considered above.

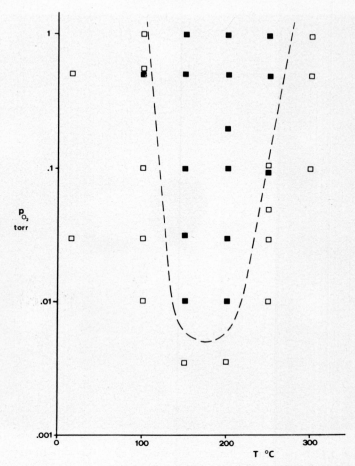

Fig. 10. Ag(111) surface. Stability range of the 4 x 4 superstructure. After exposure to oxygen for 10 minutes, full squares correspond to the superstructure presence.

Fig. 12. Ag(111) crystal. Variation of the oxygen pressure (curve 1); variation of the (00) spot intensity for the clean surface (curve 2); and variation of the (00) spot intensity for an oxygen-covered surface (curve 3); all during temperature raising.

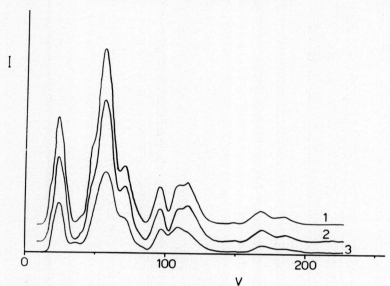

Fig. 11. Ag(111) surface. (00) spot intensity versus voltage, at 5° incidence. (1), (2) and (3): as in figure 6.

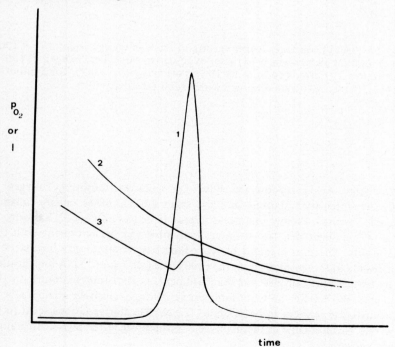

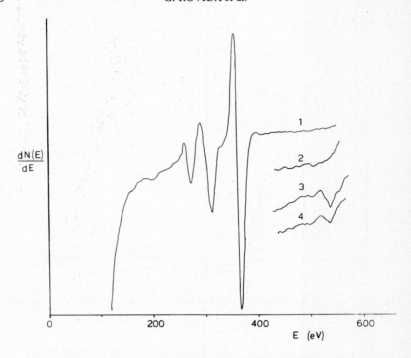

Fig. 13. Ag(111) surface. Auger spectra in the high-energy region ($E_p = 1000$ eV): (1) and (2) clean surface; (3) after exposure to 10^{-1} Torr oxygen at 150°C for 5 minutes; (4) after exposure to 10^{-1} Torr oxygen at 500°C for 3 minutes. Curves (2), (3) and (4) were obtained with higher sensitivity (3x).

Low-energy Auger spectra are shown in figure 14. Adsorbing oxygen at 150°C, large-intensity variations, and also slight shifts in the secondary emission peaks, are obtained. Energy loss peaks appear to be less influenced. Exposing at 500°C, only a slight decrease in intensity is observed, in agreement with the LEED results. The measurement of work-function variations give qualitatively similar results to those obtained on the (100) face (see figure 15). A long heating at high temperature, or else ion bombardment, is necessary to return to the initial value. We believe the work-function variation, exclusively attributable to the chemisorbed oxygen, to be of about 0.2 eV. Greater variations seem to be mainly due to internally diffused oxygen, as observed after high temperature oxygen exposure.

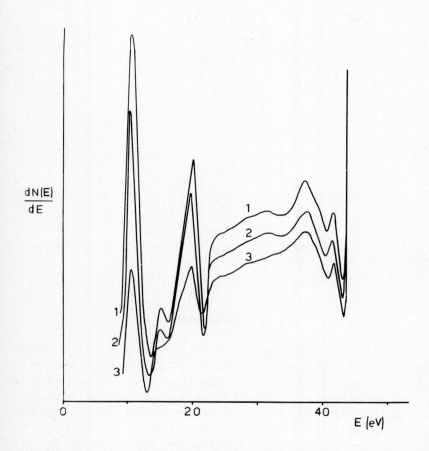

Fig. 14. Ag(111) surface. Low-energy secondary-emission spectra. $E_p = 50$ eV. (1), (2) and (3): as in figure 6.

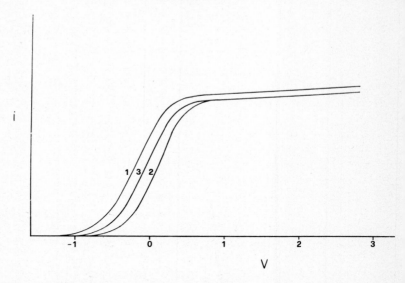

Fig. 15. Ag(111) surface. Retarding potential curves. (1), (2) and (3): as in figure 7.

IV. DISCUSSION

Under our particular experimental conditions we were only able to study the oxygen remaining on the silver after a return to ultra-high vacuum. We were able to identify at least two distinct forms of this oxygen. One, obtained by exposure at temperatures from about 50 to 250°C, is desorbed with an activation energy of about 35-45 kcal/mole. It seems to be very sensitive to surface conditions (presence of sulfur and tellurium and, at least for certain faces, the oxygen itself below the surface) and shows a behaviour that depends on the face being considered, producing a 4 × 4 superstructure on the (111) face and faceting on the (100) and (110) faces. This form of adsorbed oxygen seems to cause a work-function variation of no more than 0.3 eV.

The other form seems to be oxygen present below the surface. Probably, it does not cause structural variations of the silver surface but modifies certain electronic properties, which are then reflected in the behaviour of the surface. In particular this oxygen, which can be eliminated only by prolonged high temperature heating, greatly influences the behaviour of the (100) face. When present in sufficient quantity it seems to inhibit the faceting that is observed on the pure (100) surface and the chemisorption of the first oxygen form. It does not however, noticeably affect the properties of the (111) face. Besides, it causes a work function increase, which, when this oxygen form is present in large quantities (for example, after treatment in oxygen at 500°C) reaches a value of 0.6 eV or even greater.

This oxygen form causes only relatively small variations in the LEED intensities and in the secondary emission spectrum of silver. The oxygen Auger peak is slightly less intense with respect to the one that is obtained when the surface is covered with adsorbed oxygen. This internal oxygen should be in the dissociated state, as has been demonstrated by the studies conducted on the solubility of oxygen in silver (Steacie and Johnson, 1926). Work-function increases, related to the two oxygen forms, are in agreement with results obtained by Rudnitskii et al. (1968).

As far as chemisorbed oxygen is concerned, i.e. that which desorbs at lower temperatures, we believe that it is undissociated and that a partial negative charge on it cannot be very large.

The fact that we are not dealing with ions is indicated by the relatively small work-function variation (of the order of 0.3 eV), even in conditions of maximum coverage. On the other hand, the existence of a unit mesh with the notable dimensions of that present in the 4 x 4 superstructure, can hardly be justified by the existence on the surface of atomic oxygen at maximum coverage. Besides, atomic oxygen adsorption should cause a strong rearrangement of the surface silver atoms. This re-arrangement is not indicated by our LEED results.

REFERENCES

Bagg, J. and Bruce, L. (1963). *J. Catal.* **2**, 93.
Czanderna, A. W. (1964). *J. Phys. Chem.* **68**, 2765.
Maglietta, M., Pratesi, F. and Rovida, G. In press.
Rudnitskii, L. A., Shakovskaia, L. I., Kulkova, N. V. and Temkin, M. I. (1968). *Dokl. Akad. Nauk SSSR* **182**, 1358.
Sandler, Y. L. and Durigon, D. D. (1965). *J. Phys. Chem.* **69**, 4201.
Steacie, E. W. R. and Johnson, F. M. G. (1926). *Proc. Roy. Soc. London A* **112**, 542.

STUDY OF CHEMISORPTION OF HYDROGEN AND ETHYLENE ON NICKEL BY LEED AND MAGNETIC METHODS

G.-A. MARTIN, G. DALMAI-IMELIK and B. IMELIK

Institut de Recherches sur la Catalyse, C.N.R.S., Villeurbanne, France

I. INTRODUCTION

Recently, low-energy electron diffraction (LEED) thermodesorption (T.D.) and other physical methods, has provided useful information about adsorption phenomena. Particularly, it has been shown that adsorption of gases often takes place differently on various crystallographic planes. However, these results cannot be easily compared with those obtained on powders for which the surface is made up of numerous and unknown crystallographic planes.

In the laboratory (Martin *et al.*, 1970), nickel-on-silica catalysts showing preferential crystallographic orientation have been prepared which makes such a comparison possible. For this purpose, we undertook the study of adsorption of hydrogen and ethylene on nickel-on-silica catalysts by magnetic methods. High magnetic field saturation measurements were preferred to classical low-field technique which give in some cases ambiguous conclusions. The results obtained by magnetic methods on supported nickel powders are compared in this paper with LEED and thermodesorption data on monocrystals.

II. EXPERIMENTAL

MATERIALS

Two nickel-on-silica samples were prepared by reduction in a hydrogen flow of a basic nickel silicate, $Ni_3(OH)_4 Si_2 O_5$ (nickel antigorite) and of a compound prepared by impregnation. The nickel antigorite was obtained after a hydro-thermal treatment at $350°C$ during one week of a stoiechiometric mixture of very pure and finely divided SiO_2 and $Ni(OH)_2$. Its structure is derived from the structure of $Ni(OH)_2$ where some OH are replaced by SiO_4 tetrahedrons. The nickel crystallites issued from reduction of the nickel antigorite show an orientation with the 110 axis perpendicular to the silica substrate. This does not mean that only 110 faces are present on the surface of Ni crystallites. A preliminary study by electron microscopy shows the formation of twins and some crystal facets are developed. The impregnated compound was obtained by adding SiO_2 (aerosil) to a solution of $Ni(NO_3)_2$ 6 NH_3. After filtration, the complex adsorbed on the surface of the support was destroyed by evaporating ammonia, and the nickel hydroxide precipitated. After washing and drying, a very pure product was obtained. We did not observe any orientations of the nickel crystals nor well crystallized forms. The mean diameter measured by magnetic methods is 60 Å.

APPARATUS AND PROCEDURE

The saturation magnetization M_s of nickel catalysts during adsorption of gases is measured by the extraction method at 300, 77 and 4.2°K in magnetic fields up to 20 k Oe. M_s may be related to the quantity of adsorbed gas by the equation

$$M_s = M_{s0} - 0.25\, k\alpha v,$$

M_{s0} is saturation magnetization at zero coverage, v the quantity of adsorbed gas expressed in ml · NTP, k a constant nearly equal to one, depending upon the temperature measurement and taking into account the variation of M_s with temperature α is the change of magnetization, expressed in Bohr magnetons (β), due to the adsorption of one molecule of gas. From α, the number of bonds n between the metal and the adsorbed gas may be calculated if we assume that nickel atoms in interaction with the adsorbed molecules cease to participate to the collective magnetism. The number of metallic atoms which are demagnetized by this process is equal to n, and n is obtained by dividing α by the magnetic moment of the nickel atom expressed in Bohr magnetons $n = \alpha/0.606$. n is easily calculated from the slope of the curve M_s versus v by the equation

$$n = \frac{6.63\, M_s}{k\ \ v}$$

The heat of adsorption of hydrogen on these catalysts is measured in a Calvet calorimeter. Experiments of low-energy electron diffraction were performed on Ni monocrystals at different crystal temperatures: −10 to 500°C. After adsorption, the monocrystal is heated and gases desorbed analysed by mass spectrometry.

III. RESULTS

ADSORPTION OF HYDROGEN

The saturation magnetization measured at $4.2°K$ of the Ni/SiO_2 issued from antigorite, decreases linearly when the quantity of hydrogen adsorbed at $300°K$ increases (figure 1). It can also be seen that magnetic adsorption and desorption

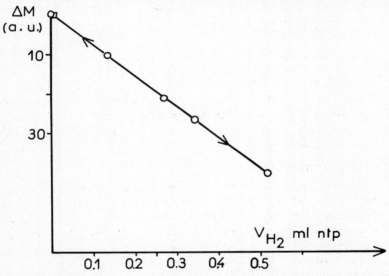

Fig. 1. Variation of the saturation magnetization measured at $4.2°K$ (arbitrary units) versus volume of adsorbed hydrogen at $300°K$ in ml·NTP. The nickel content is about 35 mg.

isotherms are identical. Similar curves are obtained when saturation magnetization is measured at 77 and $300°K$. The same magnetic isotherms are observed when hydrogen is adsorbed on Ni/SiO_2 issued from impregnation. The values of α and n relative to these systems are listed in table 1 (average of three determinations).

Table 1: α measured at three temperatures, relative to the adsorption of H_2 on the two nickel-on-silica catalysts, and number of bonds n between H_2 and Ni.

Ascendancy	T ads. (°K)	α meas. at $300°K$	α meas. at $77°K$	α meas. at $4.2°K$	n
Antigorite	300	1.73	1.56	1.47	2.43
Impregnation	300	1.40	1.30	1.27	2.09

The error on α is estimated at ∓ 0.05. A decrease of α with temperature is observed, but up to now no satisfactory explanation for this behaviour has been proposed. n has been calculated after the variations of magnetization measured at $4.2°K$; at this temperature, magnetization of the smallest particles is taken into account. The two calculated n values indicates that the adsorption is dissociative ($n \geqslant 2$) and that the hydrogen atom is bound to more than one nickel atom on the sample issued from antigorite.

Heats of adsorption of H_2 on the two samples can be compared in figure 2.

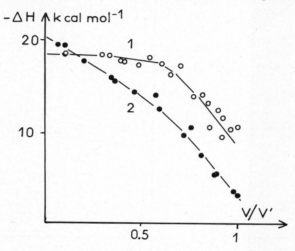

Fig. 2. Heat of adsorption of H_2 on Ni–SiO$_2$ issued from antigorite (curve 1) and from impregnation (curve 2).

Differential heat is plotted versus v/v', where v is the quantity of adsorbed gas and v' the v value when equilibrium pressure is 10 Torr. Compared with bibliographic data (Bond, 1962) relatively small initial adsorption heats are observed (-18 and -20 kcal $\cdot$ mol^{-1}). High initial heat is generally related to the presence of imperfections or high-index crystallographic planes, as shown by Eley and Norton (1970). We can conclude their absence in our system. This observation may be explained by the purity of samples and the high temperature of reduction ($620°C$). The dependence of the heat of adsorption of hydrogen on surface coverage for the two samples is different. $-\Delta H_{ads}$ decreases linearly when coverage of nickel issued from impregnation increases whereas the adsorption heat observed on the other sample seems to be practically independent of coverage for a large range of pressure. These experiments suggest either a strong neighbour interaction (Eley and Norton, 1970), or a smaller mobility of hydrogen on nickel obtained by reduction of antigorite as suggested by Beek *et al.* (1950).

H_2 adsorption has been followed on (111) (100) and (110) faces of Ni monocrystals from 0 to 60 Langmuir (Bertolini and Dalmai-Imelik, 1969; May, 1970). At room temperature a LEED pattern is observed only on 110 face. It corresponds to a 1 x 2 unit cell on the surface. When H_2 is desorbed by thermal treatment, only one maximum is observed at the coverage we are working. The temperature of this maximum is different on the three faces: 85°C on (111), 60°C on (110) and 90°C on (100).

ADSORPTION OF ETHYLENE, ETHANE AND HYDROGEN-ETHYLENE

Magnetic experiments were performed only on $Ni-SiO_2$ catalysts prepared by impregnation. Variations of saturation magnetization measured at 77°K, during adsorption of ethylene at various temperatures (−78, −20, +25°C) can be

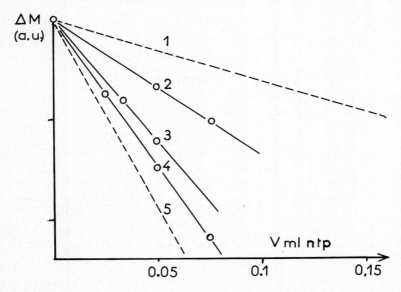

Fig. 3. Variation of saturation magnetization of $Ni-SiO_2$ during adsorption of C_2H_4 at −78°C (curve 2), −20°C (curve 3) and +25°C (curve 4). Curve 1 is relative to hydrogen adsorption and curve 5 to C_2H_6.

compared in figure 3. Surface coverages are small, to avoid physical adsorption and self hydrogenation (equilibrium pressure smaller to 10^{-2} Torr). α and n values calculated after the slopes of the magnetic isotherms are listed in table 2.

The parameter α is roughly independent of the measurement temperature. Even at low temperature, the number of bonds is always higher than 2, which indicates that adsorption of ethylene on nickel is dissociative, since the value of α for associative form ($Ni-CH_2-CH_2-Ni$) is 2.

Table 2. α *for ethylene adsorbed at* -78, -20 *and* $+25°C$, *and number of bonds* n *between adsorbed species and nickel.*

T ads °C	α meas. at 300°K	α meas. at 77°K	α meas. at 4.2°K	n
-78	–	2.8	2.7	4.6
-20	–	4.5	4.4	7.4
$+25$	6.6	6.1	6.15	10.1

These results were confirmed by adsorption of ethylene at $-78°C$ followed by progressive heat treatment up to 550°C. Variations of α with temperature are plotted in figure 4. Three adsorption states are observed: at low temperature

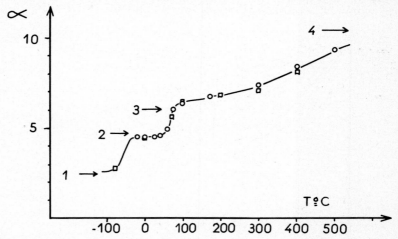

Fig. 4. Variation of α relative to C_2H_4 with temperature. Arrow 1 corresponds to $n = 4$, arrow 2 to $n = 8$, arrow 3 to $n = 10$ and arrow 4 to a complete dissolution of C in Ni.

where $n \simeq 4$, from -20 to $+40°C$ where $n \simeq 8$, and above $+75°C$ where $n \simeq 10$. For $n = 10$, we think that a complete cracking occurs, according to the reaction

$$C_2H_4 + 10Ni \rightarrow 2Ni_3C + 4NiH$$

At higher temperatures, α increases in a continuous way. The superficial nickel carbide, thermodynamically unstable, is decomposed and carbon is dissolved in the bulk nickel. Four peripheral electrons of C are then going to the unfilled d-shell of the metal. The decrease in magnetization due to a complete dissolution of two carbons is equal to $\alpha = 8$ Bohr magnetons. If the magnetic effect which is caused by the adsorption of four hydrogen is added (4 x 0.6 = 2.4), $\alpha = 10.4$. This theoretical value is not far from the observed one at 500°C ($\alpha = 9.3$).

Various models can be proposed to explain $n = 4$ (a, b, c) and $n = 8$ (d, e).

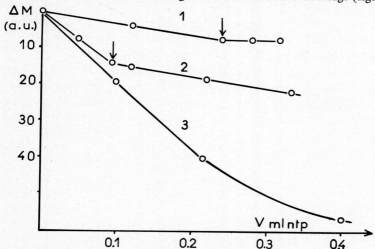

Adsorption of ethane on Ni/SiO$_2$ occurs at room temperature. The number of bonds formed, calculated from magnetic isotherm at low coverage (figure 5),

Fig. 5. Variation of saturation magnetization during adsorption of H$_2$ then C$_2$H$_4$ (curve 1), adsorption of C$_2$H$_4$ then H$_2$ (curve 2), and adsorption of C$_2$H$_6$ (curve 3).

is 12. This value is in good agreement with the complete cracking of the molecule according to the equation

$$C_2H_6 + 12Ni \rightarrow 2Ni_3C + 6NiH$$

Another set of experiments was performed at 25°C (figure 5). The surface is first covered with C$_2$H$_4$ and H$_2$ is then admitted: the magnetic effect is the same as if H$_2$ is adsorbed on a bare surface. No interaction between preadsorbed

C_2H_4 and subsequently adsorbed hydrogen is observed. However if C_2H_4 is admitted on a nickel surface precovered with H_2, the magnetic effect is different. α observed for C_2H_4 on a bare surface of nickel is 6.1 (table 2), whereas it is equal to 0 on a surface precovered with hydrogen. There seems to be interaction between preadsorbed H_2 and C_2H_4 coming from the gas phase. To explain $\alpha = 0$, two reactions are possible

$$CH_2 = CH_2 + NiH \rightarrow Ni-CH_2-CH_3$$
$$nCH_2 = CH_2 + NiH \rightarrow Ni(CH_2)_{2n-1}-CH_3 \quad \text{(polymerization)}$$

In these equations, the number of bonds between Ni and adsorbate is unchanged so that $\alpha = 0$.

We have performed with LEED and thermodesorption the same experiments on Ni monocrystals (Dalmai-Imelik and Bertolini, 1970; Dalmai-Imelik et al., 1971) and comparable results to powder samples were obtained. On the 3 crystal faces we observed the cracking of ethylene and formation of acetylenic compounds and H atoms adsorbed on Ni. Depending upon the contact time and crystal faces, some reactions may occur more or less rapidly between adsorbed H atoms and ethylene in gas phase. H atoms are replaced by acetylenic residues and more complicated radicals are formed. For short exposure at about $-10°C$ we may assume that the adsorption occurs according to b model.

Two peaks of desorption of H_2, one corresponding to H bonded to the nickel and the other to the cracking of the CH groups, are observed. At higher temperature around $25°C$ we find always the same peaks of desorption and according to the magnetic results we assume an adsorption in the following manner ($n = 8$).

```
        H           H
        |           |
H       C           C       H
|      /|\         /|\      |
Ni   Ni Ni Ni    Ni Ni Ni   Ni
```

At about $150°C$ or higher (depending on the crystal face) we observe a LEED pattern corresponding to a nickel carbide on the surface without adsorbed H. For temperatures between 400 and $450°C$ graphite is observed on the (110) face at high coverage. Finally, at temperatures above $500°C$ appears a (1 × 1) structure after dissolution of carbon in the bulk of the Ni monocrystals.

For the adsorption of ethane we have similar results on monocrystals for short exposure. We observe chiefly the desorption of hydrogen bonded to the nickel.

When the surface is first covered by hydrogen we observe a fast modification of the LEED pattern corresponding to a reaction between adsorbed H and C_2H_4 in the gas phase. H atoms are replaced by the acetylenic or more complex carbonated hydrogen compounds. On the contrary no fast variation of the LEED pattern is observed when the monocrystals have adsorbed ethylene on the surface previous to the adsorption of H_2.

IV. DISCUSSION

INTERPRETATION OF α AND MAGNETIC ISOTHERMS

The interpretation given in this paper was first proposed by Selwood (1962) to explain low-field measurements. This hypothesis is in good agreement with the theory of the de-metallization of surface atoms during adsorption (Sachtler and Van der Planck, 1969). Likely values of $n = 2$ for H_2, 10 for cracked C_2H_4 and 12 for cracked C_2H_6 can be considered as a good verification of this theory.

ADSORPTION OF H_2

Ehrlich (1966) and Tucker (1966) suggested that on (110) face, H is bound to two Ni atoms (Ni_2H) and that a close-packed row of Ni atoms is slightly displaced in respect of their clean surface position, giving a 1 × 2 pattern. Our magnetic measurements indicate that on $Ni-SiO_2$ issued from antigorite, H is bound to more than one Ni atom and we can conclude that on this sample the (110) faces are abundant.

If Tucker's model is admitted, H must be frozen to support Ni displacements (May, 1970). Mobility of hydrogen on the (110) face should be smaller than on other faces. Our calorimetric measurements are in agreement with this view: H is probably less mobile on the sample issued from antigorite, where 110 faces are rather abundant. However, these results can be also explained by a strong neighbour interaction and new work is under way from which we hope to have more conclusive evidence for this special problem.

ADSORPTION OF ETHYLENE, ETHANE AND $H_2-C_2H_4$

A certain disagreement exists in bibliographic data about adsorption of ethylene on nickel. First low-field magnetization measurements (Selwood, 1962) as well as infra-red spectroscopy (Erkelens and Liefkens, 1967; Sheppard *et al.*, 1970) suggested an associative adsorption ($Ni-CH_2-CH_2-Ni$) when adsorption occurs at room temperature. Our LEED and thermodesorption experiments performed on (111), (110) and (100) faces have shown that adsorption is dissociative. High field magnetic measurements confirm this view. The discrepancy between our results and infra-red data is probably due to low surface coverage used in magnetic measurements in respect of those generally used in infra-red spectroscopy. Indeed, recent experiments reported by Wösten, and

Erkelens (1970) show that no CH band is observed in infra-red at low coverage. These authors have also done low field magnetic measurements and their results are in good agreement with our observations. Moreover, for monocrystals it is also possible to observe the formation of $C_x H_y$ groups due to self hydrogenation if the contact time is increased. Finally, it is important that the dissociative chemisorption of hydrocarbons on Ni is also confirmed by a field emission study performed recently by Whalley *et al.* (1970). Hence, it seems that at low surface coverage ethylene is completely cracked, whereas at higher coverages self hydrogenation may occur.

Results obtained for Ni monocrystals and powders for adsorption of ethane and for the subsequent $H_2-C_2H_4$ or $C_2H_4-H_2$ adsorption are also very similar and show once more that monocrystals and Ni powders behave in the same manner.

From the standpoint of catalysis, all our experiments suggest that the hydrogenation of C_2H_4 should be governed by a Rideal mechanism. However, these results have been obtained at low coverage, at high coverage other chemisorbed species may exist and other reaction paths adding their effect to the main Rideal mechanism are possible.

In conclusion, monocrystals and supported polycrystalline nickel samples behave in a similar way. In all cases, most of chemisorbed ethylene is cracked already at temperatures as low as room temperature. Saturation measurements on orientated nickel-on-silica catalysts suggest that adsorbed atom H may be bound to two nickel atoms on 110 plane; this is compatible with LEED results on monocrystals.

REFERENCES

Beek, O. and Cole, W. A., Wheeler, A. (1950). *Disc. Farad. Soc.* **8**, 314.

Bertolini, J. C. and Dalmai-Imelik G. (1969). "Structure et Propriétés des Surfaces des Solides". C.N.R.S., Paris.

Bond, G. C. (1962). "Catalysis by Metals". Academic Press, New York and London.

Dalmai-Imelik, G. and Bertolini, J. C. (1970). *C. R. Acad. Sci. Ser. C.* **270**, 1079.

Dalmai-Imelik, G., Bertolini, J. C. and Imelik, B. (1971). *J. Chim. Phys.*

Ehrlich, G. (1966). *Annu. Rev. Phys. Chem.* **17**, 297.

Eley, D. D. and Norton, P. R. (1970). *Proc. Royal Soc. Ser. A* **314**, 319.

Erkelens, J. and Liefkens, Th. J. (1967). *J. Catal.* **8**, 36.

Martin, G. A., Renouprez, A., Dalmai-Imelik, G. and Imelik, B. (1970). *J. Chim. Phys.* **67**, 1149-1160.

May, J. W. (1970). *Advan. Catal. Relat. Subj.* **21**, 151.

Sachtler, W. M. H. and Van der Planck, P. (1969). *Surface Sci.* **18**, 62-72.

Selwood, P. W. (1962). "Adsorption and Collective Paramagnetism". Academic Press, New York and London.

Sheppard, N., Avery, N. R., Morrow, B. A. and Young, R. P. (1970). "Symposium on Chemisorption and Catalysis". London.

Tucker, C. W. Quoted as a private communication by Ehrlich (1966).

Whalley, L., Davis, B. J. and Moss R. L. (1970). *Trans. Faraday Soc.* **576**, vol. **66**, 3143.

Wösten, W. J. and Erkelens, J. (1970). "International Symposium on Heterogenous Catalysis" (Roermond).

AUTHOR INDEX

Numbers followed by asterisks are those pages on which references are listed.

A

Abon, M., 246, 249, 259*
Adams, R. O., 313, 315, 325, 328*
Alexander, C. S., 279, 290*
Anderson, J., 222, 226*, 245, 246, 249, 252, 258, 259*, 395, 404*
Anderson, P. W., 216, 226*
Andronikashvili, T. G., 71, 72, 73, 84*
Aristov, N. G., 9, 15*
Armand, G., 59, 70*
Armstrong, R. A., 16, 17*, 19, 21, 29, 31, 32, 32*, 36, 38, 47*
Astakhov, V. A., 9, 10, 12, 15*, 16
Aston, J. G., 86, 98*
Audrieth, L. F., 294, 296*
Avery, N. R., 441, 443*
Avgul, N. N., 9, 15*

B

Bacigalupi, R. J., 111, 125*
Baer, Y., 359, 368*
Baessler, H., 326, 328*
Bagg, J., 419, 420, 431*
Bailyn, M., 220, 226*
Baker, B. G., 24, 32*
Baker, F. A., 236, 243*
Ballu, Y., 408, 410, 415*
Barnes, M., 86, 98*
Barrer, R. M., 12, 15*
Barton, G. C., 381, 390*
Bashara, N. M., 377, 378*
Beach, N. A., 327, 328*
Beaufils, J.-P. A., 50, 56*, 65, 68, 70*
Becker, G. E., 224, 226*, 359, 368*, 377, 379*
Becker, J. A., 184, 188*
Beek, O., 436, 442*
Behrens, H., 342, 342*
Bell, A. E., 397, 399, 404*
Bellardo, A., 16, 17*, 20, 32*
Bennett, A. J., 369, 377, 378*

Bering, B. P., 4, 5, 6, 12, 13, 15*
Bergmark, T., 359, 361, 368*
Berkner, K. H., 208, 212*
Bertolini, J. C., 437, 440, 442*
Bezus, A. G., 72, 84*
Bird, R. B., 69, 70*
Bishop, H. E., 384, 390*
Blakely, J. M., 346, 357*
Blyholder, G., 288, 290*, 377, 378*
Böhmer, H., 210, 212*
Bohn, G. K., 360, 368*
Bond, G. C., 245, 259*, 436, 442*
Boreskov, G. K., 331, 332, 334, 343*
Born, M., 185, 188*
Boyd, M. E., 85, 97, 98*
Bradshaw, A. M., 277, 278, 286, 288, 290*
Brennan, D., 292, 296*, 304, 311*
Bretz, M., 99, 110*
Breuer, H. D., 313, 314, 325, 327, 328*
Brodd, R. J., 55, 56*
Bruce, L., 419, 420, 431*
Buckman, A. B., 377, 378*

C

Callcott, T. A., 369, 378*
Carter, G., 227, 242*
Cerofolini, G. F., 36, 47*
Chen, J., 161, 168*, 363, 367*
Chernen'kova, Yu. L., 72, 74, 84*
Chesters, M. A., 278, 289, 290*
Chubb, J. N., 138, 139, 140, 147*
Chumburidze, T. A., 71, 73, 84*
Clampitt, R., 208, 210, 212*
Clausing, R. E., 297, 311*
Clavenna, L. R., 391, 394, 400, 402, 404*
Coad, J. P., 384, 390*
Cochrane, H., 31, 32, 32*, 33, 46, 47*
Cole, W. A., 436, 442*

SUBJECT INDEX

Absolute rate theory, 228, 330
  temperature dependence of transmission coefficient, 234
Activation energy for desorption (See also: Desorption, Desorption rate, Desorption spectra)
  dependence on surface coverage, 394
  H_2 from Ni, 239
  He from solid Ar, 146
  He from solid Kr 146
  He from solid Xe, 146
  O_2 from Ag, 420
Active carbon
  adsorption isotherm of cyclohexane, 10
  characteristic curve for the adsorption of benzene, 8
Adsorption isotherm equations (See also: Dubinin-Radushkevich equation, Isotherm of adsorption)
  Dubinin-Radushkevich, 19, 35
  Hill-de Boer, 26
  Hobson, 34
  Ross and Olivier, 19, 22
  two-dimensional virial isotherm, 24, 58, 88
Alkali metal atoms
  electron tunnelling and surface ionization on Mo, 175-176
  surface charge in adatoms, 186
  surface lifetimes on polycrystalline Mo, 172
  surface mobility on Mo, 183
Ammonia
  field emission microscopy on Mo, 248-251
  intermediate formation of NH_x species in decomposition on Mo, 258
  non-dissociative adsorption on Mo, 249

Ammonia—*continued*
  work function variation during adsorption and desorption on Mo, 247-250
Anderson's Hamiltonian, 217-219
  Hartree-Fock theory of Anderson's Hamiltonian, 221
Anelastic scattering
  dependence of anelastic scattering on the Debye temperature of the solid, 158
  of Ar beams from Ag(111) and Au(111), 160
  of Ar beams from Ni(111) and Pt(111), 160
  of Ar beams from W(110), 160-161
  of He beams from diamond, 167
  of Kr beams from Pt(111), 157
  of Ne beams from Ag(111) and Au(111), 159
  of Ne beams from Ni(111) and Pt(111), 159
  of Ne beams from W(110), 159
  trapping probability in anelastic scattering, 158
Auger electron spectroscopy
  changes in transition probabilities, 360, 365
  coaxial cylindrical electron spectrometer in, 361
  energy shift in metal Auger electrons, 362
  identification of adsorbed states by, 359
  in revealing carbon contamination, 287
Auger spectra
  Ag-O_2 system, 428-429
  C in various chemical forms, 364
  CO on Ta(100), 365
  CO on W(112), 362
  Mo(111) surface, 383
  O_2 on Mo(111), 384